STEM of Desire

Cultural and Historical Perspectives on Science Education *Research Dialogs*

Series Editors

Catherine Milne (*New York University, USA*)
Kathryn Scantlebury (*University of Delaware, USA*)

VOLUME 8

The titles published in this series are listed at *brill.com/psed*

STEM of Desire

Queer Theories and Science Education

Edited by

Will Letts and Steve Fifield

BRILL
SENSE

LEIDEN | BOSTON

Cover illustration: From ATOM-r *Kjell Theøry* (2017). Performers (from left to right): Christopher Knowlton, Justin Deschamps, Leonardo Kaplan and Judd Morrissey. Photograph © Grace DuVal

All chapters in this book have undergone peer review.

The Library of Congress Cataloging-in-Publication Data is available online at http://catalog.loc.gov

ISSN 2589-6199
ISBN 978-90-04-33105-1 (paperback)
ISBN 978-90-04-36998-6 (hardback)
ISBN 978-90-04-33106-8 (e-book)

This book is printed on acid-free paper and produced in a sustainable manner.

CONTENTS

Acknowledgments vii

List of Figures ix

Illyria 1
Charlotte Boulay

Prolegomenon: Queer Theories and STEM Education 3
Steve Fifield and Will Letts

1. I : Snow Queen :: "Nigger" : (School) Science 41
Francis S. Broadway

2. Queering Science Education without Making Too Much Sense 59
Steve Fifield and Will Letts

3. Beyond Nature Talk: Transforming Environmental Education with Critical and Queer Theories 89
blake m. r. flessas and Timothy D. Zimmerman

4. Wonder in the Science Classroom 109
Andrew Gilbert and Emily M. Gray

5. Teaching Queering Physics: An Agenda for Research and Practice 125
Helene Götschel

6. What Does Queer Theory Have to Do with Teaching Science in Elementary Schools? 147
Kristin L. Gunckel

7. Queering STEM Learningscapes 161
Joe E. Heimlich

8. What's in a Name? Reflections on Learning and Teaching in Central Texas 177
Tommye Hutson

9. Strange Precipitate: How Interest in Science Produces Different Kinds of Students 191
Kathryn L. Kirchgasler

10. What Makes *Girls* and *Boys* So Desirable? STEM Education beyond Gender Binaries 209
Michelle L. Knaier

11. Children, Nomads, and Queering: Desire and Surprise in a Wiggly World 223
Sheri Leafgren and Scott Sander

12. Inviting the Mess: A Children's Museum's Transgressive Tactics for Unleashing Play 243
Anna MacDermut and Adrian Zongrone

13. Thinking Like a Fox: Queering the Science Classroom When Teaching about Sex and Sexuality 255
Michael J. Reiss

14. Exhibiting Doctors and Nurses: Queering Professional Education in a Medical Museum 269
Cecilia Rodéhn

15. Camping Science Education: A Trip to Camp Wilde and the Queer Nature of Nature 289
Nicholas Santavicca, Jesse Bazzul and Stephen Witzig

16. Towards a Queer Curriculum of Infinity 307
James Sheldon

17. Yearning, Learning, and Earning: The Gritty Ontologies of American Engineering Education 319
Amy E. Slaton, Erin A. Cech and Donna M. Riley

The Bargain 341
Charlotte Boulay

Index 343

ACKNOWLEDGMENTS

We've worked on this book for more than four years, while between us we've lived in five cities and three countries on two continents. We're grateful to our awesome contributors for their steadfast commitment to the project and for the passion that shows through in their writing. Thanks to series editors Kathryn Scantlebury and Catherine Milne for inviting the book (and persistently nudging it along!), and to Peter de Liefde, Michel Lokhorst, Evelien van der Veer, and Jolanda Karada at Brill Sense. Our appreciation to Lisa DeBoer for producing the book's index.

Special thanks to Anatomical Theatres of Mixed Reality (ATOM-r) for the book's cover image, taken at the International Museum of Surgical Science in Chicago by Grace DuVal (see imss.org/kjell). ATOM-r is a provisional collective exploring forensics, anatomy, and 21st-century embodiment through performance, language, and emerging technologies. We're indebted to Grace and performers Chris Knowlton, Justin Deschamps, Leonardo Kaplan, and Judd Morrissey for permission to use the photograph. The image is from *Kjell Theøry* (2017), an augmented reality performance juxtaposing Alan Turing's theory of morphogenesis with algorithmic mutations of Guillaume Apollinaire's 1917 play *Les Mamelles de Tirésias* (*The Breasts of Tiresias*), a genderfluid spectacle for which the author invented the word *surrealism* (see atom-r.com). We salute Jorge Lucero for connecting us to ATOM-r!

Will: I acknowledge with gratitude Charles Sturt University and specifically Lesley White, Executive Dean of the Faculty of Arts and Education, for her support and for allowing me the time and space to pursue this never-ending passion. To Sandra Harding and Nancy W. Brickhouse, who were there at the beginning of this work, I acknowledge and thank you for the inspiration, motivation, mentoring, and friendship. Huge thanks to Brett Hartmann and our boys for sharing me with this project and enduring the absences—and especially for unstinting love and support. And most of all, deep, heartfelt thanks to Steve Fifield for being an amazing intellectual travelling companion over decades of adventures, occasional mischief, and lots of laughs. Steve, I'm in awe of your ideas, inspired by your writing, and so very grateful for our friendship—intellectual and otherwise.

Steve: My work on this book was aided immeasurably by adjunct faculty appointments in the University of Delaware School of Education that allowed access to the university's research library. Thanks to Zoubeida Dagher, Danielle Ford, Chrystalla Mouza, Carol Vukelich, and other colleagues in the School of Education for supporting those appointments. At important points in my work life as a grant writer, Marisa Wigglesworth, then at The Franklin Institute Science Museum, and Patricia Brooke, at the Oregon Museum of Science and Industry, helped me craft day jobs that accommodated this project in off-hours. I couldn't have asked for more supportive bosses. I may have never turned to queering science education without

Lee Swain's curiosity about his experiences and his generosity in sharing what he learned. Will Letts, my memory, though perhaps not yours, is that 20 years ago you catalyzed our subsequent work together during a conversation at Iron Hill Brewery. You're the best intellectual instigator I know, and a loving friend. And deep gratitude to Kit for sharing our life with contributors to this volume, virtual guests in our home who couldn't leave until Will and I finished our work.

FIGURES

1.1.	Odi et Amo Angelic Angelica Elessar (photograph © L. Sue Rooney-Flynn)	52
9.1.	A developmental sequence of science education aims (Downing, 1925, p. 83)	196
14.1.	The brown-skinned mannequin (photograph by Patrik Klingborg)	270
14.2.	The Museum of Medicine in Uppsala (photograph by Patrik Klingborg)	273
14.3.	Representation of a medical doctor in the downstairs exhibition (photograph by Patrik Klingborg)	274
14.4.	Augusta Strömberg's artwork in the exhibition on Ulleråker psychiatric hospital (photograph by Patrik Klingborg)	276
14.5.	The history of nursing on display (photograph by Patrik Klingborg)	278
15.1.	Birchbark sketch	292
15.2.	Chalk outline on cabin floor	297

CHARLOTTE BOULAY

ILLYRIA

At dawn a charm of finches fighting
in the privet, swooping low over spilled seed,
wings tazzing the grass, rustling like folding
newspaper, like time. All morning the rowdy
flutter and chirp, both presence and absence.
Ask, *who can command this coming?* They flash
over the roofs of thought, twittering into
the emptiness, the weather of what's gone.
You think, one more small thing could break you.
Relinquishment, spent under the covers
while all else continues: dawn pales
over and over, and with it the birds argue
bitterly, scritching at the sill. Each month
building an edifice in the body, a sculpture of hope.
I thought when finished it would break into flight.
The radio calmly lists losses: houses, lovers,
glaciers, coins. Impossible to hold everything.
Outside the magnolias burst their velvet buds,
cluster in thick pink clouds then fall
and slick the ground with sweet rot. Tomorrow,
let it happen: shipwreck, crumpling, sink.
After the storm, sit up against the pillows
in the doomed light and look outside. See
the parcel of brown linnets visited by an exotic,
her breast snow white, head splashed with yellow
and blue. She is washed up, unrecognized as herself.
She knows nothing of the posters on every tarred
telephone pole searching for the lost cockatiel,
please call if you have seen Edmund. Best bird ever.
She asks, *what country, friends, is this?* She whistles
a new world, sings a snatch of jingle, a jazz lick.
She alights in the thicket and speaks her name.

 | DOI:10.1163/9789004331068_001

ABOUT THE AUTHOR

Charlotte Boulay's poetry has been published in the *New Yorker*, the *Massachusetts Review*, the *Boston Review*, the *Michigan Quarterly Review*, *Beloit Poetry Journal*, *The Pinch*, *Cyphers*, and other journals, and has appeared on the *PBS NewsHour* poetry blog and on NPR's *Morning Edition*. Her book, *Foxes on the Trampoline*, was published by Ecco Press/HarperCollins in 2014. She worked at the Franklin Institute science museum in Philadelphia, Pennsylvania (USA), for five years. She is currently director of development at the Miquon School.

STEVE FIFIELD AND WILL LETTS

PROLEGOMENON

Queer Theories and STEM Education

She asks, *what country, friends, is this?* She whistles
a new world, sings a snatch of jingle, a jazz lick.
She alights in the thicket and speaks her name.

—Charlotte Boulay ("Illyria," this volume)

One may start by acknowledging first that teaching and learning are basically acts of desire.

—Michalinos Zembylas (2007, p. 333)

Any educator should consider the dynamics of sexuality as central to the capacity for human curiosity, for living a vital intellectual and social life, and for our capacity to attach passionately to knowledge, other people, and life projects.

—Deborah Britzman (2000, p. 36)

Queer loosely describes a diverse, often conflicting set of interdisciplinary approaches to desire, subjectivity, identity, relationality, ethics and norms.

—Noreen Giffney (2009, p. 2)

Desire is always surpassing the frames established for it, and a queer politics of desire allows us to become open to what exists beyond the discursive frameworks that have been established for these experiences.

—Catriona Mortimer-Sandilands and Bruce Erickson (2010, p. 37)

INTRODUCTION

STEM of Desire: Queer Theories and Science Education investigates and instigates entangled relations of STEM education, sex, sexuality, gender, and manifold desires using queer theories and allied perspectives to advance constructive critique, creative world-making, and (com)passionate advocacy.[1] Science, technology, engineering, and mathematics (STEM) education entails disparate settings, subjects, objects, relationships, practices, affects, and knowledges that blend people and things in restless naturecultural (re)productions.[2] Institutionalized regulatory processes of naming,

 | DOI:10.1163/9789004331068_002

categorizing, normalizing, and explaining, such as STEM education, necessarily induce transgressions via infidelities, slippages, resistances, improvisations, and perverse repurposings that both define and undermine what's normal and natural, and what's not. Queer theorizing uncovers and incites paradoxical relations of the normal and strange in ways of knowing, feeling, and being as they manifest in and through sex, sexuality, gender, and polymorphous desires. Queering attends to far-reaching desires and odd (im)possibilities in processes that normalize and naturalize (and abnormalize and denaturalize) particular manifestations of human subjects, society, culture, and nature.[3]

Like any cultural practice, queer theorizing (re)produces its own normalizations. These have included persistently abrading dominant sexualities, genders, and desires without taking up hopeful world-making with similar passion. Further, with nourishing roots in textual criticism, queer theorizing has been prone to disregarding—to failing to imagine material possibilities for—the animated matter of nature in/as culture. We encouraged contributors to this volume to queer these and other habits of queer theories by leavening their critiques of STEM education with hope, renewal, and appreciation for feral agencies that transgress and transmute "nature" and "culture."

For more than two decades, occasional research on STEM education and queer theories has appeared in scattered venues near and beyond disciplinary margins. This intersectional scholarship isn't the normal occupation of either STEM educators or queer theorists. But rather than normalizing this work, a goal of *STEM of Desire* is to present and elicit a host of abnormal perspectives that are among the "indispensable interior exclusion[s]" (Fuss, 1991, p. 3) of STEM education and queer theories as typically conceptualized and practiced. It's a queer cliché to say that queer theory is so queer it can't be defined. *Definition* gestures to situated and interested claims about how things are (un)related to each other. Of course, the issue isn't that queer theory is undefinable, but that it won't settle into a single, all-encompassing definition. "The wanderings of sexuality," writes Deborah Britzman (2000), "signify excess, more than we may consciously ask for or even want" (p. 33). "Queer attachments work both to celebrate the excess of life and to politicize the sites at which this excess is eradicated" (Mortimer-Sandilands & Erickson, 2010, p. 37). Queer theorizing is excessive in its abundant definitions and its orientations to the ways things exceed themselves (which is to say that when we attempt to view a thing or being in exquisite ontological isolation, as if self-sufficient, it fails to attain itself—only with/as others do things come into being). Queering attends to promiscuously cocreative intimacies in which things touch each other, recoil, and touch again even as they occupy, and are occupied by, myriad others in relation to which they arise.

To negotiate this excessiveness, queer theorists (like all theorists) are partial to some material, enacted, and interpretive dimensions more than others. As participants in and interpreters of STEM education's manifold desires, we must critically interrogate and connect with partial perspectives—not to transcend them but to cultivate and appreciate their generative tensions. STEM education is a regime of order even as the impossibility of its faithful reproduction undermines foundational dualisms like nature/culture. Phenomena of interest to queer theorizing overlap with, though may be differently construed by, other lived and analytical perspectives, including gay, lesbian, bisexual, trans, feminist, gender, sexuality, psychoanalytic, crip, "race," postcolonial, class, language, and so on. Queering supports intersectional analyses while recognizing that dimensions of lived experience, cultural production, and analysis are already so entangled and entangling that none stand outside of intersection. Queering notes and perverts the desire to construct intersectional selves and inquiries just so by seeking situated, embodied, interested, accountable, and incompletable—in Donna Haraway's (1988) sense, more *objective*—partial perspectives.[4]

As instigators of this collection, our core concern (which we don't assume all our contributors share) isn't to counter systemic inequities faced by sexual and gender minorities in STEM education. It isn't to increase the representation of sexual and gender minorities in STEM fields, nor to raise the voices and visibilities of LGBTQQI+[5] learners, educators, scholars, scientists, technologists, engineers, and mathematicians. And it isn't to queer STEM curriculum by integrating LGBTQQI+ issues and role models. Certainly, queering STEM education as variously conceived in this volume can critically rework and incorporate these objectives and approaches. Many of the chapters in this collection engage and inform these important concerns. But this book isn't about advancing LGBTQQI+ inclusion in STEM education. Instead, our core concern is what and how desiring subjects and STEM education might be(come) in relation to each other and the world. We hope that "the pedagogy at work [here] is one where the desire for knowledge interferes with the repetition of both heterosexual and lesbian/gay [LGBTQQI+] normalization" (Luhmann, 1998, p. 141). We further hope that the curriculum relations at work here interfere with habitual normalizations of STEM and STEM education.

In this book you will encounter various conceptions and practices of queering STEM education that are in conversation though not necessarily agreement. Disrupting habituated practices of knowing, feeling, and being in and with STEM education arouses passions. Things are wonderfully and terribly implicated, especially when we wish to mark their uniqueness, self-sufficiency, agency, and/or (un)naturalness. We think of queer theories less as tools to tease out distinct valences of people and STEM education (e.g., sexuality, gender, desire, "race," class, ability, etc.) than as *orientations* (Ahmed, 2006b) for approaching and eliciting responses from STEM education and beings of all sorts that coarise and are caught up with/in it. But don't take our word for it. We suggest you approach particular definitions of queering STEM education as provisional orientations, not as determinative demarcations of proper and improper queering and STEM education.

We next offer contemplations on *orientation*, *implication*, *identity*, and *normativity* to variously ground and unfix views of queer theorizing and STEM education. Then we sketch a mud map of the territory that to date has been visited by bringing queering to STEM education.[6] To conclude we offer impressions of each chapter that preview the multifarious perspectives in this collection. We wish to support divergent conceptions and practices of queering STEM education through our necessarily constrained visions of it. Constraint offers creative leverage, something to converse with, refuse, or repurpose. We invite you to pursue critical, creative, and virtuous world-making by reading playfully, subversively, and with a "desirous intent of being" (Alexander, 2018, p. 277).

CONTEMPLATIONS

See endnote 7 for several excellent books that address the development and application of queer theories.[7] Familiarity with these foundational sources will support critical readings of *STEM of Desire*, but this book can be approached on your own terms. Here are several framing concepts to keep in mind and in motion as you negotiate this collection.

Orientation

> A queer phenomenology might involve an orientation toward what slips, which allows what slips to pass through, in the unknowable length of its duration. In other words, a queer phenomenology would function as a disorientation device; it would not overcome the "disalignment" of the horizontal and vertical axes, allowing the oblique to open up another angle on the world. (Ahmed, 2006b, p. 172)

In their magisterial taxonomy of qualitative social science research, Yvonna Lincoln, Susan Lynham, and Egon Guba (2018) classify queer theory as a paradigm in its own right and as an interpretive perspective within and across the paradigms of critical theory and constructivism. Taxonomy invites ambiguity: every carefully drawn boundary is another distinction to blur. For instance, calling queer theory a *paradigm* is too master-narrative for our tastes; instead, as just noted, we think of queer theorizing as enacting *orientations* in and toward the world that locally and temporarily, but persistently, reorient the world. Sara Ahmed (2006b) writes that "orientations shape what bodies do, while bodies are shaped by orientations they already have, as effects of the work that must take place for a body to arrive where it does" (p. 58). Imagine Ahmed's "bodies" expansively: bodies of knowers, bodies of knowledge, sensate bodies, institutional bodies, nonliving bodies, worlds of bodies.[8] Bodies matter and so do orientations: they enable and constrain. "Orientations are about the direction we take that puts some things and not others in our reach" (p. 56).

Thus oriented, we imagine how queer theories enact deconstructive, critical, and reparative orientations toward (im)possibilities for bodies of all kinds under normalizing regimes of sex, sexuality, gender, and manifold desires. Queer theories deconstruct when they mark and question the discursive and material reproduction of foundational structures of (non)sense-making and being. Queer theories are critical when they trouble (im)possibilities for ideology, power, and social justice. And queer theories at their most hopeful are reparative when they queer their own habitually paranoid critiques—which advance a "hermeneutics of suspicion and exposure" (Sedgwick, 2003, p. 140)—with an easy openness to surprise, improvisation, and renewal.[9]

Implication

> At its simplest queer theory is a critical inquiry into the alignments of sex, gender and desire that are in the service of normative forms of heterosexuality, the heteronormative, that saturates the social and cultural order. (Graham, 2014, p. 6)

> Sexuality is not easily separable from that which it is supposedly not. (Cohen, 2013, p. 151)

> Constantly pushing beyond its own limits, the sexual always has less to do with interlocking genitalia than with interlocking webs of discourse. (Azzarello, 2008, p. 146)

Sexuality isn't everything, but its residues are everywhere. In a founding work of queer theorizing, *Epistemology of the Closet*, Eve Kosofsky Sedgwick (1990) positions the homo-/heterosexual binary as indispensable to analyses of modern Western culture:

> An understanding of virtually any aspect of modern Western culture must be, not merely incomplete, but damaged in its central substance to the degree that it does not incorporate a critical analysis of modern homo/heterosexual definition. (p. 1)

Drawing on and inflecting Michel Foucault in *The History of Sexuality*, Sedgwick mobilizes sexuality not as a hulking mass at the core of modern Western culture but as kinetic, distributed, and often unmarked meanings and practices that are in "distinctively privileged relation to our most prized constructs of individual identity, truth, and knowledge." "The language of sexuality," Sedgwick continues, "not only intersects with but transforms the other languages and relations by which we know" (p. 3). Here is an alluring paradox of queer theorizing: Foregrounding sexuality alerts us to distal, diffuse, and yet robust relations of desire that are neither simply sexuality nor unrelated to it.

Even concise definitions of queer theory, like the preceding epigraph from Mark Graham, suggest that sexuality exceeds itself—it can't accommodate all the relations from which it arises and those it conditions. Consider Annamarie Jagose's (1996) oft-cited encapsulation of queer:

> Broadly speaking, queer describes those gestures or analytical models which dramatise incoherencies in the allegedly stable relations between chromosomal sex, gender, and sexual desire. Resisting that model of stability—which claims

> heterosexuality as its origin, when it is more properly its effect—queer focuses on mismatches between sex, gender, and desire. (p. 3)

There is plenty to chew on in "the allegedly stable relations between chromosomal sex, gender, and sexual desire" for STEM educators interested in troubling heteronormative formulations of genetics, anatomy, physiology, and behavior, for instance. Holding formulations of sex and sexuality in school science curricula to rigorous standards of scientific accuracy and objectivity should be a pressing concern for queer theorizing, because so much of what is taught about sex and sexuality in the name of science misrepresents what we know to be true (Lemke, 2011; Reiss, this volume, Chapter 13). Notice, however, in Jagose's definition the shift, perhaps incidental, though to us enticing, from "sexual desire" in the first sentence to "desire" in the second. We read this unqualified desire as beckoning queer theorizing beyond critiques of the science (education) of sex, gender, and sexuality (as crucial as those critiques remain!) to the many, typically unmarked desires that infuse and animate all of STEM education, including desires aroused and denied by it. To reiterate, the animating desires in and around STEM education are neither simply sexuality nor unrelated to it.

In the first pages of her rewarding introduction to queer theory, Nikki Sullivan (2003) explicitly calls out this wider realm of queer theorizing:

> Rather than focusing narrowly on sexuality and/or sexual practices, [this] book aims to consider critiques of normalizing ways of knowing and of being that may not always initially be evident as sex-specific. ... This sort of approach is crucial, it seems to me, if we are to understand the broader significance of Queer Theory and the extensive range of ways in which notions of sexuality and gender impact—at times implicitly—on everyday life. (p. vi)

Similarly, for Steven Seidman (1996) the concern of queer theory isn't sexuality narrowly construed but the routine sexualization of social and cultural organization. Queer theory is the

> study of those knowledges and social practices that organize "society" as a whole by sexualizing—heterosexualizing or homosexualizing—bodies, desires, acts, identities, social relations, knowledges, culture, and social institutions. (p. 13)

Because residues of sexuality are strewn about, queer theoretical concerns can be "somewhere to the side of [sexuality], tangentially or contingently related or even rather unrelated to it" (Sedgwick, 2003, p. 132). Queering turns toward disciplined and feral desires of material, personal, social, cultural, and other (imagine what others there may be!) (im)possibilities near and far from what we *normally* think of as sexuality—including, if we dare, motley naturecultural conglomerates like STEM education.

Queer theorizing arose as critiques of heteronormative sexuality, but this is a contingent foundation, not an essence. The proposition here is that the scope of queer theorizing ought to be vast and discipline-resistant because the sexual and nonsexual are constitutively entangled. Cultural domains and relations conventionally understood as nonsexual—like most of STEM education—are imbued by and imbue normative (and so ab-normative) sexuality. Graham (2014) suggests we consider the notions of *implicate* and *explicate order* to imagine the pervasive, and often diffuse, nature of heteronormativity in terms of loose networks rather than rigid cultural or analytical hierarchies. The "overdetermined entanglement" of the heteronormative in cultural implicate order "enables us to look at different phenomena from multiple angles" (p. 6). Graham continues:

> One of the qualities of an implicate order is that it performs a kind of origami that brings into propinquity phenomena that in the explicate order appear distant, unrelated and sometimes even antagonistic. … To explicate the (hetero) normative is to unfold the implications of this order across a range of social and cultural phenomena. (p. 6)

This suggests that when we begin with sexuality and trace it through the implicate order of STEM education, sexuality may diffuse and suffuse into relations of desire that bear little resemblance to what we ordinarily think of as sexuality, even as these diffuse desires carry robust residuals of normative (hetero, homo, whatever) sexual formations and discourses. Conversely, when we bring queering sensibilities to inquiries around "nonsexual" topics in STEM education, we may discover (and incite) unimagined implications in and for the sexual order.[10] Such queer theorizing presses us to interrogate commonsense distinctions like "sexual" and "nonsexual" as *effects* of normalizing regimes that orient us to see and participate in the world in some ways and not others. While STEM topics like genetics, anatomy, physiology, and behavior are ripe for queering, queering explorations are needed *all across* STEM education to expose, abrade, and generatively repurpose normative sexualities and desires that are implicit in topics and practices that may seem like (and perhaps are) strange concerns for queering.

How might we imagine sexuality that exceeds itself in/as manifold desires that are implicit in the bodies, subjectivities, objects, and cultures of STEM education? Consider, for instance, how Britzman (2000) works with Sigmund Freund's view that

> sexuality is other to education, to cure, and to the confines of authoritative knowledge, even as sexuality is the condition of and cause for our capacity to learn, to seek pleasure in both the making and unmaking of knowledge, and to craft erotic relations from the strangest places. (p. 34)

Erotic, indeed, and not merely "sexual." Here sexuality broadens to eros, the "original urge," "the push that comes before representation" (p. 33). Early in a human life, the workings of eros—"the urge that wants satisfaction and the desire for pleasure" (p. 38)—"are precocious, arriving before knowledge and understanding" (p. 33). In this sense, says Britzman, "sexuality is the first condition for human curiosity and hence the first condition or force of learning" (p. 38). Later, as self-identified selves, we come to intimately inhabit, but never really to understand, the "intractableness of Eros" (p. 33), the murky depths and dividedness of conscious and unconscious being(s) in which "our desires slip between the fault lines of recognition and misrecognition, [and] our urges divide in the strange calculus of ignorance and knowledge" (p. 33).

If eros exposes division within, it also integrates mind-body through the sensuality of embodied knowing. "The urge of sexuality … is made from the desire to touch and to be touched by people, by ideas, and by living" (p. 38). Kristin Gunckel (this volume, Chapter 6) orients bodies and embodied cognition toward sexuality reimagined as exploring, engaging, contacting, and intervening in the world:

> The physical, emotional, and psychological pleasure derived from engaging in this quest pushes the body to move, interact with, and connect to the physical world, and it is from these interactions that knowledge is constructed. In remobilizing sexuality in this way, Britzman unchains sexuality from normalizing links to acts of sex and sets free the notion that sexuality is the underlying driver of humans' insatiable quest for knowledge and understanding. … Thus, knowledge of the world may come from bodily experiences in the world, but the experiences bodies have are inspired and guided by bodily desires. (Chapter 6)

Eros, then, enlarges possibilities for thinking about and queering sexuality in STEM education as "a bodily desire to satisfy curiosity" (Chapter 6).

For a different (in other words, queerly complicit) perspective on desire in excess of sexuality that we might bring to queering STEM education, consider Michalinos Zembylas's (2007) reading of Gilles Deleuze and Felix Guattari's notions of desire. Here desire is an "autonomous and productive … *flow* that is continuous and is always *becoming*" (pp. 335, 336, original italics). Formulated in part in resistance to Freudian and other psychoanalytic models of desire, Deleuzo-Guattarian desire isn't a drive to fill an unfillable psychic void, but a productive force of expanding possibilities; not the stuff of psychocognitive fantasies, but of material, sociopolitical relations; not an expression of divided subjectivity, but ceaseless synthetic becomings. "In other words, according to this notion, desire is not a libidinal but an *existential* force, … an ontological and epistemological process of becoming" (p. 338, original italics). Deleuzo-Guattarian desire is queering in affect and effect: For instance, it expresses and is expressed by agential assemblages that include and exceed human being(s). This opens the productivity of desire and the reach of queering STEM education to nonhuman being(s) (see Giffney & Hird, 2008), to "life [that] reads and rewrites itself" (Kirby, 2011, p. xi), and to vibrantly inanimate matter (see Bennett, 2010; Cohen, 2013).

Zembylas (2007) uses Deleuzo-Guattarian desire to theorize a pedagogy of desire:

> The claim of pedagogy of desire is that through the mobilisation and release of desire, teachers and students make available to themselves the forces of pleasure and the powerful flows of desire, thereby turning themselves into subjects who subvert normalized representations and significations and find access to a radical self. … A pedagogy of desire dissociates itself from the instrumental purposes of education in order to embrace neglected aspects of learning and teaching such as joy, pleasure, happiness and transgression. (pp. 332, 340)

Sheri Leafgren and Scott Sander (this volume, Chapter 11) trace STEM pedagogies of desire in children's and educators' wiggly walks through cities, schools, and curricula. They eschew standardized STEM inquiry skills and invite us to consider STEM inquiry in/as assemblages of being(s) and becoming(s) through which desires flow. Consider, for instance, a teacher and several 4- to 5-years-olds strolling through a city, gently intending to pay attention. But to what? The paranoid pedagogue in us imagines pointless wandering and bad surprises (Sedgwick, 2003). Then "a child noticed a plant—a flower—growing from the pavement. This flower held lovely significance: 'That plant wants to tell us it has perfume,'" said the child (Leafgren & Sander, this volume, Chapter 11).

> Desire happens without positing a desiring agent or desirable object; it is emergent in relations and not the property or act of any one being or thing. This gives desire back to the relational universe of being—being scent, flower, city, human. (Chapter 11)

No longer a disciplined way of knowing for children who age-appropriately regulate desire, STEM inquiry is desire-becoming-inquiry immanent in the assemblage children + teacher + city + pavement + flower + scent. "In their work together, the children and the teacher as part of desiring assemblages relax contentedly into the moment, becoming a communion of mutual curiosity and wonder—a collaborative inquiry of the world" (Chapter 11). A collaborative inquiry *of* the world isn't inquiry executed by stable, coherent, self-contained subjects upon passive objects in the world, but the world (re)producing and differentiating, immanently coming to know and feel and express in/as transient, heterogeneous assemblages of desire.[11] That's pretty queer(ing).

Identity[12]

> Queer seeks to disrupt the discrete, fixed locations of identity by understanding sexuality and its meanings not as a priori or given but as constructed, contingent, fashioned and refashioned, and relational. (Talburt, 2000, p. 3)

> [Queer] signifies the messiness of identity, the fact that desire and thus desiring subjects cannot be placed into discrete identity categories, which remain static

> for the duration of people's lives. Queer thus denotes a resistance to identity categories or easy categorisation, marking a disidentification from the rigidity with which identity categories continue to be enforced and from beliefs that such categories are immovable. (Giffney, 2009, pp. 2–3)

> An important step in answering my colleagues' questions about what queer theory has to do with anything is to set them straight about what a queer inquiry into education is not. It is not a project of assimilating LGBTQ people into schools. That is a worthy project and one in which I participate, but that is not the focus of queer theory as a lens of critique. (Gunckel, this volume, Chapter 6)

Queering STEM education as we imagine it is about how normative and strange sexes, sexualities, genders, and desires shape and are shaped by (dis)identifications in and with STEM education. By applying the poststructuralist credo that "there is no true self that exists prior to its immersion in culture" (Sullivan, 2003, p. 41)—and by tantalizing *culture* with the notion that perhaps it has been *nature* all along (Kirby, 2008, 2011)—queering critically and creatively engages how STEM education makes identity, for better and worse, (im)possible. This queering orientation troubles assumptions and incites subversive reworkings of STEM education equity and inclusion projects that reproduce normative LGBTQQI+ identity politics. Which *isn't* to say that queering versus identity politics is an unambiguous binary (i.e., queering/identity politics) that permits clear choices between opposites.

Queer theorizing deconstructs normalized and naturalized asymmetrical formations like male/female, man/woman, masculine/feminine, heterosexual/homosexual, straight/gay, cis/trans,[13] innate/chosen, normal/strange, natural/unnatural, mature/immature, knowledge/ignorance, thinking/feeling, culture/nature, science/nonscience, and so on. Disrupting normative binaries undermines claims to stable identities and other demarcations that carry inequalities baked into binary oppositions. Those pursuing queer theorizing can get caught up in dualistic thinking about binaries, contrasting the queering deconstruction of identity with identitarian perspectives (including gay, lesbian, bisexual, trans, and others) that speak for individuals and groups that have been disregarded and silenced by heteronormativity and other dominant discourses. Rather than simply thinking of binary thinking as bad, it seems our relationship with it must be ambivalent and ambiguous, a sort of nondualistic abiding with dualism.[14]

This ambiguity renders queering paradoxically alluring and confounding to those of us who want to disrupt normalizations around sexes, sexualities, genders, and desires while advancing social justice concerns that are commonly framed around tidy identity categories that express the very normalizations we are trying to move beyond. A good deal of queering work in education, including work in this volume, is entangled in and exploring ways to productively negotiate this potentially generative mess. For example, in *Queer Inclusion in Teacher Education: Bridging Theory, Research, and Practice*, Olivia Murray (2015) writes, "to be *queer* is to have a sexual orientation, gender identity, and/or gender expression different from that acceptable in mainstream society" (p. xii, original italics). Being "inclusive of queerness" means that educators and institutions address "*queer issues*," which are "points of interest surrounding the claimed identities, experienced oppression, and resistances enacted by queer people and straight allies" (p. xiii, original italics). In this framing, to queer a practice or institution means to meaningfully and transformatively address "queer issues."

Murray's definition of *queer* in some ways resembles the positionality-not-positivity vision of queer identity advocated by David Halperin (1995), which we embrace in the next section. For example, as what is acceptable in mainstream society shifts, so too might Murray's *queer* fluidly mark new ways of being *unacceptable.* We suggest that the persistence of the unacceptable is acceptable, in fact, desirable, because we seek marginal positions from which to trouble and subversively rework what is normally acceptable. But in conventionally successful queer inclusion projects, wouldn't queer (as that which is unacceptable) disappear by assimilation into the acceptable? Further, as Murray elaborates "queer inclusion," *queer* functions as a catchall for people identifying (and being placed by institutions) as LGBTQQI+. In contrast, our interest in this volume is to cultivate productive marginalities that respond to normalizing shifts in what is acceptable by seeking virtuous possibilities that are, in conventional terms, less-than-acceptable ways of knowing, feeling, and being in STEM education. We are trying to resist the normalizing pull in/on STEM education of indelible, knowable, institutionalized identities.

The queering we wish to elaborate in *STEM of Desire* isn't limited to concerns that are the primary providence of queer people and straight allies. Our view (articulated by many before us) is that when *queer* is used to collect disparate arrays of people and issues, the tendency is to (a) reduce the heterogeneous collective to a dominant and homogenizing core, thus "obliterate[ing] the situatedness of individuals" (Morris, 1998, p. 276), and/or (b) encourage flight from the overly broad category of queer by proliferating more finely drawn and *exclusive* categories that harden into identities based on claims to stable qualities of being (hence the "+" in LGBTQQI+). We imagine queering to mark productive, critical practices that may precede and exceed being queer or a queer. Thus, *STEM of Desire* attempts to turn from queer as noun, and even as adjective (though this is often irresistible), toward queer as verb, recognizing that queer the noun remains just to the side, still in/at play, but not central.

In *Educators Queering Academia: Critical Memoirs*, sj Miller and Nelson Rodriguez (2016) use queer as a marker for LGBTI+ categories *and* for postmodern, queer theoretical critiques that strongly challenge stable identity categories and claims, including those marked by various versions of the moniker LGBTQQI+. "We recognize," they write, "that embedded in this usage are theoretical and political tensions between identitarian and non-identitarian approaches to thinking about gender/sexual identities and categories and their intersections with other categories of social difference" (p. xv). They accept and leverage these tensions to bring into conversation "a broad range of identities, experiences, and approaches to teaching and research that ultimately aims to interrogate and disrupt what is perceived as (hetero and cis)normative structures within academia" and, we add, broader cultures (p. xvi).

We recognize similar tensions in our past and continuing struggles to imagine *queer* inclusively as marking lives and standpoints from which critique might arise without homogenizing and essentializing the undisciplined diversity we are attempting to (re)cognize and embody.[15] We all do identity work. Identity is a life craft, a worthy, troublesome, and impossible effort to negotiate sense-making about one's self in light of one's and others' excessive being(s) and becoming(s). The point isn't to deny identity but to practice becoming selves with curiosity, creativity, wonder, and humility by accepting identities as convenient (and inconvenient) stories enfolding mysterious realities of human being(s)—stories and mysterious realities that require intentional practices to unpack, engage, revise, and release. In this spirit we appreciate Miller and Rodriguez's mingling of queer as identity with queering as critical practice:

> Our use of queer embraces the freedom to move beyond, between, or even away from, yet even to later return to (myriad) identity categories that support the perpetual reinvention of the self within the context and concerns of an antihegemonic queer politics and praxis. (p. xvi)

To further unpack these tensions, we briefly consider the difficult familial relations of queering as critical practice and the identity politics often associated with LGBTQQI+ equity and inclusion projects in education. We follow Dennis Carlson's (1998) helpful account of the relations of gay and lesbian identity politics and queer theories.

In the United States, gay and lesbian movements, like black civil rights and women's movements, have used identity politics to counter histories of being stereotyped, pathologized, and silenced. Identity politics pursues recognition, renaming, and re-representation for people who claim shared histories, cultures, and concerns that have been denied voice and self-representation. This politics has powerfully catalyzed certain kinds of social change. Unfortunately, identity politics easily slides into essentialism and fundamentalism that, for example, explicitly or tacitly ground "being gay" in genetics or other unspecified inborne qualities while discounting historical contingencies and embodied sociocultural realities (Allen, 1997). Equity and inclusion projects pursued through liberal identity politics take up identity categories in ways that risk reducing open-ended selves to fixed, knowable, normalized, and naturalized subjects.

While identity politics (pro)claims the ontologies of being(s), queer theories seek to deconstruct identities by tracing the trajectories of social epistemologies and phenomenologies (Ahmed, 2006b). We recognize this facile dualism, so we proceed with attention to Susan Talburt's (2000) insight that queer theories are constitutively haunted by identity. *Queer* carries identity in several guises: as a "denunciative heterosexualizing" slur (Honeychurch, 1996, p. 340), as a subversively repurposed identity label claimed by queer activists, and as a catchall synonym claimed by some and imposed upon others in the collective LGBT + other non-cis-heterosexuals. Even in our preferred formulation as a verb, *queering* leaves the mark of being a queer, like gardening makes you a gardener.[16] Making our way in the world, we all take up and are pressed into multiple, shifting, and contradictory subject positions. For example, in our own (Steve and Will's) modest efforts to undermine normalized identities, we fluidly take on and/or are ascribed (perhaps by you, right now) identities as those who wish, perhaps unwisely, to trouble the notion that people mature into identities with which they are born. Paradoxically, and yet necessarily, queering practices that unsettle identities inscribe identities back upon those who queer.

And so queering doesn't end identity work—in fact, Susanne Luhmann calls for the "infinite proliferation of new identifications"! (1998, p. 151)—but inhabits identifications as contingent, contested, storied, and temporary, even fleeting, rather than inborn essences or even stable, self-regulated productions. As a practice of self, queering pays attention to how we relate to subject positions; how we hold, refuse, confuse, and reuse them; and whether we firmly attach self to identifications or, instead, attempt to hold identifications lightly and skillfully in embodied awareness as stories that simultaneously serve and ensnare (Loy, 2010). If identity politics soothes existential angst, queer theorizing construes and celebrates personhood as "a contradictory and complex phenomenon" (Morris, 2000, p. 15). Queer theorizing, therefore, offers little on which to ground identity-based STEM equity and inclusion projects as they are *normally* conceived.

Normativity[17]

> One of the central tenets of queer theory is that all sexualities, indeed all attempts at normativity, are bound to fail because achieving normal is an impossible task. (Mayo, 2007, p. 183)

> "Queer" does not name some natural kind or refer to some determinate object; it acquires its meaning from its oppositional relation to the norm. Queer is by definition *whatever* is at odds with the normal, the legitimate, the dominant. *There is nothing in particular to which it necessarily refers.* It is an identity without an essence. (Halperin, 1995, p. 62, original italics)

> *Queer* can … be considered the intersection of queer as verb and queer as adjective: giving an account of queer highlights the radical process of reconsidering identity politics (verb) so as to describe and rearticulate the self as an ongoing negotiation working through the politics of identity (adjective). (Ruffolo, 2007, p. 257, original italics)

Early on in queer theorizing, Halperin (1995) reframed queer in a way that helps us turn from *queer* as identity to *queering* as critique. "'Queer,' …" he writes, "demarcates not a positivity but a positionality vis-à-vis the normative" (p. 62).

Halperin extends Foucault to imagine a "de-essentialized" (p. 62) queer identity as positionality by noting how the figure of "the homosexual" as developed in the West starting in the late 19th century is "an impossibly contradictory creature, not a natural reality but a fantasmatic projection" that "consolidate[s] the cultural meaning of heterosexuality by encapsulating everything that is 'other' than or 'different' from it" (p. 61). In this heteronormative and homophobic discourse, "'the homosexual' is an identity without an essence" (p. 61). In terms of liberal, humanist identity politics, this is a dehumanizing evacuation of substance. But through a subversive logic of *reverse discourse*,[18] in which liberationists appropriate and repurpose authoritative discourses to speak of and for themselves, the "contradictory and incoherent definitional content" of the homosexual is an opportunity to shift the terms of identity from substantial positivity to relational positionality (p. 61). For Halperin, "those who knowingly occupy such a marginal location, who assume a de-essentialized identity that is purely positional in character, are properly speaking not gay but *queer*" (p. 61, original italics).

Positional queer identities don't define the substance of particular kinds of people, instead they mark shifting positions, which many "kinds" of people might take up, of opposition and resistance to norms. Although beginning with the problem of "the homosexual," by shifting queer identity from positivity to positionality, Halperin radically and indeterminately extends the relational scope of and possibilities for queerness and queering. This "positionality … is not restricted to lesbians and gay men but is in fact available to anyone who is or feels marginalized because of her or his sexual practices: it could include some [straight] married couples without children, for example, or even (who knows?) some married couples *with* children—with, perhaps, *very naughty* children" (p. 62, original italics).[19]

In the spirit of our previous discussion of sexuality as an original urge to know, and of desire as productive flows, we read Halperin's "sexual practices" broadly as urge(nt) and agential ways of knowing, feeling, and becoming. Queer identities are relational positions for queering practices of opposition and resistance to norms. But queering isn't simply antinormative and anti-identitarian. Without norms to queer, there would be no queering. Don't imagine queer identities and queering as seeking positions distant from, or *merely* in opposition to, norms and normalizing processes. "Resistance to normativity is not purely negative or reactive or destructive, … it is also positive and dynamic and creative" (p. 66). Queering is about strangeness that is proximal, not so distant as to be utterly inconceivable and of no sensible concern to normativity. Strangers and monsters are unsettling and intriguing because there is something familiar about them that is out of place, out of bounds but in sight. Queerness and normality, queering and normalization are in this sense intimate companions and mutually enabling conditions.

For instance, in the United States and elsewhere, recent advances in civil rights for gays and lesbians, and to lesser degrees for other sexual and gender minorities, have proceeded by normalizing proximal segments of the LGBT Other and admitting them to modestly revised versions of the hetero norm. "The norm and its negated other" writes Luhmann, "are implicated and mutually constitutive of each other" (1998, p. 151). The coconstitutive nature of the normal and abnormal gives nonheterosexual identities a foot in the door, so to speak. Because "non-straight sexualities are simultaneously marginal and central" (Luhmann, 1998, p. 144), to marginalize the Other, heteronormativity must (to its considerable discomfort and perverse fascination) hold the Other close as an "indispensable interior exclusion" (Fuss, 1991, p. 3).

Interior exclusions are necessary coconstituents of (neo)liberal cis-hetero norms. Even as some former Others move into mainstream society, transformed from interior exclusions to comfortably interior inclusions, other sexual and gender minorities, as conjured via hetero and homo normalizations, remain to serve as interior exclusions. In the meantime, proximal segments of gay and lesbian and trans populations that are, accurately or not, perceived to share the educational, socioeconomic, ethnic, "racial," and family values of the cis-hetero majority increasingly enjoy the fruits of being (more or less) normal in relation to their straight neighbors.[20] We suggest that, as in broader society, this is how certain normally excluded and silenced ways of knowing, feeling, and being have begun to enter some curricula and classrooms. If LGBTQQI+ equity projects intend to include more ways of knowing, feeling, and being within modestly refigured cis-hetero norms, then assimilationist paths may address these concerns without taking on the added complexities and contradictions of queer theorizing. But in *STEM of Desire*, we seek something more complex and contradictory from the strange relations of normativity and queering.

In his introduction to queer theories, Donald Hall (2003) notes:

> 'Queer' is not chaos; it is always implicated within and impacted by surrounding definitions, norms, and rules of engagement. Yet it—*we*—can certainly question and abrade those definitions, norms, and rules forthrightly and with overt (and covert) political intent. (p. 148, original italics)

Not only does queering arise in the *context* of the norms it abrades, it infiltrates and reappropriates the very *text* of norms, enacting "a simultaneous tactical deployment of *and* critical engagement with the 'rules' of conventional behavior" (p. 7, original italics). With gratitude to Audre Lorde for spinning a rich trope we further twist, queering is, in part, about taking up and—sometimes subtly, sometimes radically—repurposing the master's tools.[21]

David Ruffolo (2007) elaborates a vision of working with and reworking the complex and intimate relations of normativity and queering. He begins by explicitly calling out the reorientation in queer theory from *identities* to *identifications*. Identities are "fixed and stable articulations of the collective 'I'," which are at best idealizations—that is, impossibilities (p. 256). Identifications are "mobile and fluid negotiations among momentary subjects," which emphasize the temporary position of social subjects whose subjectivities are continually in process (p. 256). This "shift from *identities* to *identifications* highlights the negotiation of differences, rather than similarities, among subjects" (p. 255, original italics). Ruffolo echoes Dennis

Sumara and Brent Davis (1998) when they write, "queer theory asks that the form of curriculum and the relations of pedagogy be appropriated as spaces to interpret the minutiae of differences among persons, not merely among categories of persons" (p. 198).

Ruffolo integrates theorizing from Foucault (*disciplinary power*), Althusser (*interpellation*), and Butler (*accounts*) on subjectivation: the production of socially intelligible subjects (and so also unintelligible subjects). Arising in the cultural interface of what we conventionally understand as individuals and society, intelligible subjects form through the operation of disciplinary power and speech acts that hail subjects into being and demand that proper subjects properly account for themselves. Queer theorists have argued that normative discourses of subjectivation express ideologies that (re)produce binary and unequal subject positions like male/female, masculine/feminine, and heterosexual/homosexual. When subjects fail to properly reiterate dominant disciplinary norms, they risk being deemed abnormal, the other, or unintelligible. But we have already seen how the exclusionary logic of binaries is defeated by their own internal constitution: The dominant and subordinate terms of binaries are mutually constitutive, compelled to coexist too close for comfort.[22]

Further, the processes of subjectivation are reciprocal and indeterminant. The positions of subjects are at stake in their own reiteration of cultural norms through which subjects are constituted; reciprocally, the fate of cultural norms rests on their articulation and (re)production by imperfect subjects. The open secret of subjectivation is that despite the robust cultural reproduction of sexes, sexualities, genders, "races," classes, (dis)abilities, desires, etc., the "proper" kinds and categories of subjects that normally seem ontologically secure are in fact at risk and in motion in each and every moment of subjective be(com)ing. In the reiteration of disciplinary power, by hailing and responding to the calls of others, subjects acquire and exercise agency (whether strategic or haphazard) through inattention, error, resistance, curiosity, improvisation, wandering eros, and more that rearticulate the normative discourses from which those subjects arise. The messy subjects of subjectivation are at once reliable and capricious, replicative and creative, ensnared and free.

Ruffolo's (2007) discussion of queering straight teachers illustrates wider, radical possibilities in the relations of queering and normativity. To be intelligible, straight teachers (like all proper subjects) must account for themselves "as *coherent* 'identities'" that, if successful, *appear to be* "fixed and stable articulations of binary associations" (p. 268, original italics). To counter this normalizing intelligibility, straight teachers might queer themselves—what Ruffolo calls "giving an account of queer"—not by becoming queers, but by making the implicated, contingent figure of the *straight teacher* "queerly intelligible" (p. 268). This shift isn't from one apparently "stable signifier of normative discourses" (p. 268) (i.e., *straight*) to another (i.e., *queer*), but from reiterating the normalizing conceits of identity politics to revealing and appropriating the politics of identity "by becoming radical subjects that re/negotiate new 'norms' that are unstable, unfixed, and indefinable" (p. 269). Ruffolo offers a curious image:

> Giving an account of queer involves momentary commitments that insert queer ideologies into the crevices of normative discourses so as to rework a "new" (queer) disruptive norm that can be reproduced as a mobile and fluid identification—an identification that can never be fully knowable or completely coherent. (pp. 269–270)

This subversive appropriation of the mechanisms of subjectivation evokes relations of queering and normativity as viral, parasitic—or perhaps playfully naughty symbionts. Queering is indeed naughty—its satisfactions can be discomforting to any and all involved, including those doing the queering. We don't queer to end normalization once and for all but to subversively leverage normativity's robust (un)reliability and perturb it into new possibilities. Utopian (or dystopian) thoughts of ending normalization and the hierarchies it produces disregard the radical potential of the here and now. From contingent positions of intimate marginality, queering momentarily but persistently turns normalization back upon itself, exposes its operation, disrupts its ideologies, and elicits strange possibilities that through abjection and assimilation recondition normativity and reshape (im)possibilities for further queering engagements. Situationally co-opting the rules of normativity to expose, rework, and exceed normativity "represents queer activity at some of its most sophisticated and provocative" (Hall, 2003, p. 7).

LITERATURE MUD MAP[23]

Published work on queering STEM education has appeared sporadically since the late 1990s. We next sketch a mud map of what's been taken up in the literature to date to provisionally establish some useful points of reference and to place in relief possibilities for imagination, invention, and discovery.[24]

Early work considered how science education curricula, pedagogies, and policies might be re-viewed using queer theoretical lenses to reveal how heteronormativity is manifest in and through them. For instance, Will Letts (1999) considered how heteronormativity was apparent in elementary school science, becoming part of the hidden curriculum of the subject. Using examples of dialogue from classrooms and excerpts from curriculum materials and policy documents, he excavated how heteronormativity is discernable in these scenarios and the effects it can have if left unnamed and unexamined. Letts continued this work in 2001 by examining how seductive school science can be in attracting certain pupils and teachers (and not others) to science. Again drawing from the realms of curriculum, pedagogy, and policy, Letts advocated that "not only can a variety of texts be examined and deconstructed, but the activities and experiments that constitute the 'doing' of science can also be interrogated as part of the science lessons" (p. 270). Without such sensibilities and awareness, Letts concluded, "our science lessons are partial, science can unwittingly continue to seduce us, and we perpetuate a view of science that normalises … hegemonic heterosexual masculinity" (p. 271).

Some early work, while not explicitly queering, suggests things to come. For example, Anne-Marie Scholer's (2002) examination of the content of her university course in anatomy and physiology for students in nursing, athletic training, and physical therapy is a close deconstruction of how norms of heterosexuality can largely invisibly shape what we teach if not actively mitigated. Her article interrogated her course and her teaching of it, noting that "many of [my students] thank me for my 'stories', which help them remember concepts" (p. 83) by providing meaning and context to the facts. Scholer worked hard to situate reproduction in a much broader context of sexuality and paid special attention to female sexuality, nonheterosexuality, and the sexuality of older adults—all topics underrepresented or not even covered in the textbook she used.

Other early work examined heteronormativity in secondary school science classrooms and teacher education. For example, Steve Fifield and Howard (Lee) Swain (2002) exposed the mash-up of heteronormativity in/as biological content in learners' minds and bodies when high school students titter over terms like *phagocytosis* and *homozygous*, or protest that students of the same sex should not be paired in a classroom simulation of Mendelian inheritance. This study explored how the sense that Lee made of himself as a gay student teacher was shaped by and re-shaped his feelings for and understandings of the ontological and epistemological status of scientific knowledge. Lee became acutely aware that important realities of his life as a gay person were utterly absent from the biology content he had learned and was learning to teach. He came to believe that he could no longer trust science and the science curriculum to offer unbiased accounts of the facts of life. This article used the concept of curriculum companion meanings (see Roberts & Ostman, 1998) to illuminate "how heteronormativity is communicated in a curriculum, and how a queered science curriculum might foster more critical and humane understandings of science and self" (p. 183).

Science textbooks have been a favored object for antiheterosexist and/or queering analysis. Although not explicitly framed as queering, Michael Reiss's (1998) early study of the representation of sexuality in a selection of 15 science textbooks for 14–16-year-olds in the United Kingdom rates the books on areas of sex education to do with human sexuality, including anatomy, puberty, contraception, sexual identity, and intercourse. He found tremendous diversity across the texts, demonstrating that some are "sexist, fail to tackle personal issues to do with menstruation, ignore lesbian and gay issues and either omit or make a mess of cultural issues" (p. 147), and noting that "the 'sins' of th[e]se books are mostly those of omission rather than commission" (p. 147). Reiss maintained that although "no textbook is perfect," "justice demands that such inequalities are removed" (p. 148) so that textbooks offer more complete and accurate sex education.

Other studies of biology textbooks explicitly brought queer theorizing, variously interpreted, to bear on the science curriculum. Vicky Snyder and Francis Broadway (2004) examined eight widely adopted secondary school biology textbooks for their treatment of sexualities outside the heterosexual norm and found, perhaps not surprisingly, a profound silence. Just three of the eight books referred to sexuality outside the straight norm, and those three mentioned homosexuality only in relation to AIDS. These are the curricular silences around sexual diversity in relation to human genetics, behavior, and reproduction that the student teacher Lee (in Fifield & Swain, 2002) said made him feel excluded from the biology in school. Snyder and Broadway conceived of queering textbooks as "pointing out the silence and absence of sexuality outside of the heterosexual norm" (p. 629). This queering of absence reflects an orientation toward bringing in outsiders "who do not 'fit' into the heterosexual norm" (Snyder & Broadway, 2004, p. 618), an approach Broadway (2011, see below) would later question.

Jesse Bazzul and Heather Sykes (2011) used queer theories to analyze how a high school biology textbook widely used in Ontario addressed sexuality and gender. Like Snyder and Broadway (2004), Bazzul and Sykes noted the absence of any treatment of sexualities outside the heterosexual norm. They further pushed queering beyond marking silences by identifying heteronormativity in discussions of behavior and hormones, and in terminology that conflated sex and gender within a heteronormative framework. In a response to Bazzul and Sykes, Jay Lemke (2011) took science and science education to task for failing to live up to epistemic virtues those disciplines use to claim special privilege as producers and stewards of knowledge, minds, and bodies:

> How dare a textbook or a curriculum that claims to be both scientific and educational represent as biological fact that there are only two sexes, perfectly aligned with only two genders, which are each biologically determined to be only one or the other, and biologically determined to find one another sexually attractive, whether with reproductive results or not? (p. 288)

He noted how the nuance and complexity of how people become sexed and gendered is unaccounted for in the biology textbook in Bazzul and Sykes's investigation. Lemke's call for science and science education to live up to their professed standards of objectivity affects a paradoxical queering by demanding that scientists and educators not rest comfortably in a myth of objectivity, but that they critically account for the broader cultural beliefs that necessarily empower, advance, limit, mislead, and bias scientific knowledge. In this volume Reiss (Chapter 13) extends this call to teach better science about human sexuality.

Broadway's (2011) enrichingly difficult response to Bazzul and Sykes (2011) is insistent in bringing us back to queering curriculum, queering pedagogy and science, and queering queering. He suggested that in Bazzul and Sykes, and in Snyder and Broadway (2004), queering seems to mean adding queer (noun) identities to institutionally recognized categories of diversity, and adding queer (noun) content to already overstuffed textbooks. By 2011, including "'queer' in the hegemony" (p. 294) was not what Broadway had in mind for queering. Against the evident impulse to queer (verb) just enough to normalize queers (noun), Broadway imagined a queered curriculum as storied journeys of self, other, and the world that arc from the familiar to the strange. Queer pedagogies propel storied lives in relations of learning that put at risk illusions of a unitary, sovereign self. Broadway's queering memoir in Chapter 1 of this volume draws readers along just such an arc from familiar to strange.

Malin Ah-King's (2013) analysis of five Swedish secondary school biology textbooks in relation to animal sexual behavior found that "the textbooks offer dichotomous descriptions of males and females, and they are heteronormative in that they all describe sexual behavior in only the context of opposite-sex interactions and reproduction" (p. 75). It seems not much has changed since the first queering critiques of biology textbooks.

In terms of pedagogy and iconic manifestations of science curriculum through pedagogy, Sara Tolbert (forthcoming) notes that "dissection is often seen as a cornerstone of life sciences education, particularly in secondary school settings" (p. 2). She mobilizes queer and gender theories to illustrate how "students orient, or re/orient, to science through their teacher, their own experiences, and through the material/discursive context of school science activities" (p. 10), in this instance by engaging in dissection of preserved specimens. Tolbert demonstrates how these theoretical lenses allow us to expose and probe taken-for-granted practices and their attendant, presupposed value propositions.

When it comes to queering, K. Rands (2009b) characterizes mathematics as "the subject that dare not speak its name" (p. 182). While most of STEM remains unexplored from a queering gaze, we concur with Rands that a sacred hierarchy sets mathematics aloft as the language of nature read directly (though not without ingenuity) from the trustworthy source. (Physics, of course, fancies itself close to this mystical ideal, with other sciences in turn striving to ascend.) Rands (2009b, 2016) proposed the notion of mathematical inqu[ee]ry (later mathematical inqueery) "to deconstruct and disrupt educational norms as well as imagine new possibilities in mathematics and in the world" (2009b, p. 183). Mathematical inqu[ee]ry entails "questioning the tasks, the strategies, the very ways of thinking and doing mathematics, as well as the way mathematics is used to interpret and act in the world" (2009b, p. 186). In this volume James Sheldon (Chapter 16) brings mathematical inqu[ee]ry to infinity.

Some work has taken a holistic view of queer theorizing in and about STEM education. For example, Gunckel (2009) wrote about how "science education serves as an example site for applying queer theory and asking all sorts of questions about schools, identities, knowledge, teaching, and learning" (p. 68) in making the compelling case that "science education for a liberal democracy requires queer Science for All" (p. 72). As conceptualized by Gunckel,

> queering science education requires that science educators and researchers, teachers, and students examine how science is (hetero)sexualized, and how this helps or hinders students whose identities are constructed outside the normalized territory. ... Queering science education means inquiring into the nature of science and science knowledge, disrupting the heteronormative

structures, and opening spaces for exploring new identities and knowledge possibilities. (pp. 70, 71)

Reminiscent of earlier studies of science classroom practice, Mattias Lundin (2014) wrote about sexuality education in Sweden by examining syllabus documents and sexuality education classroom lessons for 14-year-olds. Wishing to foster science education that "engages and relates to all students" (p. 380), Lundin examined how teacher and student classroom talk "established and questioned ideas related to the hetero norm, identified the sexualisation of pedagogy and documented challenges to the hetero norm" (p. 382). We (Fifield & Letts, 2014) wrote about Lundin's article, appreciative of the ways the study traced hetero norms in the classroom and curriculum but also mindful of the reorientations entailed in bringing queer theories to bear on science education. As such we used Lundin's piece to (re)consider the relationship between queer theory and gay and lesbian studies; the border surveillance that characterizes so much of science education; the alluring call of binaries and binary thinking; the "all" in the catchcry "science for all"; and the need to queeringly engage the fullness of science in the curriculum. Our response to Lundin in turn informed the call for chapters for *STEM of Desire*.

Kerrita Mayfield (2016) practices queering as a biology teacher and teacher educator as "a method of inquiry and mode of analytical pushback against dominant constructions of legitimated curricular knowledge" (p. 121). Mayfield understands at a "cellular level" that her presence as a "tall, curvy, unabashedly brown woman, with a lush, multicolored nimbus of natural hair, a voluminous laugh, [and] an affinity for bright colors … queers narratives of power" (p. 115), but the queering "emancipation" she seeks "does not rely upon the presence and uplift of invisibilized groups" (p. 122). Instead, she suggests, through queering inquiries all students might locate their several identities in kyriarchal webs of privilege and oppression where possibilities for their own and others' bodies and knowledge coarise in/as science curriculum.

Several publications on queering environmental education have appeared, irregularly, in the last 15 years. Constance Russell, Tema Sarick, and Jackie Kennelly's (2002) prospectus argued that "queer pedagogy's insightful probing of the perils of essentialism" (p. 61) can sharpen the critical edge of environmental education's concerns around sexism and homophobia by interrogating identity, naturalization, and heteronormativity. Noel Gough et al. (2003) pressed the queering imagination in critical and fanciful directions. They conjured Camp Wilde as a "conjectural [and paradoxical] space" to "perform a queer(y)ing of environmental education research" (pp. 46, 45). Their undisciplined performance/disruption of the content, form, and identities of environmental education research(ers) is delightfully provoking—and so struck some reviewers as "unclear and confusing" (p. 61).

Perhaps reviewers weren't the only ones struggling to see the point. A decade passed before another major statement on queering environmental education, Joshua Russell's (2013) proposal for a queer ecopedagogy. If pedagogy orients subjects toward objects of inquiry and action, and ecopedagogies integrate social justice and ecological concerns, then J. Russell's queer ecopedagogies embrace a queer phenomenology (see Ahmed, 2006b) of bodily orientations toward subjects, objects, thoughts, and actions beyond conventions of "normal" and "natural." Two self-proclaimed queering articles appeared in a 2017 special issue of *The Journal of Environmental Education* on gender and environmental education. Jesse Bazzul and Nicholas Santavicca (2017) modeled how environmental educators and students can work with and rework intra-actions that entangle sex/gender and sexuality in/as material-discursive-affective things of concern using diagrams of Deleuzo-Guattarian assemblages. Diagramming assemblages is a productive (and so ethically weighty) practice of "critical engagement with unstable categories and the co-emergent nature of our shared world(s)" (p. 64). Chessa Adsit-Morris and Noel Gough (2017) stay with things Deleuzian "to touch, to invite, to dance with" possible lingering essentialisms in phenomenology taken up for queering purposes in environmental education and elsewhere (p. 72). Performing with and from difference, they "dance [queer tango, specifically] around the edges of phenomenology" (p. 69), "blurring … the boundaries of the essential(ized) body," (and, we add, the singularly embodied mind) to "grapple with the partial and processual becoming of our bodies-in-relation" (p. 73). Speaking of relations, there hasn't been much sharing among queering projects in environmental and STEM education. We trust the writing on environmental education in this volume (i.e., flessas & Zimmerman, Chapter 3; Santavicca, Bazzul, & Witzig, Chapter 15) will help to stimulate generous exchanges, less to get everyone on the same page than to discover, create, and work with generative differences.

As mentioned above, most of STEM education is unvisited by queering critiques. A good deal of (potentially) related work concerns the experiences of gender and sexual minorities, without pursuing queering as construed in this collection. Several studies have focused on LGBTQQI+ students and student retention in STEM disciplines at secondary (e.g., Gottfried, Estrada, & Sublett, 2015) and tertiary (e.g., Hughes, 2018) levels. Similarly, research in medicine (e.g., Hollenbach, Eckstrand, & Dreger, 2014), nursing (e.g., Felsenstein, 2018; Goldberg, Ryan, & Sawchyn, 2009), and engineering (e.g., Cech & Waidzunas, 2011; Hughes, 2017) has focused on attracting and retaining LGBTQQI+ students. There is also work examining how to prepare physicians to work competently with "individuals who are or may be lesbian, gay, bisexual, transgender …; gender nonconforming …; and/or born with differences in sex development" (Donald, DasGupta, Metzl, & Eckstrand, 2017, p. 345). Pressing beyond inclusion, in this volume Cecilia Rodéhn (Chapter 14) and Amy Slaton, Erin Cech, and Donna Riley (Chapter 17) bring queering analyses to medical and nursing education, and to engineering education, respectively.

THE CHAPTERS

The essays in this volume draw on queer theories to examine STEM education in schools and universities, museums and other informal learning environments, and everyday life. Subject areas include physical and life sciences, engineering, mathematics, nursing and medicine, environmental education, early childhood education, teacher education, and education standards. The chapters appear alphabetically by author. Our attempts to arrange this eclectic collection in topical sections seemed to obscure more connections than were revealed. So we trust the following aperçus will help you pick and choose among the chapters. Or take Leafgren and Sander's advice in Chapter 11: wander attentively with no predetermined destination.

We are delighted to open and close the collection with poems by *Charlotte Boulay* that speak of (im)possibilities for being(s) in worlds that include nature and technoscience. The poems speak of other things, too. We encourage you to revisit them as you come and go, pausing to note what, in the moment, emerges in play of the poetic and prosaic.

Francis Broadway begins with a proportional expression: (A is to B) as (C is to D). Linear. Latin school. SAT. But Broadway's terms are surprising, *non*linear. I : Snow Queen :: "Nigger" : (school) Science. We try tracking the expression from beginning to end but must stop, back up, step aside to wonder and worry, and try again. This analogical figure births syllogisms, surface relations that suggest deeper realms. Broadway cautions: "There is no vocabulary that denotes the science of which I speak."[25] So the realms aren't simply known; they are ghostly, and very real—understanding here is a presence, a felt sense. This is a warm-bodied memoir on intersectionality and transmutation. It's about a boy, descended from enslaved Americans, who loves science and literature and white men; about a science fair project on index cards in a file box; about a big black Snow Queen who sometimes worries passers-by and colleagues; and it's about salukis. These bodies and more—analogical expressions. (School) Science, it seems, offers certitudes not available in the Self, until troubling logics confound Self and Science as fantastic uncertainties that needn't be resolved to be abundantly lived and evocatively shared. Read and consider: What are the terms of your expression?

We (*Steve Fifield* and *Will Letts*) remember 20 years of intermittently working together and apart on intersections of queer theorizing and science education. We narrate touchpoints from this work: palpable zones of contact, exchange, and change among ourselves, others, science education, and queer theorizing. We are cautious of sense-making. Our retrospections fidget between autobiographical impulses for "a singularity of [self and] story," and queerer virtues of "multiple, conflicting, and even odd and abnormal … stories and identities" (Miller, 1998, p. 369). We trace (re)orientations in our work: from queer identity to queering standpoints and positionalities, from visibility to epistemology, from LGBTQ inclusion to entangling pedagogies of desire. Plenty of nonsense remains. Unaccountably, we fancy science education. "But the object of our longing is even more distributed and bizarre than science (education)," in ways we feel but can't explain—which makes sense, in another sense.

blake flessas and *Timothy Zimmerman* combine critical and queer theories to reimagine environmental education that "leads to liberation, reparation, and emancipation for all of nature, including all human beings." They integrate micro- and macro-level perspectives on social justice and environmental harm to reclaim environmental education's radical roots. The authors decry the field's mainstream inclination to adopt depoliticized ideologies and to neglect macro issues like, say—capitalism! Early visions of queering environmental education carry some of this assimilationist legacy in calls for LGBT normalization and inclusion. And although *ecopedagogy* (Kahn, 2010) uses critical theories to address macro issues, flessas and Zimmerman argue that it doesn't adequately consider individual- and group-level relations of humans and the environment like queer theorizing can support. Taking an important step, *queer ecopedagogy* (Russell, 2013) robustly queers—(dis/re)orients—humans in/and nature and eschews assimilationist identity (and educational) politics. flessas and Zimmerman now want to join queer ecopedagogy's robust queering of subjects and objects with critical pedagogy's macro perspectives on power and knowledge. This provocative revisioning concludes with design principles and practices that draw complements together—micro with macro, queer with critical, theory with practice—to craft *critical queer environmental pedagogies*.

Andrew Gilbert and *Emily Gray* use wonder and an affective turn in queer theorizing to queer the "discursive narratives/practices/identities associated with [school-based science education] and to challenge science to enact the ideals it claims to promote regarding wonder." "Wonder," writes Jeffrey J. Cohen (2013), "is irruption and eruption, outside coming in and inside moving out at once, altering immediately the world's seeming placidity. To wonder actively is to open oneself to the world made strange, to the world's queerness" (p. 153). We hear a lot about *curiosity* in science education. By *wonder* Gilbert and Gray mean something else. Curiosity is cognitive, its trajectory toward knowledge, understanding, resolution. Wonder is aesthetic, emotional—it arcs toward sensibilities of beauty, awe, humility, wisdom. This distinction, though too tidy, helps us appreciate the fine particularities that manifest in/as science (education) when the fruits of curiosity deepen wonder, when awe accents understanding, and when empirical inquiry meets eros. Lived passionately, in fullness, science (education) is intimate(ly) in nature. Neoliberal agendas, in contrast, pursue passions that enroll science (education) in contests of economic fitness. Taking a stand, Gilbert and Gray call us to science (education) more wonderful than merely "seeking information and data when what is on offer is communication and entanglement" (Haraway, 2008, pp. xxv–xxvi).

Helene Götschel's "teaching queering physics" is about teaching students "to critically question and transcend norms around gender, sexuality, desire, narratives, cultures, facts, and other powerful social categories that shape hegemonic presentations of physics and physicists." In her university course on classical mechanics for first-semester engineering students, Götschel draws on gender, queer, and new materialism theories to develop "competencies of queering" as integral to scientific literacy for students and future practitioners of science and engineering. In this chapter she illustrates teaching that queers physics narratives, culture, knowledge, and representations. Her lectures bring a critical edge to accounts of the "underrepresented or silenced in the history of physics" by "using these stories to reflect on the processes and mechanisms of othering" in physics. She is encouraged by physics workplace diversity initiatives but wishes they embraced LGBT+ and other underrepresented physicists more as "enriching source[s] of diversity for scientific development" than as "vulnerable victims." Turning to queering physics knowledge, the "entanglement of matter and meaning" in quantum physics is now well-rehearsed (if not widely appreciated; see Barad, 2007) but remains little studied in other areas of physics. This makes further work by Götschel and others even more beguiling! Pedagogical representations of physics typically reinforce, and so can be reoriented to forcefully queer, physics knowledge and culture. For instance, Götschel pricks the thin skin of heteronormativity and sexism by using a video of a male figure skater to illustrate angular momentum, which reliably elicits squirmy giggling from many male students in her lectures. Götschel's pedagogies press students to observe, investigate, and own their entanglements in physics culture and knowledge as performative, material-discursive practices.

Kristin Gunckel's colleagues and graduate students have often asked her, What does queer theory have to do with teaching science in elementary schools? Gunckel's thoughtful response addresses queering science education standards, science teaching and learning, and teacher preparation. She begins by clarifying: Queering isn't only about LGBTQ-identified people or the worthy goal of LGBTQ inclusion, nor about bringing sex to elementary classrooms. Instead, "to queer something is to ask what it is not, to disrupt normal constructions, and to release the possibilities that normal constructions hide." For example, in the United States, *A Framework for K–12 Science Education* (*Framework*) and the *Next Generation Science Standards* (NGSS) weave together core disciplinary concepts, crosscutting concepts, and scientific practices. Missing from this useful braid, says Gunckel, is the strand of desire: living sinews of body-minds as desirous cognitive agencies seeking "a feeling for the organism." Such pedagogies of desire in elementary school are discouraged by "notions of childhood innocence and ignorance [that] conspire to cocoon children in suffocating webs that isolate sexuality from learning." As we discussed earlier, this *sexuality* references impulses for embodied knowing. Pleasurable (not synonymous with "fun") science teaching and learning "satisfy a deep curiosity," not necessarily with ready-made answers but by visiting "uncharted territory where ambiguities, contradictions, and uncertainty exist." Rounding out her response, Gunckel considers elementary school teacher identities that hold loving science apart from loving children. Love needs fantasy (see Berlant, 2014), as do teaching and learning science. Reading Gunckel, we're struck by how queering elementary teacher preparation mobilizes fantasy—quite real and impactful—as a medium of identifications that might reconcile love of self, of children, and of science in classroom practices that nurture "visceral connections to phenomena."

Joe Heimlich takes us out of school and into lifelong, life-wide, life-deep STEM learning with a focus on informal learning environments.[26] Heimlich rouses us from default-mode learning to become critically conscious and purposeful about STEM-related learning across all aspects of our lives. He introduces the notion of *learningscapes*: "spaces of living through which the many pathways of an individual's experiences, memberships, identities, and activities run and are interpreted." These are the paths we knowingly or obliviously travel (which throw into relief those we don't travel) through potentially vast ecosystems of learning. Queering learningscapes "is an aspirational, intentional practice" of critical reflection and action on how we move through life as learners, a "dis-settling of the unconscious and bringing the self as a conscious, desiring learner through life to the experiences comprising the learningscape." Queering the learningscapes of LGBT people may be an especially fruitful way to reveal and counter heteronormative mediators of everyday STEM learning. Informal STEM learning institutions, says Heimlich, should offer all visitors experiences that incite and support queering learningscapes. This edgy proposition reorients typically risk-averse institutions and fun-loving visitors toward learningscapes less travelled, perhaps less comfortable, quite possibly more fun, and certainly more educative.

"How is being so smart as a woman gonna help you when Jesus comes back?" a hairdresser asked *Tommye Hutson*. The things people say in beauty shops! Teaching science for 35 years (22 in rural Texas), Hutson has stories to tell about naming, expectations, and fitting in, or not. Her tales trace the implicatedness of self, friends, family, school, community, science, sex, gender, sexuality, and more through quotidian scenes that, as we dwell in and upon them, reveal the *extra* in *ordinary*. She exposes, troubles, hopes, and, when possible, repairs. Hutson allows us to ponder the entanglements of an anxious grandfather of a young man: "Could it be that arsenic in them wells? What about it, Doc? Could that make 'em go gay?" And of Clay, an eighth grader who reads "with the same intensity that cedar trees burn—voraciously and tenaciously," whose father says it's time to leave school, to be a man. And of a two-year experiment with single-sex seventh-grade science classes—supposedly about improving academic achievement but really about scheduling weightlifting for football players. Hutson writes, "the single-sex classroom experience had given [students] opportunities to think about their identities … with fewer concerns about being proper girls and boys." It later seemed that more of those students than usual came out as lesbian or gay. Or was it arsenic in the water?

Kathryn Kirchgasler's comparative historical analysis of *interest in science* draws on contemporary U.S. standards documents (i.e., NGSS and *Framework*) and *General Science Quarterly*, a science education journal from the early 20th century. She deftly integrates history of science education, science studies, and queer theorizing to examine how interest in science "has repeatedly crystallized around notions of difference in sex and race."[27] In the early 20th century, differences in interest weren't directly measured but "inferred based on [explicit, evolutionary] theories of race and sex differences." Accordingly, white Anglo-Saxon Protestant males were suited to science at its most refined and abstract, while "the 'mongrel lot of pupils of all races'" (Woodhull, 1917, p. 224), might be expected to develop a modest "appreciation for science," especially its practical lessons concerning "physical, mental, and sexual hygiene." Later in the century, new survey and statistical techniques established interest in science as a measurable matter of fact. Once quantitatively established, the numbers drop away leaving an "invisible attitude in an individual's mind, one that can be probabilistically inferred from a student's gender and racial classification." With interest in science thus depoliticized (i.e., seeking "individual-level explanations for gender and racial disparities in science"), "the sociopolitical dilemma of how to redress science education's exclusions is replaced with an agenda of attitudinal change." Kirchgasler worries that NGSS instructional case studies may "serve as primers, training the 'classificatory eye' (Foucault, 1995, p. 147) of the teacher to read particular expressions of interest in science as a sign of a certain 'kind' of student: one with greater or less talent." Such are the hazards of using scientific theories and evidence to support differentiated instruction when science carries unexamined prejudices about different kinds of people. Historicizing the tools and perspectives we inherit is a necessary strategy for engaging the sociopolitics of access to and success in science learning. With the benefit of historically informed hindsight and reflection on the present, Kirchgasler presses us to ask, "What other purposes might animate a science education freed from the injunction to respond to presumed differences in students' interests and needs for rigor versus relevance?"

Complementing Kirchgasler's analysis, *Michelle Knaier* ponders how NGSS perpetuates a heteronormative cisgender binary by marking "girls" as a nondominant group with particular needs and interests. NGSS (2013) "Case Study 5" features instructional approaches it says work especially well for girls. Knaier doesn't doubt the general efficacy of the student-centered pedagogies NGSS prescribes, but she does question who counts as a girl and why particular science pedagogies should work especially well for them. What, after all, "is the difference between a 'girl' interest and a 'boy' interest?" Would NGSS's prescriptions hold for "girls who present masculine traits [and] boys who present feminine traits"; trans girls and boys; genderqueer, intersex, and questioning students? If gender is salient in science for all, then two genders clearly won't do. Riffing on NGSS's language, Knaier cheekily asks if we might "purposefully interest and engage the [genderqueer students] in the class" or adopt "a strategy that enhances [androgynous students'] confidence in science"? Perhaps (see Rands, 2013)—but the intersectional implications (gender + sexuality + "race" + ethnicity + class + ability + etc.) of identity-reductionist pedagogies boggle the mind. "Curriculum," Knaier wisely advises, "should not be packaged in little pink, blue, or purple boxes—and … neither should our students." The alternative, Knaier says, isn't ignoring gender but "curbing the *desire* for gender conformity" (original italics) in order to recognize in *all* students the complex realities and possibilities of gender and STEM.[28]

We earlier highlighted *Sheri Leafgren* and *Scott Sander's* account of Deleuzo-Guattarian desire in the richly wriggly paths of children's everyday STEM-related inquiries. The learning spaces in which children and their caring adults move, or are stuck, are typically striated by sharply delimited objects, meanings, practices, and ways of being human. Being educated means growing (up) used to and embodying these structures. Grown-up ways can enable what are viewed within narrow registers as effective actions, while (and perhaps precisely because) they preempt so many other possibilities. Not being all grown up, Leafgren and Sander move through striated spaces like 5-year-olds who are attuned to playful possibilities between and across structures, thus effecting a local "smoothing of the striations of State pedagogy … [that recovers] possibilities foreclosed by straight-ahead looking and straight thinking." Educators at all levels feel the pull, and often seek the comfort, of developmentalist standards and best practices that define sanctioned learning trajectories from the not-fully-realized human to, as Slaton, Cech, and Riley (this volume, Chapter 17) put it, proper "personhood." A grown-up teacher tells kindergartners to ignore strange noises coming from the roof of their school building; a queering teacher attuned to the generative risks of walking children's "nomadic paths" accompanies them outside to investigate the noises and to test their conjectures (is it a witch? a rabbit? a witch chasing a rabbit?). Such "queering," say Leafgren and Sander, "honors children as complete and real and fully human beings—albeit human beings who often do human being differently than older beings tend to do."

In an essay that resonates with Leafgren and Sander's, *Anna MacDermut* and *Adrian Zongrone* elaborate a recurring theme in this collection: the intimate, even playful, relations of STEM learning norms and transgressions. They focus on children's museums while keeping schooling in sight as a generative counterpoint. "Play …," they write, "allows desires to be fulfilled and imaginations to take control" of learning in which "no single thought or idea creates the rest; no conclusion is an ending; no person is the sole possessor or producer of knowledge." They show how play lyses boundaries and recombines subjects and objects, and how it extends rhizomatically, making connections "among entities that are presupposed to be dissimilar." Play performs counter-heteronormative epistemontologies that mess up patrilineal family trees of conceptual development; play assembles rhizomatic affinities of choice and chance in which ideas proliferate, though "one idea is not inherently the root of other ideas." Conventional associations are stubborn and reconstitute following playful disruptions. But they return changed, invested with irreverent possibilities in and by the wayward body-minds of learners. Learners too are re-formed through transgression. MacDermut and Zongrone demonstrate how by crossing limits and returning, again and again, children's museums and young learners (and adults who aren't all grown up) can unleash the "transformative power of play, a recreational science."

Michael Reiss writes, "It is increasingly the case that conventional teaching about human sex and sexuality is poor science, … outdated fodder … served up in bland science textbooks and standardized approaches to teaching human reproduction" (and reproduction, after all, doesn't exhaust human sex and sexuality).[29] *Poor science*, says Reiss, not just bad teaching. In a queering twist, this is an epistemological critique of science teaching that disregards the radical norms that science professes: empiricism, reason, objectivity (look around, in STEM education and beyond—these are radical norms!). Reiss considers three distinctive (though not mutually exclusive) aims for the science curriculum. Under each, he argues, queering how human sex and sexuality are conventionally presented in the science curriculum strengthens science *and* curriculum. If science education is to promote flourishing lives, then "much richer, less binary account[s] of sex and gender" are needed. If science education's aim is to instill powerful knowledge and wise dispositions toward it, then "troubling what otherwise would remain unwarranted presumptions about how humans exist and behave" is necessary. And a science curriculum for social justice should examine the suffusion of the science of human sexuality in and by moral, political, and economic discourses. In the post post-Enlightenment—and in light of scientific virtues variously reenvisioned by Donna Haraway, Sandra Harding, Evelyn Fox Keller, Helen Longino, and others—might queering science education make once-/still-radical Enlightenment values new again?

Cecilia Rodéhn joins nursing and medical students on guided tours of Uppsala University's Museum of Medicine and draws on Ahmed's (2006a, 2006b) work to examine "how the exhibition space orients teaching, how volunteers and curators create lines of orientation when teaching, and how that teaching orients students." In a cluttered exhibition hall, medicine occupies the most space and is presented through its techniques and technologies as the domain of "emotionally restrained" white men at work "liberating the population from unwanted diseases and disabilities." Histories of nursing, in contrast, are in the corners and back of the room where women are depicted as emotionally engaged (though chaste) heterosexuals, surrogate mothers (in the old-fashioned sense), and natural caregivers. In another exhibit the psychiatric hospital that originally occupied the museum building is framed as home to an extended family, with the caregiver-to-patient relationship like that of parent to child. The disabled or mentally ill body is, like the child's, ambiguously embraced as needing correction and often a danger to itself and others. Rodéhn illustrates how teaching could queer this and other museums by "disorienting heteronormative lifelines and desires" that run through the exhibits. For example, echoing Reiss's (this volume, Chapter 13) call for better science, better (and so dis-/re-orienting) histories of nursing would note that some women formed loving, same-sex relationships in nursing, not being restricted either to unattached lives committed to nursing or to leaving the profession to marry a man. Along these and other lines of (dis)orientation, Rodéhn considers how teaching in medical museums (even and especially heteronormative ones!) can open new possibilities for students, and for the practices of nursing and medicine.

In 2003, Noel Gough, Annette Gough, and several friends established "Camp Wilde, an imaginary intellectual space dedicated to alleviating 'the irony deficiency that is a hallmark of so many academic texts' (McWilliam, 1999, p. x) by queer(y)ing the earnestness of much environmental education research (and perhaps provoking some subversive laughter)" (p. 45). Now *Nicholas Santavicca, Jesse Bazzul,* and *Stephen Witzig* send Chase, Dirk, and Henri (who aren't the authors nor other than the authors), to Camp Wilde "as queer(er)s … to rediscover nature's queer performativity … and maybe even to develop a queer-ethical-political-scientific performativity of the researcher/observer—performativities that allow productive assemblages to come forward." That's serious business. But like truths exceed facts, serious business exceeds sobriety. Camp Wilde arouses slippery narratives, ambiguous classifications, uncertain bodies, and fresh-cut realities. And earth worms—what's with the worms? This text breaches fictional worlds of reality and imagination to inquire into relations of nature and environmental and science education as they are and as we dare them to be(come). Reimagining worlds demands seriously ethical games of truth or dare. It's your turn—Chase, Dirk, and Henri are waiting: truth or dare?

James Sheldon extends mathematical inqu[ee]ry (see Rands, 2009b) to infinity by reviewing its treatment in an undergraduate mathematics textbook and by illustrating a queer curriculum of infinity. Mathematical inqu[ee]ry "move[s] beyond a politics of inclusion and representation to uncovering and then challenging the mechanisms that produce the mathematical canon, … [including] take[ing] seriously the libidinal dynamics that underpin all mathematics." Infinity, Sheldon says, is one of several mathematical topics with "a certain queer potential, aesthetic, or history."[30] These are "good starting points for mathematical inqu[ee]ry … into why these [queer] topics are left out [of the curriculum] and how they get 'straightened' when presented in a formulaic, formalized, and/or rote algorithmic fashion." Sheldon relates how the authors of an "unconventionally conversational" undergraduate textbook begin a chapter on cardinality (a foundational topic for infinity) with poetical and comic references that evoke "haunting amazement and campy humor." But the text promptly moves "straight" into the usual litany of "propositions, theorems, and proofs." With a queer curriculum of infinity in mind, Sheldon reflects on a course they taught in which "infinity formed the common thread" through a heterodox blend of mathematical, historical, philosophical, and literary perspectives. Such inqu[ee]ry seeks to liberate the "discrete, bounded objects" of conventional mathematics into "queer bodies of knowledge [that] captivate, engross, and puzzle in a similar fashion to the way that literal bodies do."

Scene: A middle school hallway in middle America. A poster handwritten by students on a teacher-assigned theme declares: "To have grit you must push through tough times and fight advirsity" (sic). Grit is having a moment in education. *Amy Slaton, Erin Cech,* and *Donna Riley* bring a process-ontological view to grit in U.S. engineering education to illustrate how the commonsense notion that students and subject matter are distinct discretely obscures the coproduction of proper personhood and proper knowledge.[31] "This is a loop in which doing what is understood/felt/intuited to be good engineering not only enfolds unmarked assumptions about eligible bodies but re-produces eligible and ineligible bodies and desires as the very practices of engineering." The authors take up grit, recently renovated from folksy idiom to meritocratic creed, to parse the "selective logic" by which "*who counts as engineering* … enact[s] neoliberal distributions of power with exquisite precision" (original italics). Grit extends self-efficacy by insisting that "satisfaction is to be attained *by way of* the difficulties produced by a degenerate and daunting world" (original italics). Metricized as the Grit Scale, it becomes a *technology of desire* that assesses "sturdiness in the face of discomfort or challenge" in order to "produce, measure, sort, transform, and/or discard learners." The "self-reproducing character of U.S. engineering knowledge" in which grit and other (de)merits circulate corrupts familiar inclusion-as-reform projects because "that which is defined as engineering performs the work of discrimination." Engineering's paranoia ("there must be no bad surprises"; Sedgwick, 2003, p. 130) "disallows not merely technical uncertainty and any deep questioning of what the engineer should do next, but also strange selves." We imagine that the authors' passionate sturdiness in the face of these suspicious prohibitions lies off the Grit Scale, which can't fathom the desires animating their de-/re-engineering of ontologies in engineering education.

LOOKING AHEAD

Unlike many edited collections, this one isn't a snapshot of a well-developed field of scholarship—it is, rather, catalytic. Many of the contributors to this volume hadn't previously worked extensively with queer theories. They used this project to think in new ways about their lives and scholarship, to scrutinize and cross familiar and unfamiliar boundaries. The results of their personally and professionally gutsy (not to mention gritty!) explorations leave us hopeful about further inquiries that use queer theories and allied perspectives to reimagine and reenact STEM education. There is, of course, much that isn't here, including, for example, queering perspectives on education in computer science, informatics, and tech-heavy technosciences. Further, we want to see additional critical studies of STEM education subject matter and practices that integrate history, philosophy, and sociology of technosciences with queer theorizing. We also look forward to more work that takes up trans analytical perspectives in and on STEM education. During the development of this volume, we were in touch with several individuals who were still formulating projects from trans perspectives, and we are eager for such work to appear.

NOTES

1 By *gender* we mean, more or less, "the cultural meanings that the sexed body assumes" (Butler, 1990/1999, p. 10) through stylized enactments and readings of those enactments, while we also maintain that the sexed body isn't merely a material given but an indeterminate mash-up of "nature" and "culture." See Stimpson and Herdt (2014) on the many lives of the word "gender." And see Glasser and Smith's (2008) criticism of education research and policy (including much of gender-equity work in STEM education) for synonymizing gender and sex, which in effect reduces gender to biological sex. We appreciate Roughgarden's (2004) inflection of gender with posthumanist materialism that

extends gender to nonhumans or, turning that around, integrates humans in wider empirical inquiries on diversity, sex, and gender: "*Gender is the appearance, behavior, and life history of a sexed body*" (p. 27, original italics). Finally, we prefer "queer theories" to "queer theory." As Hall (2003) says, "There is no 'queer theory' in the singular, only many different voices and sometimes overlapping, sometimes divergent perspectives that can be loosely called 'queer theories'" (p. 5).

2 See Haraway (2003) on *natureculture* and Barad (2007) on *intra-action* as the entangling way of naturecultures.

3 We generally don't use "queer(s)" to refer to the collective lesbian, gay, bisexual, transgender, queer, questioning, intersex, and other nonheterosexual and/or gender-nonconforming (LGBTQQI+) people in contrast to what must be, in turn, *nonqueer* (gender-conforming and heterosexual) people. That said, some contributors to this volume do use "queer(s)" as an umbrella term for LGBTQQI+. We think the human condition is pretty queer, but we don't want to obscure difference by calling everyone (a) queer. (One could, we suppose, rescue difference by imagining that we're each queer in our own special way!) We're more interested in *queering* as discombobulating (ab)normalizing desires by inquiring into why some people/beings/things/practices are seen as strange.

4 "The knowing self is partial in all its guises, never finished, whole, simply there and original; it is always constructed and stitched together imperfectly, and *therefore* able to join with another, to see together without claiming to be another. Here is the promise of objectivity: a scientific knower seeks the subject position, not of identity, but of objectivity, that is, partial connection. There is no way to 'be' simultaneously in all, or wholly in any, of the privileged (i.e., subjugated) positions structured by gender, race, nation, and class [etc.]" (Haraway, 1988, p. 586, original italics).

5 Lesbian, gay, bisexual, transgender, queer, questioning, intersex, etc.

6 "Mud map" is Australian slang for a map or set of directions, often drawn on the ground with a stick, to roughly portray the territory.

7 Jagose (1996) is an influential early synthesis of queer theory that reflects its founding concerns with sexuality and gender. Sullivan (2003) remains one of the best book-length introductions to queer theory, in part because it frames queering more widely than conventional understandings of sexuality. Hall (2003) is an engaging introduction pitched to those who want to take up queer theorizing in literary analyses, where so much queering was first elaborated. Wilchins (2004) remains a fresh and accessible take on sexuality, gender, and identity. Turner (2000) is a deep dive into the intellectual history of queer theory.

8 This broad notion of bodies is in the vicinity of Deleuze and Guattari's *body without organs*: "any organized structure such as government, a body, and the universe" that arises and functions in/as the continuous flow of productive desire (Zembylas, 2007, pp. 336–337).

9 Sedgwick's appropriation of the term *reparative* in relation to *queer* (another appropriation) lives and theorizing is cleverly discomforting and enabling. See Sedgwick (2003) and Love (2010) for more on reparative readings in queer theory.

10 Morris's (1998) notion of *queer sensibility* informs our sense of queering as an *orientation* toward reading and writing the world.

11 "Acknowledging the agency of the world in knowledge makes room for some unsettling possibilities, including a sense of the world's independent sense of humor. ... We are not in charge of the world. We just live here and try to strike up noninnocent conversations by means of our prosthetic devices" (Haraway, 1988, pp. 593, 594). See Barad (2007), Bennett (2010), and Kirby (2011) for notions of a playful world making (sense of) itself.

12 Portions of this subsection and the next previously appeared in *Cultural Studies of Science Education*, "[Re]considering Queer Theories and Science Education," volume 9, 2014, pages 393–407, Steve Fifield and Will Letts.

13 See Enke's (2012) critical reflections on *cisgender*.

14 Kirby (1999) cautions against too easily acting "as if oppositional logic is an error to be corrected" and notes that "the identity of an error will always infect its correction, as the latin root *errare*—to wander, to be uncertain—suggests" (p. 28). And so, although reading binaries dualistically threatens to collapse possibilities for knowing and being into dominant and subordinate poles with, perhaps,

a linear continuum between them, the terms of binaries needn't be strictly dualistic. For example, we can repurpose binaries as mobile containers of complex relations that, with nudging, expand into multidimensional (im)possibilities. Unfortunately, taken-for-granted dualism rules in postpositivist applications of the continuum model of binaries in, for instance, qualitative and quantitative scales that fix a subject's or group's position and track "development" in relation to standardized states of knowing, feeling, and being. In this volume see Kirchgasler's (Chapter 9) historical analysis of interest in science, and Slaton, Cech, and Riley's (Chapter 17) cautionary tale of the Grit Scale.

15 In Chapter 2 we reflect on our efforts to turn from queer(s) to queering.

16 "Reading constructs the reader as well as the text. ... Reading queerly creates a queer reader" (Morris, 1998, pp. 276–277).

17 See Colebrook's (2008) compelling vision of queer(ing) and normativity that differs from what we present here. In Colebrook's Deleuzian-inspired account, queer theory isn't

> theory that sets itself against normativity through either a recognition of another group of bodies or through a destabilization and negation of norms. Rather the theory would set as its task the notion of the Idea as a problem: how have such relations and terms emerged, what – given effected relations – might have occurred otherwise; what are the forces of potentiality hidden in our experienced encounters? (p. 31)

Ruffolo's (2009) similarly Deleuzian "Post-Queer Considerations" complements Colebrook's view.

18 See Foucault (1978/1990, p. 101).

19 See Mayo's (2007) discussion of queer queers, straight queers, and intermittent queerness.

20 See Duggan's (2002) discussion of the *new homonormativity*.

21 See https://collectiveliberation.org/wp-content/uploads/2013/01/Lorde_The_Masters_Tools.pdf

22 "Binary logic undoes its truths even as it affirms them, so that an effective way to displace and intervene into what appears to be a repressive mono-logic is to consider its essential perversity" (Kirby, 1999, p. 28).

23 Portions of this section previously appeared in *Cultural Studies of Science Education*, "[Re] considering Queer Theories and Science Education," volume 9, 2014, pages 393–407, Steve Fifield and Will Letts. © Springer Science+Business Media Dordrecht 2014. With permission of Springer.

24 In this section we include selected publications from journals and books, not conference papers, unpublished manuscripts, or work from other venues.

25 In this section quotations are from the chapter being discussed, unless noted otherwise.

26 For more on STEM teaching and learning in museums, see MacDermut and Zongrone (Chapter 12) and Rodéhn (Chapter 14) in this volume.

27 See Slaton, Cech, and Riley (this volume, Chapter 17) for a related analysis of *grit* as a measurable personal attribute that mediates affinity for STEM.

28 See Rands's (2009a) discussion of gender-complex teacher education.

29 "As the anthropologist Carole Vance once cautioned, we should be wary of regarding 'reproduction' as the 'core' of gender and sexuality" (Stimpson & Herdt, 2014, p. 6).

30 See Sheldon and Rands's (2013) list of mathematical concepts that are good starting points for queering.

31 Götschel (this volume, Chapter 5) makes a similar point: "In my teaching … I keep in mind that processes of selection, experiment, observation, and labeling create at the same time physical matter on a macroscopic scale *and* future engineers with physical knowledge."

REFERENCES

Adsit-Morris, C., & Gough, N. (2017). It takes more than two to (multispecies) tango: Queering gender texts in environmental education. *The Journal of Environmental Education, 48*(1), 67–78.

Ah-King, M. (2013). Queering animal sexual behavior in biology textbooks. *Confero, 1*(2), 46–89.

Ahmed, S. (2006a). Orientations: Toward a queer phenomenology. *GLQ: A Journal of Lesbian and Gay Studies, 12*(4), 543–574.
Ahmed, S. (2006b). *Queer phenomenology: Orientations, objects, others*. Durham, NC: Duke University Press.
Alexander, B. K. (2018). Queer/quare theory: Worldmaking and methodologies. In N. K. Denzin & Y. S. Lincoln (Eds.), *The Sage handbook of qualitative research* (5th ed., pp. 275–307). Los Angeles, CA: Sage Publications.
Allen, G. (1997). The double-edged sword of genetic determinism: Social and political agendas in genetic studies of homosexuality, 1940–1994. In V. A. Rosario (Ed.), *Science and homosexualities* (pp. 242–270). New York, NY: Routledge.
Azzarello, R. (2008). Unnatural predators: Queer theory meets environmental studies in Bram Stoker's Dracula. In N. Giffney & M. J. Hird (Eds.), *Queering the non/human* (pp. 137–157). Aldershot: Ashgate.
Barad, K. (2007). *Meeting the universe halfway: Quantum physics and the entanglement of matter and meaning*. Durham, NC: Duke University Press.
Bazzul, J., & Santavicca, N. (2017). Diagramming assemblages of sex/gender and sexuality as environmental education. *The Journal of Environmental Education, 48*(1), 55–66.
Bazzul, J., & Sykes, H. (2011). The secret identity of a biology textbook: Straight and naturally sexed. *Cultural Studies of Science Education, 6*, 265–286.
Bennett, J. (2010). *Vibrant matter: A political ecology of things*. Durham, NC: Duke University Press.
Berlant, L. (2014). Love. In C. R. Stimpson & G. Herdt (Eds.), *Critical terms for the study of gender* (pp. 250–270). Chicago, IL: University of Chicago Press.
Britzman, D. (2000). Precocious education. In S. Talburt & S. Steinberg (Eds.), *Thinking queer: Sexuality, culture, and education* (pp. 33–59). New York, NY: Peter Lang.
Broadway, F. S. (2011). Queer (v.) queer (v.): Biology as curriculum, pedagogy, and being albeit queer (v.). *Cultural Studies of Science Education, 6*, 293–304.
Butler, J. (1990/1999). *Gender trouble: Feminism and the subversion of identity*. New York, NY: Routledge.
Carlson, D. (1998). Who am I? Gay identity and a democratic politics of the self. In W. F. Pinar (Ed.), *Queer theory in education* (pp. 107–119). Mahwah, NJ: Lawrence Erlbaum Associates.
Cech, E. A., & Waidzunas, T. J. (2011). Navigating the heteronormativity of engineering: The experiences of lesbian, gay, and bisexual students. *Engineering Studies, 3*(1), 1–24.
Cohen, J. J. (2013). Queering the inorganic. In E. H. Yekani, E. Kilian, & B. Michaelis (Eds.), *Queer futures: Reconsidering ethics, activism, and the political* (pp. 149–163). Farnham: Ashgate.
Colebrook, C. (2008). How queer can you go? Theory, normality and normativity. In N. Giffney & M. J. Hird (Eds.), *Queering the non/human* (pp. 17–34). Aldershot: Ashgate.
Donald, C. A., DasGupta, S., Metzl, J. M., & Eckstrand, K. L. (2017). Queer frontiers in medicine: A structural competency approach. *Academic Medicine, 92*, 345–350.
Duggan, L. (2002). The new homonormativity: The sexual politics of neoliberalism. In R. Castronovo & D. Nelson (Eds.), *Materializing democracy: Toward a revitalized cultural politics* (pp. 175–194). Durham, NC: Duke University Press.
Enke, A. F. (2012). The education of little cis: Cisgender and the discipline of opposing bodies. In A. Enke (Ed.), *Transfeminist perspectives in and beyond transgender and gender studies* (pp. 60–77, 210–217). Philadelphia, PA: Temple University Press.

Felsenstein, D. R. (2018). Enhancing lesbian, gay bisexual and transgender cultural competence in a midwestern primary care clinic setting. *Journal for Nurses in Professional Development, 34*(3), 142–150.

Fifield, S., & Letts, W. (2014). [Re]considering queer theories and science education. *Cultural Studies of Science Education, 9*, 393–407.

Fifield, S., & Swain, H. (2002). Heteronormativity and common sense in science (teacher) education. In R. M. Kissen (Ed.), *Getting ready for Benjamin: Preparing teachers for sexual diversity in the classroom* (pp. 177–189). Lanham, MD: Rowman & Littlefield.

Foucault, M. (1978/1990). *The history of sexuality, Vol. 1: An introduction* (R. Hurley, Trans.). New York, NY: Random House.

Fuss, D. (1991). Inside/out. In D. Fuss (Ed.), *InsideOut: Lesbian theories, gay theories* (pp. 1–13). New York, NY: Routledge.

Giffney, N. (2009). Introduction: The 'q' word. In N. Giffney & M. O'Rourke (Eds.), *The Ashgate companion to queer theory* (pp. 1–13). Farnham: Ashgate.

Giffney, N., & Hird, M. J. (Eds.). (2008). *Queering the non/human*. Aldershot: Ashgate.

Glasser, H. M., & Smith III, J. P. (2008). On the vague meaning of "gender" in education research: The problem, its sources, and recommendations for practice. *Educational Researcher, 37*(6), 343–350.

Goldberg, L., Ryan, A., & Sawchyn, J. (2009). Feminist and queer phenomenology: A framework for perinatal nursing practice, research, and education for advancing lesbian health. *Health Care for Women International, 30*(6), 536–549.

Gottfried, M., Estrada, F., & Sublett, C. (2015). STEM education and sexual minority youth: Examining math and science coursetaking patterns among high school students. *The High School Journal, 99*(1), 66–87.

Gough, N., Gough, A., Appelbaum, P., Appelbaum, S., Doll, M. A., & Sellers, W. (2003). Tales from Camp Wilde: Queer(y)ing environmental education research. *Canadian Journal of Environmental Education, 8*(1), 44–66.

Graham, M. (2014). *Anthropological explorations in queer theory*. Farnham: Ashgate.

Gunkel, K. L. (2009). Queering science for all: Probing queer theory in science education. *Journal of Curriculum Theorizing, 25*(2), 62–75.

Hall, D. E. (2003). *Queer theories*. New York, NY: Palgrave Macmillan.

Halperin, D. (1995). *Saint Foucault: Towards a gay hagiography*. New York, NY: Oxford University Press.

Haraway, D. (1988). Situated knowledges: The science question in feminism and the privilege of partial perspective. *Feminist Studies, 14*(3), 575–599.

Haraway, D. (2003). *The companion species manifesto: Dogs, people, and significant otherness*. Chicago, IL: Prickly Paradigm Press.

Haraway, D. (2008). Foreword: Companion species, mis-recognition, and queer worlding. In N. Giffney & M. J. Hird (Eds.), *Queering the non/human* (pp. xxiii–xxvi). Aldershot: Ashgate.

Hollenbach, A. D., Eckstrand, K. L., & Dreger, A. (Eds.). (2014). *Implementing curricular and institutional climate changes to improve health care for individuals who are LGBT, gender nonconforming, or born with DSD: A resource for medical educators*. Washington, DC: Association of American Medical Colleges.

Honeychurch, K. (1996). Researching dissident subjectivities: Queering the grounds of theory and practice. *Harvard Educational Review, 66*, 339–355.

Hughes, B. E. (2017). "Managing by not managing": How gay engineering students manage sexual orientation identity. *Journal of College Student Development, 58*(3), 385–401.

Hughes, B. E. (2018). Coming out in STEM: Factors affecting retention of sexual minority STEM students. *Science Advances, 4*, 1–5.

Jagose, A. (1996). *Queer theory: An introduction*. New York, NY: New York University Press.

Kahn, R. (2010). *Critical pedagogy, ecoliteracy, and planetary crisis: The ecopedagogy movement*. New York, NY: Peter Lang.

Kirby, V. (1999). Human nature. *Australian Feminist Studies, 14*(29), 19–29.

Kirby, V. (2008). (Con)founding 'the human': Incestuous beginnings. In N. Giffney & M. J. Hird (Eds.), *Queering the non/human* (pp. 35–54). Aldershot: Ashgate.

Kirby, V. (2011). *Quantum anthropologies: Life at large*. Durham, NC: Duke University Press.

Lemke, J. (2011). The secret identity of science education: Masculine and politically conservative? *Cultural Studies of Science Education, 6*, 287–292.

Letts, W. J. (1999). Making "boys" and "girls" in the classroom: The heteronormative nature of elementary school science. In W. J. Letts & J. T. Sears (Eds.), *Queering elementary education: Advancing the dialogue about sexualities and schooling* (pp. 97–110). Lanham, MD: Rowman & Littlefield.

Letts, W. J. (2001). When science is strangely alluring: Interrogating the masculinist and heteronormative nature of primary school science. *Gender & Education, 13*(3), 261–274.

Lincoln, Y. S., Lynham, S. A., & Guba, E. G. (2018). Paradigmatic controversies, contradictions, and emerging confluences, revisited. In N. K. Denzin & Y. S. Lincoln (Eds.), *The Sage handbook of qualitative research* (5th ed., pp. 108–150). Los Angeles, CA: Sage Publications.

Love, H. (2010). Truth and consequences: On paranoid reading and reparative reading. *Criticism, 52*(2), 235–241.

Loy, D. (2010). *The world is made of stories*. Somerville, MA: Wisdom Publications.

Luhmann, S. (1998). Queering/querying pedagogy? Or, pedagogy is a pretty queer thing. In W. F. Pinar (Ed.), *Queer theory in education* (pp. 141–155). Mahwah, NJ: Lawrence Erlbaum Associates.

Lundin, M. (2014). Inviting queer ideas into the classroom: Studying sexual education from a queer perspective. *Cultural Studies of Science Education, 9*(2), 377–391.

Mayfield, K. (2016). Queering the inquiry body: Critical science teaching from the margins. In sj Miller & N. M. Rodriguez (Eds.), *Educators queering academia: Critical memoirs* (pp. 113–124). New York, NY: Peter Lang.

Mayo, C. (2007). Intermittently queer. In N. M. Rodriguez & W. F. Pinar (Eds.), *Queering straight teachers: Discourse and identity in education* (pp. 183–200). New York, NY: Peter Lang.

Miller, sj, & Rodriguez, N. M. (2016). Introduction: The critical praxis of queer memoirs in education. In sj Miller & N. M. Rodriguez (Eds.), *Educators queering academia: Critical memoirs* (pp. xv–xxiii). New York, NY: Peter Lang.

Morris, M. (1998). Unresting the curriculum: Queer projects, queer imaginings. In W. F. Pinar (Ed.), *Queer theory in education* (pp. 275–286). Mahwah, NJ: Lawrence Erlbaum Associates.

Mortimer-Sandilands, C., & Erickson, B. (2010). Introduction: A genealogy of queer ecologies. In C. Mortimer-Sandilands & B. Erickson (Eds.), *Queer ecologies: Sex, nature, politics, desire* (pp. 1–47). Bloomington, IN: Indiana University Press.

Murray, O. J. (2015). *Queer inclusion in teacher education: Bridging theory, research, and practice*. New York, NY: Routledge.

NGSS Lead States. (2013). *Next generation science standards: For states, by states.* Case study 5: Girls and the *Next Generation Science Standards* [June 2013 version]. Retrieved from http://www.nextgenscience.org/sites/default/files/%285%29%20Case%20Study%20 Girls%206-14-13.pdf

Rands, K. (2009a). Considering transgender people in education: A gender-complex approach. *Journal of Teacher Education, 60*(4), 419–431. doi:10.1177/0022487109341475

Rands, K. (2009b). Mathematical inqu[ee]ry: Beyond 'add-queers-and-stir' elementary mathematics education. *Sex Education, 9*(2), 181–191. doi:10.1080/14681810902829646

Rands, K. (2013). Supporting transgender and gender-nonconforming youth through teaching mathematics for social justice. *Journal of LGBT Youth, 10*(1–2), 106–126.

Rands, K. (2016). Mathematical inqueery. In N. M. Rodriguez, W. J. Martino, J. C. Ingrey, & E. Brockenbrough (Eds.), *Critical concepts in queer studies and education: An international guide for the twenty-first century* (pp. 183–192). New York, NY: Palgrave Macmillan.

Reiss, M. J. (1998). The representation of human sexuality in some science textbooks for 14–16 year olds. *Research in Science & Technological Education, 16*, 137–149.

Roberts, D. A., & Ostman, L. (1998). Preface. In D. A. Roberts & L. Ostman (Eds.), *Problems of meaning in science curriculum* (pp. ix–xii). New York, NY: Teachers College Press.

Roughgarden, J. (2004). *Evolution's rainbow: Diversity, gender, and sexuality in nature and people.* Berkeley, CA: University of California Press.

Ruffolo, D. V. (2007). Giving an account of queer: Why straight teachers can become queerly intelligible. In N. M. Rodriguez & W. F. Pinar (Eds.), *Queering straight teachers: Discourse and identity in education* (pp. 255–276). New York, NY: Peter Lang.

Ruffolo, D. V. (2009). Post-queer considerations. In N. Giffney & M. O'Rourke (Eds.), *The Ashgate companion to queer theory* (pp. 379–394). Farnham: Ashgate.

Russell, C. L., Sarick, T., & Kennelly, J. (2002). Queering environmental education. *Canadian Journal of Environmental Education, 7*, 54–66.

Russell, J. (2013). Whose better? [Re]orienting a queer ecopedagogy. *Canadian Journal of Environmental Education, 18*, 11–26.

Scholer, A.-M. (2002). Sexuality in the science classroom: One teacher's methods in a college biology course. *Sex Education, 2*, 75–86.

Sedgwick, E. K. (1990). *Epistemology of the closet.* Berkeley, CA: University of California Press.

Sedgwick, E. K. (2003). *Touching feeling: Affect, pedagogy, performativity.* Durham, NC: Duke University Press.

Seidman, S. (1996). Introduction. In S. Seidman (Ed.), *Queer theory/sociology* (pp. 1–29). Cambridge, MA: Blackwell.

Sheldon, J., & Rands, K. (2013). Queering, en-gendering, and trans-forming mathematics and mathematics education. In M. Martinez & A. Castro Superfine (Eds.), *Proceedings of the 35th annual meeting of the North American Chapter of the International Group for the Psychology of Mathematics Education.* Chicago, IL: University of Illinois at Chicago. Retrieved from http://files.eric.ed.gov/fulltext/ED567771.pdf

Snyder, V., & Broadway, F. (2004). Queering high school biology textbooks. *Journal of Research in Science Teaching, 41*, 617–636.

Stimpson, C. R., & Herdt, G. (Eds.). (2014). *Critical terms for the study of gender.* Chicago, IL: University of Chicago Press.

Sullivan, N. (2003). *A critical introduction to queer theory.* New York, NY: New York University of Press.

Sumara, D., & Davis, B. (1998). Telling tales of surprise. In W. F. Pinar (Ed.), *Queer theory in education* (pp. 197–219). Mahwah, NJ: Lawrence Erlbaum Associates.

Talburt, S. (2000). Introduction: Some contradictions and possibilities of thinking queer. In S. Talburt & S. R. Steinberg (Eds.), *Thinking queer: Sexuality, culture, and education* (pp. 3–13). New York, NY: Peter Lang.

Tolbert, S. (forthcoming). Queering dissection: "I wanted to bury its heart, at least". In C. Taylor, C. Amade-Escote, & A. Abbas (Eds.), *Gender in learning and teaching: Feminist dialogues across international boundaries*. London: Routledge.

Turner, W. B. (2000). *A genealogy of queer theory*. Philadelphia, PA: Temple University Press.

Wilchins, R. (2004). *Queer theory, gender theory: An instant primer*. Los Angeles, CA: Alyson Books.

Zembylas, M. (2007). Risks and pleasures: A Deleuzo-Guattarian pedagogy of desire in education. *British Educational Research Journal, 33*, 331–347.

ABOUT THE AUTHORS

Steve Fifield is an affiliated assistant professor in the School of Education at the University of Delaware (USA). He previously worked at the University of Delaware as a faculty and staff member in science teacher education and program evaluation. He has also been a grant writer at The Franklin Institute Science Museum in Philadelphia, Pennsylvania, and the Oregon Museum of Science and Industry in Portland, Oregon. He's currently a paraprofessional in special education services. Steve has a PhD in curriculum and instruction (with a minor in studies of science and technology) and an MS in ecology and evolutionary biology from the University of Minnesota–Twin Cities. He's interested in cultural studies of science and science education, including intersections with queer theoretical and Buddhist conceptions and practices of subjectivity and (non)self.

Will Letts is associate dean academic in the Faculty of Arts and Education at Charles Sturt University in Bathurst, NSW, Australia. His research interests include the cultural studies of science and science education, especially with respect to sex, gender, and sexuality; interrogating subjectivities in ECE/teacher education; and the enactment of pedagogical documentation in tertiary settings. He has loved learning about and doing science since he was very young. That's what led to him becoming a high school science teacher and a lecturer in science and technology. You can reach Will at wletts@csu.edu.au

FRANCIS S. BROADWAY

1. I : SNOW QUEEN :: "NIGGER" : (SCHOOL) SCIENCE

INVITATION[1]

The title of this chapter is a didactic, linear statement of its aim with the hope of normalizing and generalizing the meaning readers create through their transactional readings of the text. In other words, with the question of hope as Hope—one of the Pandora's evils (Hesiod, 2004)—the title postulates and structures the meaning of the chapter by explicating, granting, and privileging authorial meaning. Valuing to queer rather than to normalize, I would like to queer the way the title of the chapter—I : Snow Queen :: Nigger : (school) Science—is ordinarily read, in terms of the definition of an analogy, as *I am to a Snow Queen as a Nigger is to (school) Science*. However, queerly, the association and synonym relationships within the title might give it the meaning (a) I am a Snow Queen and (b) a Nigger is (school) Science. Furthermore, the title queerly posits the following relationships: (c) I am a Nigger and (d) a Snow Queen is (school) Science, as well as (e) I am (school) Science and (f) a Nigger is a Snow Queen.

INVOCATION

My intention when writing on queer theory and science education is to write a text within which I queer, but always the writings tend to end as extremely normative, if not prophetically didactic. The mantra for my writing is "to dismantle the dominant educational narrative in which one passes from ignorance to knowledge about both self and other" (Miller, 1998, p. 372), and "to challenge categorical thinking, and [to] foster critical consciousness" (Sears, 1999, p. 5), all the while remembering that "under certain circumstances failing, losing, forgetting, unmaking, undoing, unbecoming, not knowing may in fact offer more creative, more cooperative, more surprising ways of being in the world" (Halberstam, 2011, pp. 2–3). This chapter is no different, other than I am different. One of the differences within this chapter, to which I will hold myself in many of its parts, is I will act like I know something and show that what I have been studying has left an imprint on me to some degree. I want to own my studying. This leaves me open to not being understood, to being misunderstood, to be unwise, and to need to say, "I didn't think about that." To queer or to write queerly "is supposed to be scary, and it is okay to be afraid" (J. Cruze, personal communication, July 1, 2017).

 DOI:10.1163/9789004331068_003

Here, within this chapter, is an autobiographical narrative. Heeding the caution of Delany (1996) that "you do not need an identity to become yourself; you need an identity to become like someone else" (p. 19), this chapter is authored by a queer, often spoken and written as gay, who is queering queer theory (Broadway, 2011), who is questioning "becom[ing] like someone else," and who is explicating his desires within the context of schools (Pinar & Grumet, 1976) and schooling (Wolcott, 2002). Maybe if the author, as queer, is seeking or needing an identity, then the identity would be male (biological sex), homosexual (sexual orientation), man (gender identity), masculine (gender roles), same-sex (sexual behavior), gay (sexual identity) based on the bifurcated heteronormative typology of Sears (1999, p. 5). Furthermore, in order to allow the reader, who needs not be queer or to queer, to pick and choose, to change and reshape the identity of the author as a member of the science education community as his primary social affiliation (Heath, 2004, p. 160), the author would be a middle school and high school science and mathematics teacher who taught in two "historically white [Southern] elite K-12 independent schools [in Atlanta, Georgia (see Wilson & Frank, 1983) and Columbia, South Carolina] established prior to the landmark 1954 *Brown v. Board of Education* decision and 'considered the most privileged and prestigious of private schools'" (Purdy, 2016, p. 62), and who also taught at an independent school in Michigan. The author is a science educator, an early-childhood teacher educator, a multicontent-area teacher educator, a curriculum theorist, et cetera. When all is said and done, and with the assumption that the reader self-identifies as a member of the science education community writ large, this memoir is a skeptical critique of science education through autobiographical insights that enable the author to be queer and to embody, if not to expose, his desire and desires. In short, this chapter uncovers and investigates the *his*tory (sic) of one worker within science education who wears the public mask of a science educator while questioning the desires behind the undefinable private mask of a monstrous stranger.

If the last paragraph was the *I*, then the *Nigger*, if implicit, was not explicit. Always conflated with biological sex, sexual orientation, gender identity, gender roles, sexual behavior, and/or sexual identity are race and class. Although there is a racialized science (Harding, 1993), in terms of the Nigger, there is no vocabulary that denotes the science of which I speak. Much as there is a science that is conducted, honored, and utilized by those who accept the label "scientist," and a "school science," there is no "Nigger science." I posit that the cultural sciences, such as indigenous science, are only science because they are compared to and measured by the science of "scientist"—the science immortalized in school science textbooks, research journals, conference proceedings, literature reviews, et cetera. In other words, cultural sciences seek an identity: to be like something else. Therefore, I talk about a racialized science that does not seek to be like someone/thing else; hence, I must add to that which implies identity—Nigger.

In addition to the endnote explication of Nigger, at this point I want to define Nigger through the metaphorical lamentation, "Why did God make me an outcast and a stranger in mine own house?" (Du Bois, 1903, p. 3). The Nigger is a stranger, if not a monster, whereupon a stranger is neither insider nor outsider, "neither us nor clearly them, not friend and not enemy, but a figure of ambivalence who troubles the border between us and them" (Phelan, 2001, pp. 4–5); and a monster, a "body [that] literally incorporates fear, desire, anxiety, and fantasy (ataractic or incendiary), giving them life and an uncanny independence" (Cohen, 1996, p. 4). This aforementioned worker within science education, the person that causes people to cross the street, to jaywalk, when approaching, cannot summarize who I am without mentioning the number of teacher preparation colleagues, usually white heteronormative females, who after years of working with me admit that they were "afraid" of me when we first met; nor without mentioning the students who viewed me as a stranger, if not a monster, being their first and often only Nigger teacher. Thus, this chapter is a stranger- (if not a monster-) comes-to-town narrative that takes the form of the regressive stage: "Observ[ing] oneself functioning in the past[,] … one takes special notice of one's past life-in-schools, with one's past life-with-schoolteachers and one's past life-with-books and other school-related artifacts" (Pinar & Grumet, 1976, p. 8); [this is] "an autobiographical strategy by means of which one may understand the nature of one's life in schools, and the functions of school in one's life" (Pinar, 1975/1994, p. 60).

Through a queer memoir and transactional reading of *A Separate Peace* (Knowles, 1959/1975), the first sections of a monstrous Snow Queen stranger story that intends upon queering science education (i.e., this chapter) chronicle the rejection of the humanities, literature specifically, by a 10th grader who embodied the conflation of race, class and sex. The second section, after an introduction and description of the term *Snow Queen*, argues that the creation of a heteronormative identity, albeit the queer Snow Queen, is the rejection of the monstrous stranger and the rejection of the Nigger in order to be the Scientist: white, masculine, and elite. A Snow Queen, metaphorically, is a person who loves/does science who is "baptiz[ed] into white citizenship and simultaneously excommunicate[ed] from the black freemen" (Brady, 1976, p. 70). A Snow Queen

> symbolizes a bridge between the worlds of black and white [and] incorporates the concept of the black middle class and the ghetto-dwelling freedman. While wearing the citizen mask, intellectual achievement allows him mobility in white-dominated society. While wearing the freedman mask, [he] connect[s] with natural identity rooted in the spontaneous, emotional black culture. (p. 74)

In other words, if and only if the Nigger forfeits being a Nigger and acts white, "excel[s] in academic arenas which traditionally have been defined as the prerogative of white Americans" (Fordham & Ogbu, 1986, p. 177), the Nigger—through the hidden curriculum of obedience, compliance, and complacency and "policies and practices that produced and reproduced knowledge that simultaneously reproduced inequality" (Romero, 2017, p. 213)—is not lynched. The Nigger be, becomes, a scientist.

A READING OF *A SEPARATE PEACE*

In the summer between my eighth- and ninth-grade years, I read *A Separate Peace* (Knowles, 1959/1975) as part of my summer reading list. I was a reader who

> brings to the work personality traits, memories of past events, present needs and preoccupations, a particular mood of the moment, and a particular physical condition [and] carries on a give-and-take with the signs of the page. … As the text unrolls before the reader's eyes, the meaning made of the early [signs] influence what comes to the mind and is selected for succeeding signs. (Rosenblatt, 1938/1985, pp. 30–31, 26–27)

Reading is what I wanted to do, all that I wanted to do. I did not know that (school) reading was primarily about uncovering and articulating authorial intention. I was often told that I did not know what the author meant! I worked that summer reading *A Separate Peace* with visions of white men, naked at least to the waist, dancing in my dreams, my moments of acting like I was dreaming, when I could close my eyes and make the world that I wanted to see, in which I wanted to be.

> But what interested [me] more than words, or all the little things that were said and done between long pages of the book [I] read, were the things that were *felt* there, and [I] said to [my]self that in [my] life [I] would feel things too, all the things [I] had ever want[ed] to, because [I] would not be afraid. … [I] would find out what it was like to be alive. (Morgan, 2004, p. 57, original italics)

I loved Phineas,

> [a] student who combined calm ignorance of the rules with a winning urge to be good, who seemed to love the school truly and deeply, and never more than when he was breaking the regulations, a model boy who was most comfortable in the truant's corner. (Knowles, 1959/1975, p. 16)

And it took me many years to not try to love only Phineas even when someone loved me, when someone wanted me, when someone was not repulsed by me—although I might as well have been like James Baldwin, ugly with

> the face of Satan … two great eyes, and a broad, low forehead, and the triangle of his nose, and his enormous mouth, and the barely perceptible clef in his chin, which was … the mark of the devil's little finger. (Baldwin, 1952/1985, p. 27)

Phineas loved Gene. "Phineas had thought of [Gene] as an extension of himself" (Knowles, 1959/1975, p. 171). Phineas needed Gene:

> Phineas was shocked at the idea of my [Gene's] leaving. In some way he needed me. He needed me. I was the least trustworthy person he had ever met. I knew that; he knew or should know that too. I had even told him. I had told him. But there was no mistaking the shield of remoteness in his face and voice. He wanted me around. (p. 100)

Phineas needed to love Gene and needed Gene to love him:

> "I hope you're having a pretty good time here. I know I kind of dragged you away at the point of a gun, but after all you can't come to the shore with just anybody and you can't come by yourself, and at this period in life the proper person is your best pal." He hesitated and then added, "which is what you are," and there was silence on his dune. (p. 40)

I needed Phineas to love me, but I wanted to discuss *A Separate Peace*. I wanted to discuss the love between Gene and Phineas: "'No, I don't know how to show you, how can I show you, Finny? Tell me how to show you. It was just some ignorance inside me, some crazy thing inside me, something blind, that's all it was'" (p. 183). I wanted to discuss me—"I alone was a dream, a figment which had never really touched anything" (p. 178)—to discuss love, to discuss how to bring a peace, separate from the war outside, to the war inside me.

I made meaning of *A Separate Peace* as a fan. Fans are "active producers and manipulators of meanings[,] … readers who appropriate popular texts and reread them in a fashion that serves different interests" (Jenkins, 1992, p. 23). I found the romance, love, between Gene and Phineas. Additionally, as a textual poacher (Jenkins, 1992) who "takes away only those things that are useful or pleasurable to the reader" (p. 24) and recognizes

> the erotic friendship between Gene and Finny, the homophobic motivation for Gene's panic reaction to Finny's affection, the significance of the loss of the boys' passionate friendship, or the way in which Gene's social contribution and maturity are coded in gendered terms (Tribunella, 2011, p. 125),

I would learn the moral of the tale: FAGGOT DIES! Gene's

> maturation throughout the novel represents his movement away from an effete intellectualism and "adolescent" homoerotic relationship. His "moral" progression involves abandoning the queer possibility and accepting a hegemonic and necessarily heterosexual masculinity that adolescent readers of the novel are tacitly encouraged to emulate and valorize. (p. 125)

COMMENTARY ON THE READING OF *A SEPARATE PEACE*

In order to connect *A Separate Peace* to science education, first, I need my reading of it to denote that I would not be successful at school reading where success at school is being assimilated and acculturated as elders—those who determine what is of worth and should be transmitted to the young (Pinar, 2004)—have determined. This could have also been illustrated by my memorable, but forgotten, undergraduate reading of and writing about *The Structure of Scientific Revolutions* (Kuhn, 1970) in the philosophy department's philosophy of science course, which earned me a D (as I did submit a paper on time and about the meaning I made of the text and which, as you could guess, was not what the instructor expected and wanted). Again, I did not master the hidden curriculum of obedience, compliance, and complacency and "policies and practices that produced and reproduced knowledge that simultaneously reproduced inequality" (Romero, 2017, p. 213).

Secondly, my rising-ninth-grader reading of *A Separate Peace* explains my queerness, oft writ gayness, specifically a Nigger man who desires and has sex with men, white men. But my reading of *A Separate Peace* was not the reading that I (and many others) would voice. I share the above story in order to give a reading that would never be the reading in any English classroom including, if not especially, at the school where I was assigned to read *A Separate Peace*, a school that embodied "the prefect-fagging system, [where] 'senior' boys are given a major role in governing the school, wielding discipline, and carrying responsibility" (Nash, 1961, p. 14), wherein personal relations were based on "cruelty, arrogance, and servility" (p. 19)—the training of the elite, the aristocracy, the colonialist, the imperialist. If I continue to read as a Nigger, covertly a queer Nigger man, then I would not be successful in school—the continuation of being a failure in school which is what landed me in said all-boys boarding school.

Therefore, the reading of *A Separate Peace* leaves two questions: Where do I go to be successful at school? and Where do I go to find white men, if not to be white? Where could I, the Nigger, forfeit being a Nigger and act white to "excel in academic arenas which traditionally have been defined as the prerogative of white Americans" (Fordham, 1988 p. 177)? Simply answered, I went to the sciences, school Science.

MEMORY INTERLUDE: PRACTICING TO BE A SCIENTIST

Although I have assigned, evaluated, and graded and will assign, evaluate, and grade numerous science fair projects, my first, and my only, science fair project was in a grey, three-by-five (5.5 inches by 3.5 inches by 2.5 inches), steel card file box.

> A grey, three-by-five (5.5 inches by 3.5 inches by 2.5 inches), steel card file box sat on a table. If this was today, I would say that to the right of the grey, three-by-five file box there stood a white trifold display board with a description of a science project, and to the left of the grey, three-by-five file box was a white trifold display board with a description of a science project. Alone, naked, stood a grey, three-by-five file box. Date: seventh grade. Zoom in on the grey, three-by-five file box. Open the grey, three-by-five file box. Look inside the grey, three-by-five file box. See the three-by-five cards in the grey, three-by-five file box. Notice that there are no buff-colored manila dividers. Observe the three-by-five, unruled index cards organized based on the American Kennel Club alphabetic organization, rather than on the world's traditional and classic alphanumeric order. Heed the breed of dog, the breed's country of origin, and the height and weight of the breed handwritten on a three-by-five, unruled index card. A grey, three-by-five (5.5 inches by 3.5 inches by 2.5 inches), steel

> card file box containing three-by-five, unruled index cards was my seventh-grade science fair project.

By seventh grade I did things that earned the comment, "You should be a scientist." I wanted to be a veterinarian. Are not veterinarians scientists? Veterinarians need to know the different breeds of dogs. I need to name all the parts of the dog if I was to be a veterinarian. It was the same as when I, without being asked, named all the books of the Old Testament and all the books of the New Testament years ago when I was going to be a preacher man, "this man full of The Word, with The Power and The Light" (Rylant, 1986, p. 8) who "grips the pulpit and gazes raptly into the air over the heads of the congregation as if he can see the Savior there; his voice swelling with holy thunder" (Grimsley, 1995, p. 1) and "drawing on Scripture, tortuously interpreted, to support these statements" (Bell, 1992, p. 184) as if he "speak[s] with that authority which could only come from G-d" (Baldwin, 1952/1985, p. 94). I did not include the names of all the parts of dogs because I was learning the scientific names of all the bones in the dog skeleton. The three-by-five (5.5 inches by 3.5 inches by 2.5 inches), steel card file box was to demonstrate my science, my knowledge of dog breeds. Furthermore, the three-by-five (5.5 inches by 3.5 inches by 2.5 inches), steel card file box represented my "strong, positive feelings of self-efficacy in science[,] … [my comfort] with being identified as a scientist[,] … [and my] definitive plans to attend college and study science or engineering" (Dublin, Sigman, Anderson, Barnhardt, & Topkok, 2014, pp. 171–173).

COMMENTARY ON THE READING OF *A SEPARATE PEACE*, CONTINUED

The reading of *A Separate Peace* was not in isolation. At the same time, I was engaging in doing science, albeit school Science. Parallel to learning to read was learning science and mathematics at which, if memory is correct, I excelled. Science, in school, was the naming of things, vocabulary, or learning as placing objects into long-term memory (Freire, 1970/2000).

> This boy said to me, "See that bird standing on the wheat there? What's the name of it?" I said, "I haven't got the slightest idea." He said, "It's a brown-throated thrush. Your father doesn't teach you much about science."
>
> I smiled to myself, because my father had already taught me that that doesn't tell me anything about the bird. (Feynman, 1969, p. 316)

Tenth grade was when I formally declared, to myself, silently, that I wanted to be a scientist and abandoned school reading because "I wanted to know which way was right and which way was wrong" (Rich, 2014). Tenth-grade biology became a place, "a sense of self and a sense of being in the world" (Whitlock, 2007, p. 2), to demonstrate what I had become, who I am, and the meaning science. I gave myself an identity. I needed an identity. It was in 10th grade that science became my identity, who I was to become and to be.

School science, as place, necessitates science as historical, cultural, and political as well as sexed, gendered, classed, and raced (Pinar, 2001). Through running from, avoiding, and disavowing my race, my class, and my sex, I adopted and became school Science as “an incomplete being in a sea of possibility” (Phelan, 2001, p. 129), studying, joining, and embodying a school Science which would be “the personal, social, and ecological dimensions of where [I] live [my] lives (He & Ross, 2012, p. 6).

Again implicit, but not explicit, in the description of the rejection of a Humanities identity and the adoption of a school Science identity is school science (and mathematics) as the place where I could be successful because being a Nigger was not necessary. However, being color-blind, oft writ white, was required. I remember the pre-eighth-grade Nigger social and cultural dictum to acknowledge being a Nigger, but in order to be “successful,” especially in school, I needed to become white. Being a man who had sex with men was not necessary, no matter that the Hall of the Sciences and Engineering was the home to men. To be successful in school Science, I did not need the lived experiences of authors, I did not need the lived experiences of others, I did not need myself. I just needed to know the facts, to be objective. Additionally, I, as a scientist, could become heteronormative, white, masculine, and elite—all the things that this 10th-grade reader of *A Separate Peace* wanted to be.

SNOW QUEEN

Propositions

First, Snow Queen is a metaphor for “becoming white” predominately as part of the autobiographical memories of a person who is labeled science educator. Second, science, as an ontology and epistemology, is White. For example, in the call for increasing underrepresented peoples in science, science is opening opportunities, especially for me and Niggers, to be psychologically and culturally the overrepresented majority, oft writ White. Last, socially, politically, and psychologically, Snow Queen and science are metaphors in that they both resemble the same object of desire, Whiteness. If a Snow Queen is one who desires White, then the Nigger who is desiring science is desiring White. In other words, a Snow Queen, metaphorically, is a person who loves/does science. Thus, the result of queering science is to say that school Science is a means to making oneself White, if only performatively.

Definitions

> According to Green (2010), the term *queen* means a gay man. However, Baker (2002, p. 49) suggests that the homosexual use of *queen* stems instead from the older word *quean*, meaning harlot. Unlike *queen*, which is used for women of high rank, *quean* has connotations of low status and eventually became associated with homosexuals. (Hamilton, 2014, p. 163)

The connotation of "queen" is masquerading, "embrac[ing] gender fluidity by performing a separate identity that they could put on and take off" (Taylor & Rupp, 2004, p. 120). In other words, being, becoming, a queen is performative, and "understood not as a singular or deliberate 'act,' but rather, as the reiterative and citational practice by which discourse produces the effects that it names" (Butler, 1993, p. 2). In defining queen, the effect is "the jester" of the gay community (Simmons, 2014, p. 631) or the stereotype of heteronormative hegemonic womanhood, hyperfeminine (Hillman, 2011). For example, "'Walking down the runway, I'm feeling super sexy, and flirty, and I'm just having a good time out there. It's a fishy outfit. I look really pretty. And I look glamorous.' This reveals one example of the benefits of speaking like a true queen" (Simmons, 2014, p. 645).

Who is a Snow Queen? "Super-clean prim and proper middle-class men totally isolated from the African American community, including other black gay men, and only capable of being attracted to white men[,] … the gay black man as a racially isolated 'snow queen,'" (Ross, 2013, p. 679). "A gay man of color who desires white men" (Muñoz, 1997, p. 94) "who almost exclusively desire other white men" (p. 99), and "who has been locked out of the elite white gay male sphere of influence" (p. 100). Furthermore, as the "educated Negro [who] is compelled to live and move among his own people whom he has been taught to despise" (Woodson, 1933/1990, p. 5), a Snow Queen is telling, "I'm surrounded by white folks," you know. "I love them, and they love me" (X, 1965, n.p.). Clinging to

> the collection of myths to which white Americans cling: that their ancestors were all freedom-loving heroes, that they were born in the greatest country the world has ever seen, or that Americans are invincible in battle and wise in peace, that Americans have always dealt honorably with Mexicans and Indians and all other neighbors or inferiors. (Baldwin, 1996, p. 101)

Me as Snow Queen

I, as a Nigger, wrestled, dissected, and explicated the conflict between my public mask and my private face. Butterfield (1974) posits that an African American, Afro-American, Afro-american, American, Negro, black, dark-skinned person who lived in a United States that glorified and justified slavery, peonage, segregation, and lynching (Woodson, 1933/1990; Pinar, 2001) had

> to learn to wear a mask to seem as if he fits the owner's concept of 'the nigger.' At the same time he has an identity of his own that must be hidden, because it is a threat to the slave system. In effect, he maintains a double identity and shifts between the two according to the occasion. If the mask slips, the [slave] does not suffer existential crisis, but arrest, capture, whipping, and possible death. (Butterfield, 1974, p. 20)

Thus, as stated through the memories of *A Separate Peace*, my flight to becoming a Snow Queen is the result of my flight to being a successful student—"the colonized is elevated above his jungle status in proportion to his adoption of the mother country's cultural standards" (Fanon, 1968/2008)—as well as my flight to an identity.

However, as a metaphor, Snow Queen is

> a consciousness of a frozen "outside," of a being purely as seen by others, in the face of the lived-experience of an "inside," of a being who is able to see that he or she is seen as a being without a point of view, which amounts to not being seen as a human being. (Gordon, 2005, p. 2)

The Snow Queen wears a mask "that we fear we cannot live without and know we cannot live within" (Baldwin, 1996, p. 90) in order to ignore, to disavow, and to repudiate a private face. In other words, in order to keep from being one with neither a public mask nor a private face, and to embrace science as a place to be someone else, the Snow Queen adorns a public mask. My "success" with science has privileged the public mask immortalized as the mis-educated (Woodson, 1933) educated or the Talented Tenth (Du Bois, 1903/1996; 1948/1996): "The top 10 percent of the black race would serve as the leadership vanguard of the race, elevating the remaining 90 percent by their efforts, achievements, and advocacy" (Williams, 2009, p. 201 n.1)—albeit with a "double-consciousness, this sense of always looking at one's self through the eyes of others, of measuring one's soul by the tape of a world that looks on in amused contempt and pity" (Du Bois, 1903, p. 2).

Furthermore, the label of Snow Queen posits that the desire, if not the practice, of the Nigger to have sex with white men would permit the Nigger to become White. I, now as Snow Queen, can run from myself in order to run into the arms of white men. "I would obtain the benefits of being White if I did what Whites did" (Hayes, 2013, p. 17). But in Rodney Evans's *Brother to Brother* (Evans, McKay, Robbins, & Schoof, 2004), Eldridge Cleaver spat out at James Baldwin's question in a short exchange:

> Baldwin: What did I ever do to you?
> Cleaver: You let the white man fuck you in the ass. What does that make you? That makes you the lowest scum on the Earth. (34:07–10)

In other words, I, the Nigger who becomes the Snow Queen, can become white. At worst, I become, as the Talented Tenth, "the hypocrite Negro … [who] denies his/her Black self and the latter denies the reality of being born on non-African soil and influenced by Euro-Americans" (Crozier, 2008, p. 79)—the Negro in Black face. At worst,

> I begin to recognise that the Negro is the symbol of sin, I catch myself hating the Negro. But then I recognise that I am a Negro. There are two ways out of this conflict. Either I ask others to pay no attention to my skin, or else I want them to be aware of it. I try then to find value for what is bad—since

> I have unthinkingly conceded that the black man is the colour of evil. … I am compelled to choose an unhealthy, conflictual solution, fed on fantasies, hostile, inhuman, to reach out for the universal. (Fanon, 1968/2008, p. 197)

As being a Snow Queen, I cannot remove the black face, which is, itself, exposing the private face, exposing the Nigger who is a complicated private face of double consciousness, itself not either/or. “[D]ouble consciousness [i]s the existence of two competing souls, at war one with the other” (Williams, 2009, p. 18) or a “‘twoness’ of being an American, a Negro; two warring ideals in one dark body, whose dogged strength alone keeps it from being torn asunder” (Du Bois, 1903, p. 5).

> Two salient dimensions [of double consciousness are] a sense of inferiority that manifests in holding stigmatic conceptions toward a group in which one clearly is a member … [and] endorsing ideologies or policies that are repressive to the group in which one is a member, and therefore ultimately to oneself. (Wasserman, Clair, & Platt, 2012, p. 335)

Furthermore, “the double consciousness often emerges in the form of prejudices against groups to which one belongs” (p. 336), or in the words of a 4-year-old African American boy:

> I want to be white.
> Whites are good. If my family is white, then I would be white.
> Being black … being an African American isn’t good. That’s the reason why.
> (Lamb, 1991, p. b05, suspension points original)

As the private face of a Snow Queen, a Nigger longs to be white by means of being one of the Talented Tenth and embodying a double consciousness—oft writ a Nigger loving a white man.

Interlude: Amo, Amas Amat, Amamus, Amatas, Amant

Before the dance cards are handed to Science and Snow Queen, I recollect another memory of schooling that informs my and the Nigger’s meaning making. In my quest to be a Snow Queen as well as a motivator for my seventh-grade science fair project—a grey, three-by-five (5.5 inches by 3.5 inches by 2.5 inches), steel card file box—I have since 1977 owned, bred, shown, and judged Salukis under the banner of Odi et Amo (Figure 1.1).

> Odi et amo. quare id faciam fortasse requiris.
> nescio, sed fieri sentio et excrucior. (Catullus, 2002, p. xxxix)

What do you mean that “you love” *and* “you hate”? In my Latin III class during 10th grade, which included relearning, or maybe learning, Latin I and II for the first time, the concept of *odi et amo* fueled my desire and my lust for the White males and allowed me to begin to understand my feelings.

Figure 1.1. Odi et Amo Angelic Angelica Elessar (photograph © L. Sue Rooney-Flynn)

> I hate and I love. You ask me why?
> I do not know, but I feel it occurs and I am tortured.

Science and Snow Queen, as terms, relationally parallel *odi et amo*.

Science and Snow Queen Dancing

I do not want to spend time arguing that science is white, but rather simply to state that Science is White. If Science is an ontology and an epistemology, by extension, Science is both the embodiment of White and the White way of knowing. But both of these may be untrue if Science does not seek to have an identity—to be like something else. However, if truth be told, science desires, needs, an identity (Snow, 1959/2012). Science is the work of White people. Therefore, since the objects of desire for the Snow Queen are white men, then for me, the Nigger, to desire science is to be a Snow Queen.

As stated earlier, I did not go to the Sciences in order to fulfill my desire to be white, but to salvage my schooling—to be obedient, to be compliant, and to be complacent (Leafgren, 2016)—as well as, in the mind of a high school student, to hang out with the academically privileged (white and heterosexually-inferred) school boys, to take on the identity of being "smart" at the same time that I was refusing to accept my sexual being. This "hanging out" with white people continued throughout my schooling. In the quest for being intelligent, smart, and gifted, which Muñoz (2007) posits as coded racist descriptors, I become a successful student of the Sciences. The Nigger, the Snow Queen, garnered an identity that is raceless (Fordham, 1988) but writ White in the same sense that when we talk of America (sic), the unuttered, unarticulated, and unquestioned is White America.

I cannot get away from being a student of Science and an individual with the identity of Science—scientist—meaning the Nigger needs to become the Snow Queen: to desire White and male. Convenient for me, a major social justice movement aims to increase underrepresented people in Science. Who are the underrepresented people? Non-White people. Nonmale people. In other words, Science, as an ontology and an epistemology, is White. The request for underrepresented people does not alter or negate Science as the ontology of being White and the epistemology of knowing based on White lived experiences. Rather than being a movement to queer, "to produce a story of self and other with which one cannot identify" (Miller, 1998, p. 373) or to "challenge the determinism, essentialism, and desire for normality" (Fifield, 2012, p. 2), the ontologies and epistemologies of Science, if only defined by school Science, remain fixed and controlled while desiring to increase its number of piously devoted practitioners.

Furthermore, as the Snow Queen dances in the arms of and on the feet of Science, Science becomes a Nigger if only because of the hypodescent one-drop rule (Jordan, 2014; Plecker, 1925). Science, now a Nigger and like the Nigger, desires White, if only to maintain its identity and its dominance and supremacy, and thus becoming a Snow Queen. Thus, I, like the Nigger, is now White. I, like the Nigger, is now Science. I, like the Nigger, is the Snow Queen. I, like the Nigger (and through convoluted if not eugenic logic, the Snow Queen and Science), is all that and concurrently, queerly, a stranger and a monster. In other words, since Science, like the Snow Queen, seeks the same object of desire, White, to queer Science is to describe Science as a stranger and a monster. But before toying with Science as a stranger and a monster, I conceptualize the stranger and the monster.

As "ambiguous figures who transgress normal ways of being and knowing" (Fifield, 2012, p. 9), a stranger is an "individual of our times and civilization who tries to be permanently accepted or at least tolerated by the group which he approaches" (Schutz, 1976, p. 19). However, the stranger exists (with)in the borders of insider/outside, the Norm/the Other, invisible/omnipresent (Marotta, 2000), closeness/remoteness and threat/excitement (Levine, 1977), absorbed/excluded, and expulsion/assimilation (Karakayali, 2006). Thus, strangers dismantle, deconstruct, and destroy watertight borders that "protect the separation of 'natives' from 'aliens'—of 'us' from 'them'" (Bauman, 2010, p. 426). The stranger is a complex being, at one time being and at one time not being. The stranger does not want to *be* the Norm, does not want to *be* the Other, and does not want to *constantly become* (Kumashiro, 2000). Strangers are nonmembers (Bauman, 1995), interactive nonmembers (Simmel, 1964, 1977). "Those who do not conform to the values and world view of the host society are automatically classified as strangers" (Marotta, 2000, p. 129), whereas the Other refers to "the deprived, the weak and the destitute" (p. 122) or

> those groups that are traditionally marginalized in society, i.e., that are *other than* the norm, such as students of color, students from under- or unemployed families, students who are female, or male but not stereotypically "masculine," … students who are, or are perceived to be, queer [,] … students with disabilities, students with limited or no English-language proficiency, and students from non-Christian religious backgrounds. (Kumashiro, 2000, p. 26)

Thus, "the stranger is a non-member of one's community but one whose proximity to us undermines the opposition between 'us' and 'them'" (Vandekerckhove & van Hooft, 2010, p. xxiv).

"The monster, like myth, is a complex concept: at once abstract and concrete, and simultaneously specific and universal" (Beville, 2013, p. 7). Cohen (1996) explicates a monster as one who represents desires, attractions, and fantasies; reveals fear, desire, and anxiety; appears omnipresent, including after disappearing; threatens conformity and avoids categorization; and incorporates the Other, the Outsider, and the Beyond.

> Monsters are our children. They can be pushed to the farthest margins of geography and discourse, hidden away at the edges of the world and in the forbidden recesses of our mind, but they always return. And when they come back, they bring not just a fuller knowledge of our place in history and the history of knowing our place, but they bear self-knowledge, *human* knowledge—and a discourse all the more sacred as it arises from the Outside. These monsters ask us how we perceive the world, and how we have misrepresented what we have attempted to place. They ask us to reevaluate our cultural assumptions about race, gender, sexuality, our perception of difference, our tolerance toward its expression. They ask us why we have created them. (p. 20)

Conceptually, Science as a stranger and Science as a monster moves Science from the normal to the normal/not normal (stranger) and the normal being the not normal (monster), thus queering Science, without slipping into the politics, hazards, of identity. Strangers try to "shake off and forfeit their separate identities and to 'dissolve' in the uniform mass of 'the natives'" (Bauman, 2010, p. 429), "staying unknown … [or otherwise] something absolutely incomprehensible" (Hirschauer, 2005, p. 59). Monsters "fall outside accepted categories; threaten the existence of well-defined and distinct categories" (Holtz & Wagner, 2009, p. 13); and "elude and disturb a clear notion of identity, system and order" (Lassén-Seger, 2006, pp. 48–49). Furthermore, like a stranger and like a monster, I, desirous of Science as a Snow Queen, need a Science that does not exist unqueered—as a knowable, explicable, learnable and powerful simple identity—but rather a Science that is a monster and a stranger. A monstrous strange Science is an alternative (cf. Andersson, 2012) epistemological science that is neither a normal science nor a nonnormal science. Thus, it makes sense that I, as a Snow Queen and as a Nigger, desire a monstrous strange Science and thus must cradle, swaddle, and nourish (school) Science like Roy when "He stares at Karolis's body for a second, looks quickly over his shoulder, then runs and kneels before the body, weeping and cradling the head in his arms" (Jones/Baraka, 1964, pp. 58–59).

CODA

I gave the second iteration of this chapter, at that time titled "Becoming a Snow Queen: A Stranger in a Place Called (School) Science (Education)," to undergraduates in my course Education Equity and Excellence in Culturally Pluralistic Society with the purpose of explicitly "outing" myself during the last class session of the semester. The first question, after the uneasy silence, was, "Is this autobiographical?" I quickly answered, "Yes."

But what is autobiography? Which one of the aforementioned relationships suggested by the title moves beyond memoir to be the or an autobiographical statement restating and reinforcing the notion that (school) Science, as a thing, has its autobiography too? If anything, queering (school) Science is creating the memories which may become autobiography. In other words, looking at and into queer (school) Science memories does not create nor write the autobiography, for autobiography is relational, between me and the "other than me"—albeit person, place, or thing—that (excuse the normative binary) may turn me into a label or save me from becoming a label, someone else, such as "Nigger," "Snow Queen," or "(school) Science."

NOTE

[1] First, the use of quotation marks around words and terms throughout this chapter, including in this case "Nigger," is to problematize the self-evident word in the tradition of Ida B. Wells (Pinar, 2001, p. 27). Second, *Nigger* is used to connote descendants of enslaved black peoples of the United States as described and used by Thomas Jefferson (1787/2011) among others, including Mark Twain (Smith, 1984), rather than descendants of citizens of countries on the continent of Africa who may identify as African Americans. Third, as poorly and weakly argued by Broadway (2007), Nigger is an identity statement expressed as Negro, African American, Afro-American, Afro-american, American, black, dark-skinned person, etc.

REFERENCES

Adamczak, B. (2017). *Communism for kids*. Cambridge, MA: MIT Press.

Andersson, L. (2012). There is no alternative: The critical potential of alternative media in the face of neoliberalism. *TripleC: Communication, Capitalism & Critique. Open Access Journal for a Global Sustainable Information Society, 10*(2), 752–764.

Baldwin, J. (1952/1985). *Go tell it on the mountain*. New York, NY: Laurel.

Baldwin, J. (1996). *The fire next time*. New York, NY: Vintage.

Bauman, Z. (1995). *Life in fragments: Essays in postmodern morality*. Hoboken, NJ: Wiley.

Bauman, Z. (2010). Migration and identities in the globalized world. *Philosophy Social Criticism, 37*(4), 425–435.

Bell, D. (1992). *Faces at the bottom of the well: The permanence of racism*. New York, NY: Basic Books.

Beville, M. (2013). *The unnameable monster in literature and film*. New York, NY: Routledge.

Brady, O. E. (1976). Cultural conflict and cult ritual in LeRoi Jones's "the toilet." *Educational Theatre Journal, 28*(1), 69–77.

Broadway, F. S. (2007). To be a male, a Negro, a queer: Pedagogy in (science) education. In J. Settlage & A. Johnson (Eds.), *Proceedings of the Science Education at the Crossroads Conference* (pp. 14–15). Amherst, MA: Editors.

Broadway, F. S. (2011). Queer (v.) queer (v.): Biology as curriculum, pedagogy, and being albeit queer (v.). *Cultural Studies of Science Education, 6*(2), 293–304.

Butler, J. (1993). *Bodies that matter: On the discursive limits of "sex."* New York, NY: Routledge.

Butterfield, S. (1974). *Black autobiography in America*. Amherst, MA: University of Massachusetts Press.

Catullus, G. V. (2002). *The complete poetry of Catullus* (D. Mulroy, Trans.). Madison, WI: University of Wisconsin Press.

Cohen, J. J. (1996). Monster culture (seven theses). In J. Cohen (Ed.), *Monster theory: Reading culture* (pp. 3–25). Minneapolis, MN: University of Minnesota Press.

Crozier, K. D. (2008). The luminous darkness of Du Bois' double consciousness: Through the lens of contemporary Christian women. *Pastoral Psychology, 57*(1–2), 77–88.

Delany, S. R. (1996). Coming/out. In P. Merla (Ed.), *Boys like us: Gay writers tell their coming out stories* (pp. 1–26). New York, NY: Avon Books.

Dublin, R., Sigman, M., Anderson, A., Barnhardt, R., & Topkok, S. A. (2014). COSEE-AK ocean science fairs: A science fair model that grounds student projects in both western science and traditional native knowledge. *Journal of Geoscience Education, 62*(2), 166–176.

Du Bois, W. E. B. (1903). *The souls of Black folk*. New York, NY: Penguin.

Du Bois, W. E. B. (1903/1996). The talented tenth memorial address. In H. L. Gates & C. West (Eds.), *The future of the race* (pp. 133–157). New York, NY: Vintage.

Du Bois, W. E. B. (1948/1996). The talented tenth. In H. L. Gates & C. West (Eds.), *The future of the race* (pp. 159–177). New York, NY: Vintage.

Evans, R. (Director & Producer), McKay, J., Robbins, I., & Schoof, A. (Producers). (2004). *Brother to brother* [Motion picture]. United States: Wolfe Releasing.

Fanon, F. (1968/2008). *Black skin, White masks*. New York, NY: Grove Press.

Feynman, R. P. (1969). What is science? *The Physics Teacher, 7*(6), 313–320. Retrieved from https://doi.org/10.1119/1.2351388

Fifield, S. (2012). *Teachers as sexual strangers* (Unpublished manuscript). Retrieved from https://www.researchgate.net/publication/324276952

Fordham, S. (1988). Racelessness as a factor in Black students' school success: Pragmatic strategy or pyrrhic victory? *Harvard Educational Review, 58*(1), 54–85.

Fordham, S., & Ogbu, J. U. (1986). Black students' school success: Coping with the "burden of 'acting White'. *The Urban Review, 18*(3), 176–206.

Freire, P. (1970/2000). *Pedagogy of the oppressed*. New York, NY: Continuum.

Gordon, L. R. (2005). Through the zone of nonbeing: A reading of Black skin, White masks in celebration of Fanon's eightieth birthday. *The CLR James Journal, 11*(1), 1–43.

Grimsley, J. (1995). *Dream boy*. New York, NY: Scribner Paperback Fiction.

Halberstam, J. (2011). *The queer art of failure*. Durham, NC: Duke University Press.

Hamilton, R. (2014). Exploring the metaphorical use of colour with the historical thesaurus of English. In W. Anderson, C. P. Biggam, C. Hough, & C. Kay (Eds.), *Colour studies: A broad spectrum* (pp. 153–166). Amsterdam: John Benjamins Publishing Company.

Harding, S. (Ed.). (1993). *The "racial" economy of science: Toward a democratic future*. Bloomington, IN: Indiana University Press.

Hayes, C. (2013). Too White to be Black and too Black to be White: A journey away from Whiteness. In C. Hayes & N. D. Hartlep (Eds.), *Unhooking from Whiteness: The key to dismantling racism in the United States* (pp. 17–31). Rotterdam, The Netherlands: Sense Publishers.

He, M. F., & Ross, S. (2012). Narrative of curriculum in the south: Lives in-between contested race, gender, class, and power. *Journal of Curriculum Theorizing, 28*(3), 1–9. Retrieved from http://journal.jctonline.org/index.php/jct/article/view/378/pdf

Heath, S. B. (2004). Ethnography in communities. In J. A. Banks & C. A. McGee Banks (Eds.), *Handbook of research on multicultural education* (2nd ed., pp. 146–162). San Francisco, CA: Jossey-Bass.

Hesiod. (2004). Work and days. In A. N. Athanassakis (Trans.), *Hesiod* (2nd ed.). Baltimore, MD: Johns Hopkins University Press.

Hillman, B. L. (2011). "The most profoundly revolutionary act a homosexual can engage in": Drag and the politics of gender presentation in the San Francisco gay liberation movement, 1964–1972. *Journal of the History of Sexuality, 20*(1), 153–181.

Hirschauer, S. (2005). On doing being a stranger: The practical constitution of civil inattention. *Journal for the Theory of Social Behavior, 35*(1), 41–67.

Holtz, P., & Wagner, W. (2009). Essentialism and attribution of monstrosity in racist discourse: Right-wing internet postings about Africans and Jews. *Journal of Community & Applied Social Psychology, 19*(6), 411–425.

Jefferson, T. (1787/2011). *Notes on the state of Virginia*. Chapel Hill, NC: University of North Carolina Press Book.

Jenkins, H. (1992). *Textual poachers: Television fans & participatory culture*. New York, NY: Routledge.

Jones, L., & Baraka, I. A. (1964). The toilet. In *The baptism & the toilet*. New York, NY: Grove Press.

Jordan, W. D. (2014). Historical origins of the one-drop racial rule in the United States. *Journal of Critical Mixed Race Studies, 1*(1), 98–132. Retrieved from http://escholarship.org/uc/item/91g761b3

Karakayali, N. (2006). The uses of the stranger: Circulation, arbitration, secrecy, and dirt. *Sociological theory, 24*(4), 312–330.

Knowles, J. (1959/1975). *A separate peace*. New York, NY: Bantam Books.

Kuhn, T. S. (1970). *The structure of scientific revolutions*. Chicago, IL: University of Chicago Press.

Kumashiro, K. K. (2000). Toward a theory of anti-oppressive education. *Review of Educational Research, 70*(1), 25–53.

Lamb, Y. S. (1991, September 9). Children; race relations & pre-schoolers; it's never too early to begin building self-esteem: [Final Edition]. *The Washington Post*, p. b05. Retrieved from https://search-proquest-com.udel.idm.oclc.org/docview/307426702?accountid=10457

Lassén-Seger, M. (2006). *Adventures into otherness: Child metamorphs in late twentieth-century children's literature*. Turku: Åbo Akademi University Press. Retrieved from https://www.doria.fi/bitstream/handle/10024/4127/TMP.objres.85.pdf?sequence=2&isAllowed=y

Leafgren, S. L. (2016). *Reuben's fall: A rhizomatic analysis of disobedience in kindergarten*. New York, NY: Routledge.

Levine, D. N. (1977). Simmel at a distance: On the history and systematics of the sociology of the stranger. *Sociological Focus, 10*(1), 15–29.

Marotta, V. (2000). The stranger and social theory. *Thesis Eleven, 62*(1), 121–134.

Miller, J. L. (1998). Autobiography as a queer curriculum practice. In. W. F. Pinar (Ed.), *Queer theory in education* (pp. 365–373). Mahwah, NJ: Lawrence Erlbaum Associates.

Morgan, A. (2004). *Sons*. Brooklyn, NY: Lane Street Press.

Muñoz, J. E. (1997). "The White to be angry": Vaginal Davis's terrorist drag. *Social Text 52/53, 15*(3–4), 80–103.

Muñoz, J. E. (2007). Cruising the toilet: LeRoi Jones/Amiri Baraka, radical black traditions, and queer futurity. *GLQ: A Journal of Lesbian and Gay Studies, 13*(2), 353–367.

Nash, P. (1961). Training an elite. *History of Education Quarterly, 1*(1), 14–21.

Phelan, S. (2001). *Sexual strangers: Gay, lesbians, and dilemmas of citizenship*. Philadelphia, PA: Temple University Press.

Pinar, W. F. (1975/1994). The method of currere. In W. F. Pinar (Ed.), *Autobiography, politics and sexuality: Essays in curriculum theory, 1972–1992*. New York, NY: Peter Lang.

Pinar, W. F. (2001). *The gender of racial politics and violence in America: Lynching, prison rape and the crisis of masculinity*. New York, NY: Peter Lang.

Pinar, W. F. (2004). *What is curriculum theory?* Mahwah, NJ: Lawrence Erlbaum Associates.

Pinar, W. F., & Grumet, M. R. (1976). *Toward a poor curriculum*. New York, NY: Kendall/Hunt.

Plecker, W. A. (1925). Virginia's attempt to adjust the color problem. *American Journal of Public Health, 15*(2), 111–115.

Purdy, M. A. (2016). Blurring public and private: The pragmatic desegregation politics of an elite private school in Atlanta. *History of Education Quarterly, 56*(1), 61–89.

Rich, M. (2104, June 29). Math under common core has even parents stumbling. *The New York Times*. Retrieved from http://www.nytimes.com/2014/06/30/us/math-under-common-core-has-even-parents-stumbling.html

Romero, M. (2017). Reflections on "the department is very male, very white, very old, and very conservative": The functioning of the hidden curriculum in graduate sociology departments. *Social Problems, 64*(2), 212–218.

Rosenblatt, L. (1938/1985). *Literature as exploration*. New York, NY: Modern Language Association of America.

Ross, M. B. (2013). "What's love but a second hand emotion?" Man-on-man passion in the contemporary Black gay romance novel. *Callaloo, 36*(3), 669–687.

Rylant, C. (1986). *A fine white dust*. New York, NY: Bradbury Press.

Schutz, A. (1976). The stranger. In A. Brodersen (Ed.), *Collected papers II* (pp. 91–105). The Hague: Springer.

Sears, J. (1999). Teaching queerly: Some elementary propositions. In W. J. Letts & J. Sears (Eds.), *Queering elementary education: Advancing the dialogue about sexualities and schooling* (pp. 97–110). Lanham, MD: Rowman & Littlefield.

Simmel, G. (1964). The quantitative aspect of the group/the stranger. In K. Wolff (Ed.), *The sociology of Georg Simmel* (pp. 87–174, 402–408). New York, NY: Free Press.

Simmel, G. (1977). *The problems of the philosophy of history: An epistemological essay*. New York, NY: Free Press.

Simmons, N. (2014). Speaking like a queen in RuPaul's drag race: Towards a speech code of American drag queens. *Sexuality & Culture, 18*(3), 630–648.

Smith, D. L. (1984). Huck, Jim, and American racial discourse. *Mark Twain Journal, 22*(2), 4–12.

Snow, C. P. (1959/2012). *The two cultures*. Cambridge: Cambridge University Press.

Taylor, V., & Rupp, L. J. (2004). Chicks with dicks, men in dresses: What it means to be a drag queen. *Journal of Homosexuality, 46*(3–4), 113–133.

Tribunella, E. L. (2011). Refusing the queer potential: John Knowles's a separate peace. In M. A. Abate & K. Kidd (Eds.), *Over the rainbow: Queer children's and young adult literature* (pp. 121–142). Ann Arbor, MI: University of Michigan Press.

Vandekerckhove, W., & van Hooft, S. (2010). Introduction. In S. van Hooft & W. Vandekerckhove (Eds.), *Questioning cosmopolitanism* (pp. 105–120). Dordrecht: Springer.

Wasserman, J. A., Clair, J. M., & Platt, C. (2012). The "homeless problem" and the double consciousness. *Sociological Inquiry, 82*(3), 331–355.

Whitlock, R. U. (2007). *This corner of Canaan: Curriculum studies of place & the reconstruction of the south*. New York, NY: Peter Lang.

Williams, Z. R. (2009). *In search of the talented tenth: Howard University public intellectuals and the dilemmas of race, 1926–1970*. Columbia, MO: University of Missouri Press.

Wilson, Z. V., & Frank, J. R. (1983). *They took their stand: The integration of southern private schools*. Atlanta, GA: Mid-South Association of Independent Schools.

Wolcott, H. F. (2002). *Sneaky kid and its aftermath: Ethics and intimacy in fieldwork*. Walnut Creek, CA: Altamira.

Woodson, C. G. (1933/1990). *The mis-education of the Negro*. Trenton, NJ: Africa World Press.

X, M. (1965, February 21). *Speech at Ford Auditorium*. Retrieved from http://www.blackpast.org/1965-malcolm-x-speech-ford-auditorium#sthash.8PXRVz5m.dpuf

ABOUT THE AUTHOR

Francis S. Broadway is a professor of education in the LeBron James Family Foundation College of Education at the University of Akron, Ohio (USA). Current ideas of interest include blackfacing in children's literature, communism for kids (see Adamczak, 2017), currere as an instructional strategy for teacher education (especially moving beyond the regressive step to the progressive, analytical, and synthetic), and the community as a place called school. The odd couple, science and science education, is his context for exploring current interests and, in retrospect, what led to him becoming a high school and middle school science teacher before joining Academia, which has only confused him more. You can reach Francis at fsb@uakron.edu

STEVE FIFIELD AND WILL LETTS

2. QUEERING SCIENCE EDUCATION WITHOUT MAKING TOO MUCH SENSE

You're the perfect monster, and we've given you
a vengeful, hungry, human heart like ours.
—Charlotte Boulay, "The Bargain" (this volume)

INTRODUCTION

We've worked at intersections of science education and queer theories, on and off, for 20 years. In this chapter we organize some of what we've learned while troubling the urge to make sense around autobiographical reflections. We re-member and re-imagine what it was like to bring together science education and queer theories, such as we grasped them, beginning at a time when we knew of no similar work. Queer theories and science education are utterly entangled, but they have nothing in particular to do or not to do with each other. What emerges from their local and temporary contacts depends on the orientations of their meetings. Local and temporary, but karmic—because each dalliance conditions (im)possibilities for the next. Similarly for authors and readers—we and you are inextricable, but none of us knows what will come of our intimacies. So we're eager to share stories with others interested in the ongoing intercourse of science education and queer theories. We know you have more in mind-body to make and unmake of these tales than we imagine.

Working at intersections of science education and queer theories means laboring to locate or construe zones of exchange, and now and again settling into what we believe are such spaces to do something interesting there. *Intersection* is a fertile and flawed metaphor in cultural studies.[1] If everything is relational, can there really be anything that isn't already constitutively intersectional? In this sense science education is already implicated in the conditions of queer theorizing, and queer theorizing already arises in intimate, if nearly always unrecognized, association with science and science education.

Perhaps the point isn't that we must bring things into intersection, but that the quality of awareness we bring to things influences their ontologies. For those of us shaped by scientific, academic, and broader cultures that fetishize dividing the world into different *kinds* of things that preexist their relations, it takes mindful attention and countercultural values to engage (in) phenomena as co-constitutive, co-arising, intra-acting, and vastly implicated.[2] Even "intersection" suggests separate things

 | DOI:10.1163/9789004331068_004

coming together in space and time before heading off again on their own. Instead, we think of intersectionality as knowing and being in relational hotspots that reveal and multiply the mutual implication of things. With regard to science education and queer theories, we're interested in how relations recondition and reorient (Ahmed, 2006) by leaving residues of (im)possibility for what emerges next.

Another problem with an intersection is that you can't get very far as long as you remain there. You can explore the space from this side and that, but what's going on beyond the intersection? Are there other interesting intersections down the way? In our work we have come and gone and returned again to various intersections of science education and queer theories. Our visits have left traces of activity in and around a few intersections, and we have carried away residues that stick and make a difference even when we seem far from science education and queer theories. We have visited intersections of science education and queer theories in different contexts and guises, and later returned to find things somewhat changed, or utterly new, or pretty much as we remembered. This coming and going is an important quality of our work, though certainly not an intentional strategy, nor a wise career move.

Comings and goings lend lived experience an episodic, even fractured, quality that we typically stitch together and re-present for ourselves and others as coherent and continuous. We considered organizing this chapter as a series of signposts marking our journeys queering science education. But *signpost* suggests we see our work proceeding along clearly marked paths to and from something or somewhere. We want to resist that comforting conceit. We have settled instead on imagining *touchpoints* in our encounters with science education and queer theories. Cautious of making too much sense, but eager to elicit sensibilities, we explore areas of contact among ourselves, others, science education, and queer theories to see how they feel.

TOUCHPOINTS

> All touching entails an infinite alterity, so that touching the Other is touching all Others, including the "self," and touching the "self" entails touching the strangers within. (Barad, 2012, p. 214)

The following vignettes—fluid, retrospective (re)constructions of our thinking, teaching, presenting, and writing—came from touching into fraught and fulfilling memories that sustain and disturb us. Despite concerns about being too self-occupied, we take up our own call to action: "putting the self at risk to more fully participate in the complicities that the autobiographically coherent self denies, queers the preoccupation with self that so distracts conventional (and many queer) visions of (science) education" (Fifield & Letts, 2014, p. 399). So we offer these tales cautiously, aware of how easy it is to slip into autobiography as rational reconstruction of "truths" installed on the past to accommodate the present. We aim not to fall into that trap—we'll see how it goes. We're intent to avoid "a singularity of story," for "such singularity closes the doors to multiple, conflicting, and even odd and abnormal—queer—stories and identities" (Miller, 1998, p. 369). We write about moments that, for us, flit from seminal to run-of-the-mill and back again. We render times around which our memories are faulty or selective, our drives for self-preservation are high, and the allure, indeed the undertow, of a coherent story is almost impossible to swim against. Still, following Janet Miller (1998) we try to use

> autobiography in ways that [shift] autobiography in education from its modernist emphasis on producing predictable, stable, and normative identities and curricula to a consideration of "selves" and curricula as sites of "permanent openness and resignifiability." … Autobiography as a queer curriculum practice … cast[s] in new terms the ways in which we might investigate our multiple, intersecting, unpredictable, and unassimilable identities. (pp. 367, 368)

We begin with Will's touchpoints.

"Queer"—Noun or Verb?

Graduate school was transformative for me. Those were heady, indulgent days when a large part of what I did was read and write—in a manner I have never replicated since. I had been enrolled in a PhD program at the University of Delaware for about 18 months when colleagues approached me to present a paper in a symposium hosted by an on-campus group of lesbian, gay, bisexual, and transgender (LGBT) staff called, gorgeously, the Lavender Scholars. They sponsored an annual lecture series, and they wanted to include a session examining how the study of sexuality was changing various academic disciplines, such as fine arts, social sciences, and humanities. I was invited to present about the natural sciences. This would be my first attempt at sharing my budding formulations about the intersections of science and sexuality studies.

My paper, *Queering Science: Using Sexuality Theory and Research to Critique the Natural Sciences,* sought to "critique the natural sciences from the point of view of sexuality, and in particular from the point of view of sexual minorities" (Letts, 1995, p. 1). The critique was aimed at strengthening the natural sciences, not denigrating or eliminating them, although folks would repeatedly ask if my purpose was *really* to strengthen the natural sciences and whether such a critique could really facilitate or accomplish that. My (rather lofty) goal was to unify and codify many disparate and powerful critiques of the natural sciences, and to offer a framework for thinking about what this would/could look like.

Under the mentorship of Sandra Harding, then professor of philosophy and women's studies at the University of Delaware, my argument relied heavily on feminist standpoint epistemology. I had taken several courses with Sandra as I was crafting a minor in science and gender studies, and my work on this critique grew out of what we had been reading and discussing. As Harding (1991) wrote, standpoint theory enables us to think about "how to move from including others' lives and thoughts in research and scholarly projects to starting from their lives to ask research questions, develop theoretical concepts, design research, collect data, and interpret findings" (p. 268). My critique originated from standpoints based on sexuality—particularly the standpoints of sexual minorities. This was the first publicly shared piece where I invoked queer theory, but I was really concerned with "queers" and the standpoints of queers. It's clear to me, at least in retrospect, that I wasn't appreciating the messy schism between the identity politics of gay and lesbian studies and queer theory, and I was exploring what now looks like a muddled and internally inconsistent middle ground.

The seven-part framework I proposed centered on queer identities, using Halperin's (1995) far-reaching definition in which *queer*

> demarcates not a positivity but a positionality vis-à-vis the normative—a positionality that is not restricted to lesbians and gay men but is in fact available to anyone who is or who feels marginalized because of her or his sexual practices: it could include some married couples without children, for example, or even (who knows?) some married couples *with* children—with, perhaps, *very naughty* children. (p. 62, original italics)

As such my framework broadly encompassed who was queer. It named this broad group of people "for the purpose of acknowledging their lives by starting thought from their lives" (Letts, 1995, p. 3). I wanted both an inclusive and catchall term to include all these folks, while I also insisted that there was an incredible diversity of people included in this label and that we should resist letting it be homogenizing.

In a part of the framework called "queerly diverse," I wrote that it was

> a call to be wary of viewing this critique as an essentializing project. The use of the collective term queer is in no way meant to homogenize the enormous range of people (who outwardly may have seemingly little in common) who unite under this banner. (p. 10)

In this passage I tried to navigate between the usefulness and power of the collective noun queer and an insistence that this isn't an essentializing or homogenizing project—a delicate line to walk, indeed! In the paper I insisted that queer is a positionality rather than a positivity, but all the while I focused on the identities of people. That doesn't necessarily default to viewing identities as positivities, but it does require a bit of work to stake out how they are positionalities rather than positivities—work that, in retrospect, I hadn't completely elaborated.

I concluded the paper by proposing that standpoint theory can be a useful tool to engage and attempt to redress an epistemological crisis of the modern West, a crisis of authority that has challenged Western ways of knowing developed and refined for centuries. A queer critique of science (like the feminist, antiracist, and postcolonial critiques before) upends the traditional and much-loved conceptions of value-neutrality, rationality and objectivity.

What's clear to me, especially in retrospect, is that my earliest forays into thinking about and marshalling queer theory focused on queer as a noun. Queers were people from whose lives I felt it important to theorize. I acknowledged Deborah Britzman's (1995) formulation that

> those who seem uncomfortable think the term queer as a noun or an identity. But the queer and the theory in Queer Theory signify *actions*, not actors. It can be thought of as a verb, or as a citational relation that signifies more than the signifier. (p. 153)

And yet I was mostly conceptualizing and concerned with *queers* as people—however broadly defined—rather than, as Britzman advocated, with *queer* as a verb. Future work illustrates how I tried to pivot from the inclusion agenda largely rooted in identity politics to face a different direction, to reorient myself (see Ahmed, 2006) in order to think and theorize from people's lives. I aimed to push beyond the inclusion space. Where I did still explore an inclusion agenda (e.g., Letts, 1999b, 2001a, 2001b, 2002b), I started from queer theory to problematize more traditional understandings of identity categories and their irresistible binaries. Importantly, at this point in my thinking and writing, I hadn't yet accommodated how sex, pleasure, and the formation of sexual cultures posed profound challenges to the entrenched normative structures and practices of the natural sciences, or even schooling for that matter.

Outing My Research

From my earliest involvement in the National Association for Research in Science Teaching (NARST), which I'll use here as a proxy for the science education community (though I appreciate that the community exceeds NARST), I found freedom as a graduate student to identify and question normative structures and practices in our field. I felt what I thought at the time was a mandate to elucidate aspects of the field long hidden from view, and I relished the opportunity to work to hear silenced voices in what struck me to be, in the main, a very conservative field. I wrestled with and attempted to enact these desires in a variety of exhilarating ways.

For instance, in 1998 three colleagues and I presented at NARST in a symposium entitled Working and Learning on the Margins: Experiencing Doctoral Programs in Science Education as Non-traditional Graduate Students.[3] In my presentation (dramatically titled, I acknowledge) *But Still, Like Dust, I Rise: Reflections on Sexual Minority Issues in Science Education*, I explored what it was like for me as a graduate student "on the margins" to simultaneously "fight for the legitimacy of the questions I want to study while I study them" where the "general climate within science education ... feels inhospitable" (Letts, 1998, p. 2). This is why Maya Angelou's (1975) words in the paper's title were so evocative and ideal for me. Because in the paper I didn't explicitly state I was gay (all my friends and close colleagues already knew), the inference was that that's what made me a nontraditional graduate student. I worried that if I wrote I was gay, I would somehow fix others' sense of who I was, that I would be reduced to a label, to a caricature, for these labels "tend to lock people into fixed prescriptions for living" (Morris, 1998, p. 276). Identity politics and perhaps the politics of the closet (Sedgwick, 1990) were certainly at play in my life even as I tried to move away from them in my work!

In the paper I suggested a possible research agenda and provided some examples of what that research might entail. I wrote:

> The agenda originates with the questions, What can be seen from the margins that members of sexual minorities occupy? Is it different from the view from the center? What are the implications and consequences of such standpoints for science and for science education? I will … examine issues related to sexuality in relation to the curriculum, pedagogical practices, and people who occupy our schools. (Letts, 1998, p. 3)

I attempted to articulate a queer standpoint as "an epistemological stance … positioned against the 'positivist,' internalist epistemological ones that, explicitly or implicitly, largely continue to frame the knowledge-seeking projects about nature and society" (Harding, 1998, p. 17). The framework was intentionally and explicitly centered on "queer identities." I was interested in queer teachers and all teachers, queer students and all students, queer and all families and communities. I wondered how school science might be "another world" (Costa, 1995) for so many students (and teachers!).

In later work I took up parts of this proposed research agenda by examining the sexist and heteronormative nature of school science curricula, pedagogies, and policies (Letts, 1998, 1999a, 1999b, 2001a, 2001b), and the experiences of LGBT and other teachers and students in classrooms (Letts, 2004, 2005a, 2005b, 2006; Letts & Simpson, 2003).

Not Queer Enough

About a year after *Queering Elementary Education: Advancing the Dialogue About Sexualities and Schooling* (Letts & Sears, 1999) was published, a review of the book by William L. Leap (2000) appeared and subsequently preoccupied me for years. Pulling the book together felt like such a monumental accomplishment that I was just in the mood to celebrate. I was convinced we had broken new ground and even some of the hateful media attention the collection was garnering was a sign, I felt, that we had assembled something important, something that was getting noticed (if not carefully read by some!). The book had distracted me from working on my dissertation, so I had to have a pretty solid justification to warrant that suspended effort. And I felt I did. Leap's review was for the most part positive, but in the final paragraph he wrote,

> I have two very minor complaints: the title aside, there is very little in this collection that has to do with "queering" and/or the relevance of "queer theory" to lgbt education. Readers expecting detailed attention to these themes will be disappointed in this regard. (p. 499)

Not that queer? I puzzled. That was the whole reason we assembled the collection![4]

Although, strangely, it took me several years to own Leap's observation as on point, I eventually got there. Why did I initially perceive that remark as such a biting critique? I believe it speaks to desires and investments. My own (over)investment in the idea that we were engaged in cutting-edge, queer work, perhaps, and a loss of perspective about what good things the collection was provoking and inviting. At the time, "queer" seemed to be in the vanguard, and we were keen to push the field. But there wasn't yet enough work on queer theories and education out there (at least as we were tapping into it) to fill our collection. My own limited body of work at the time is indicative of this point! Folding queer theories into one's work or drawing upon them is not the same as a concerted project of queering, as Leap's review helped me see.

In retrospect I see *Queering Elementary Education* as a collection about (finally) working *towards* queering. Some chapters in the book did queer, as a verb, but Leap was quite correct—most of the chapters had their conceptual origins in gay and lesbian studies, not queer theorizing. It was a liminal collection attached to gay and lesbian studies and groping for queer things. I think because Steve and I were at that time working so hard to move away from (beyond?) that in our work, Leap's observation made me feel like I was caught in a gay and lesbian studies net. It was true, and thus somewhat dispiriting.

Positivities to Standpoints

At the 2000 meeting of the American Educational Research Association (AERA), Steve and I presented *Sexualities, Silence, and Science Teacher Education* (Letts & Fifield, 2000). Looking back on this now, it is clear to me that our conceptualizing sometimes exceeded our abilities to do what we aspired to do. That is, while we could articulate the construct of *queer as queering*, drawing on the extant literature and our own experiences, we struggled to find concrete examples to illustrate our points and to imagine how such conceptions would play out in the classroom, lab, or field.

We were concerned with the ways that the inclusion agenda in science education had either completely (willfully?) ignored issues of sexuality or attempted to address these issues in an add-LGBTIQQ-and-stir manner. Such scenarios "leave sexuality otherwise untheorized as a dimension of teacher practice and student learning" and demonstrate that "dominant conceptions of equity and multiculturalism reveal the modernist impulse to draw boundaries (Gieryn, 1999) and compartmentalize subjectivities" (Letts & Fifield, 2000, p. 1). In this work we wanted "to interrupt these modes of making selves and making sense by refusing stable identities and by producing new identifications that lie outside binary models of gender and sexuality" (Luhmann, 1998, p. 151).

Our project in the paper was ostensibly "to bring perspectives from these marginalized subject positions to bear on efforts to critique and improve science teacher education" (Letts & Fifield, 2000, p. 4) and to, in the process, mount a queer critique of many equity initiatives in science education. As we revised and built on text we had written a few years earlier, we wrestled to define a *queer community* without homogenizing or essentializing *queer*. We wrote that the

> notion [of] 'queer' highlights the immense diversity of the sexual minority community and necessitates that we be wary of inscribing a new essentialism. The use of the collective term queer is in no way meant to homogenize the enormous range of people who unite under this banner. ... [Our] use of the term is not intended to naturalize or essentialize the people who willingly adopt the label. Rather, it is used to help focus and direct our critique from the lives of so many that are so different and yet share a common set of oppressions due to their marginalized sexualities. (p. 5)

We saw this as an important science education project because we attempted to "[upend] jaundiced conceptions of equity work that suggest by knowing the categories a person belongs to, you then know the 'formula' for how to teach them" (p. 8). So it was an unabashedly political project that aimed to offer an alternative to the dominant (hegemonic?) equity projects in (science) education.

The invisibility of sexuality in science education troubled us because at stake were issues of epistemology. We were advocating for

> epistemologies [that would] foster a science that is more honest about the value-laden nature of its theories, methods, and results, more inclusive of what counts as knowledge, and more interested in questions, methodologies, and ways of knowing of marginalized, silenced, invisible people, including all those queer. (p. 10)

The concept of *queer standpoints* was reorienting us away from the allure of positivities as fixed, knowable identities and towards the diverse and even divergent lives and experiences of so many queers. This turn had bedeviled us, and it remains difficult to this day. The 2000 AERA paper marks a move on our part away from queer as collective noun toward queer as adjective—but a modest move, we see in retrospect. The fixed identity of queer as a positivity, or even as diverse identity, becomes relational when conceptualized as and from a queer standpoint. But the relationality we accomplished was critically compromised because we treated queer standpoints as knowable positions and points of view, rather than as processes and practices of knowing that arise in/as relational (and so always contingent and slippery) subject positions.

This epistemological point is important for a couple reasons (at least). First, by picking up on the work of feminist standpoint theorists, it goes to the nature of and possibilities for knowing in science as entailed in the lives of people who do science. This concern for diversity is about diversity of knowing (and doing and feeling) in science. This is a different equity concern than those that seem to motivate most equity projects (then and now), which aren't about pressing the possibilities for knowing in science, not about pressing the body of science, just about who can gain access to it. Alison Wylie (2004) notes that standpoint theory is "an inversion thesis: those who are subject to structures of domination that systematically marginalise and oppress them may, in fact, be epistemically privileged in some crucial respects" (p. 339). This point is about epistemology first and foremost, not representation. We weren't trying primarily to get more diverse people into science; we were trying to incite more and different science (including what nonscientists do to and with science in everyday life) from more and different ways of thinking and kinds of people.

Second, although 18 years ago we were moving away from positivities (and therefore essentialism) and toward standpoints, today we still feel caught on the horns of the queer standpoints versus queer identities dilemma. To be clear, a standpoint is not a view or "an ascribed position with its different perspective that oppressed groups can claim automatically. Rather a standpoint is an achievement, something for which oppressed groups must struggle, something that requires both science and politics, as Nancy Hartsock put the point" (Harding, 2004, p. 8). Standpoint theory still may seem invested in the notion that there are abiding kinds of people, but, importantly, it places lived and intentionally cultivated (i.e., theorized and enacted) standpoints out front rather than relying on inborn, stable essences of identity. A standpoint is about a collectively and politically constructed consciousness, not about individuals' experiences "as objects of others' observation, naming, and management" (p. 3). But they are still the standpoints of certain kinds of (oppressed) people (and so, not other kinds of people), rather than unusual ways of thinking and critique that might be taken up by, transform, and connect otherwise diverse individuals and collectives. Do queer subject positions enable queering critiques, or do queering critiques performatively produce queer subject positions? Of course, many others, including Judith Butler (1990), have already had similar thoughts:

> The foundationalist reasoning of identity politics tends to assume that an identity must first be in place in order for political interests to be elaborated and, subsequently, political action to be taken. My argument is that there need not be a "doer behind the deed," but that the "doer" is variably constructed in and through the deed. (p. 142)

What if doing science differently and thinking differently makes you a different kind of person? Perhaps doing/thinking about science differently and cultivating/acting from different standpoints can be in a dialectical relationship—mutually constituting, reconditioning, and reorienting each other.

Taking It to the Streets, Metaphorically

After lone presentations on sexuality and science education at NARST (Letts, 1998) and AERA (Letts & Fifield, 2000), we used the NARST 2000 and 2002 meetings to rally a few colleagues around this work.[5] In an echo of Leap's (2000) review of *Queering Elementary Education*, three NARST symposia across two years combined perspectives from gay and lesbian studies and queer theory. This uneasy mixture reflected important turning points and stubborn tensions in the evolution of our thinking and that of our colleagues.

Papers in these sessions were motivated by and started from different places in relation to gay and lesbian studies and queer theory. Work in gay and lesbian studies concerns the lived experiences of LGBT persons and the social, cultural, economic, and political conditions and injustices they face due to their minority status in predominantly heterosexual cultures. Queer theory shares concerns around sociocultural power relations of life in sexed and gendered bodies while troubling attempts to treat sexuality, or anything else, as a stable and knowable marker of identity. Queer theory includes unruly, deconstructivist perspectives that destabilize dominant norms around sexuality, gender, bodies, knowledge, power, and identity by imagining and inhabiting alternatives that resist being fully known or explicable. Jagose (1996) suggests that

> queer describes those gestures or analytical models which dramatise incoherencies in the allegedly stable relations between chromosomal sex, gender and sexual desire. Resisting that model of stability—which claims heterosexuality as its origin, when it is more properly its effect—queer focuses on mismatches between sex, gender and desire. (p. 3)

This move to denaturalize and uncouple sex, gender, and desire immediately opens spaces for inquiry that implicate and question identity, power, and knowledge more generally (Butler, 1990; Sedgwick, 1990). Queer theory includes perspectives that push inquiries into knowledge and identity simultaneously along multiple dimensions that intersect sexuality and gender without being contained by them.

In my NARST 2002 paper, *What is the 'Science'? and Who are the 'All'?* (Letts, 2002a), I aimed to critically deconstruct the popular reformist, equity-driven mantra "science for all." I outlined and advocated a queering of multiculturalism in science and technology education. Eschewing "laundry list" multiculturalism for a more critical examination of the assumptions that infuse conceptualizations of equity in education, queering multiculturalism extends beyond a concern for the identities of students who sit before us in classrooms. Instead, this stance explicitly uncovers and works against heteronormativity, deconstructs the powerful allure of dichotomies (such as male/female, nature/culture, knowledge/ignorance, reason/emotion, neutral/biased, homosexual/heterosexual, and trans/cis), and originates from a unique epistemological standpoint. I tried to deploy the strategy of queering not to make STEM more accessible to or comfortable for queer students (although both are worthy and important goals), but to move beyond containing and understanding identities to a critical multi-textual engagement that works to destabilize rather than fix these constructs. And I tried to shift the terms of debate to focus on epistemic issues rather than issues of representation. The piece interrogates how the broader norms around sexuality and desire shape knowledge production, validation, use, and dissemination. A version of this work moved away from science specifically to consider multiculturalism more generally in teacher education (Letts, 2002b).

In relation to science education, and concomitantly cast broadly in relation to multicultural teacher education, I hoped this work might illustrate what could be gained by shifting away from a politics of representation to focus on the epistemic affordances (and confines) of thinking from a variety of standpoints. Wylie (2004) reminds us that it's possible to operate from "a distinctive standpoint to which strategic epistemic advantage accrues, particularly in grasping the partiality of a dominant way of thinking, bringing a new angle of vision to bear on old questions and raising new questions for empirical investigation" (p. 349). I wanted to maximize what might be seen, learned, and prioritized from a range of marginalized (or at least on-the-margins) perspectives in order to have a richer, stronger sense of attending to issues of multiculturalism and equity without reverting back to identity politics. But that's not as easy to achieve as it might at first seem. I still stumbled over tensions between the notion of *queering* and the collectivity of *queers*. Advocating starting from marginalized standpoints is not the same as grappling with how that gets enacted and how you first notice and then learn the epistemic lessons that emerge from such a framing.[6] I moved away from a preoccupation with queers as a diverse group of people united under one banner, but I struggled to articulate, except in the broadest of terms, the new knowledge that starting from these standpoints could generate.

This Is Not a Pipe[7]

It was a long time between drinks. Although Steve and I have continued to discuss and debate these issues, we weren't moved to again write together about them until Kate Scantlebury invited us to respond to Mattias Lundin's (2014) article in *Cultural Studies in Science Education* (CSSE). It was while discussing our respective readings of Lundin's work that we realized there was a marvelous opportunity to (re)consider many of the ideas that had preoccupied us a decade and more before and that we had been discussing ever since. The result was Fifield and Letts (2014), in which

> we consider the uneasy, polygamous relationship between gay and lesbian studies and queer theories[,] … catalogue some of the challenges that persist in this work, and offer thoughts about how to work with and against them to enact a more just and compelling science education. (p. 393)

In some ways this is what we've been interested in doing for two decades, but our ability to articulate and envision it at play has taken a long time to incubate—and remains troubled and troubling!

In response to Lundin we write that "the issues of queering science education are more fundamental and foundational than the laudable, but limited, goal of including lesbian, gay, bisexual, transgender, transsexual, queer, questioning, and intersex students/identities/subjectivities/perspectives" (p. 399). Our most pleasurable insight in this piece came when we recognized the dualism we were unwittingly propagating between gay and lesbian studies and queer theories. We realized that in their familial relations, gay and lesbian studies and queer theories are mutually constitutive—"even as a verb, 'queering' leaves the residue of *being* queer" (p. 395). Still, we articulate our "distaste for visions of queering as assimilationist projects of liberally inclusive normalization" (p. 398), and we imagine "a queer science education as entailing inquiry into the implications and complicities (Davis & Sumara, 2000) of self, science, and nature" (p. 398). "Queer theories," we continue, "seek to make the familiar strange" (p. 400). "Queering can suggest that outsiders are in important respects already insiders, and that hetero norms are queerer than we may imagine" (p. 399). Reflecting on the limitations of our own and others' past work, we noted how Lundin's effort to put "multiple, overlapping enactments of sense-making in play, including how each shapes and is shaped by the other, counters some of the weaknesses of more isolated analyses that take texts and classroom vignettes at face value, as if passages and images from textbooks or snippets of classroom dialogue contain stable, unambiguous meanings" (p. 403).

Our CSSE article, in turn, foreshadows and motivates our explicit intention in *STEM of Desire* to coax conversations around sexuality and STEM education beyond what gay and lesbian studies has to offer by *queering* STEM education, sexuality, gender, and desire. For as we have noted,

> this puts love of learning, and learning to love, in a new light if learning is putting the self at risk by mindfully excavating and engaging the contingent relations that lie behind the illusion of an essential, autonomous self, and love is opening in awareness to the complicities of being(s). (p. 399)

We turn next to Steve's touchpoints.

Adding Sexuality—or Not

In August 1998 I moved halfway across the United States to begin my first university position as a secondary science teacher educator in the biology department at the University of Delaware. I arrived with two weeks to prepare a science teaching methods course for the first time. To my relief, Kate Scantlebury offered me her course syllabus and materials. Kate had been in the chemistry department at the university for several years, had built the secondary science teacher education program, and would be my close colleague. She organized her methods course around gender equity in science education. I adopted most of Kate's syllabus while integrating a few topics of special interest to me, including conceptual change pedagogies, and history and philosophy of science. I also thought there was an opportunity to extend the treatment of gender by adding LGBT perspectives to my syllabus.[8] Exploring equity issues around gender *and* sexuality seemed a good fit.

Then, with days until the course started, I found myself unable to convincingly and *comfortably* include LGBT issues in my syllabus. What *does* sexuality have to do with curriculum, lesson planning, instructional strategies, assessment, and so on in secondary science education? Gender equity approaches at the time had things to say about these issues in relation to girls and women in science education. But I got nowhere when I tried to pick up on gender equity in science education and extend it to sexuality.[9] My unproductive thinking about a dominant straight population and a marginalized (and mostly unacknowledged) LGBT population went something like this:

> Gender equity projects enhance science learning opportunities for girls. We can usually tell who the girls are in classrooms. If we now want to enhance opportunities for LGBT students, how can we tell who they are? Do we even need to know? If we figure that out, what are we supposed to do for them in curriculum and instruction? And why just sexuality? What about ethnicity, class, (dis)ability, and other aspects of identity? How will I justify to students this time spent on sexuality when they are worried about surviving student teaching?

I didn't have satisfying answers to any of these questions, and I didn't have time to find them. Reflecting on this two years later, I wrote, "I was afraid that I might be viewed as peddling a self-absorbed diversity favoritism that had more to do with my personal life than the practical challenges my students would soon face as student teachers. I feared for my credibility" (Letts & Fifield, 2000, p. 12).

When I moved to Delaware, I left where I had grown up and come out. I arrived where no one knew me, and I looked forward to being out right from the start. I was ready to engage with my fuller self in life, the classroom, and scholarship. I felt good about being out to students, being available as a role model. Still, my sense that there were interesting possibilities around sexuality and science education was clouded by cultural norms (which I embodied despite my styling as "out and proud") that treated the sexuality of teachers, students, and scientists as irrelevant in science education.[10] In apparent contradiction to this normativity, I conceived of sexuality as a knowable attribute of persons that exposes those with nonnormative sexualities to systemic discrimination that we should name and oppose in schooling, including in science classrooms. Although I was intrigued by the intersections of sexuality and science education, it seemed troubling and unproductive to frame this as a concern about LGBT students as a distinct and marginalized population in science classrooms. I later came to appreciate normative (hetero, homo, and otherwise) anxieties as surface features of deeper stuff worth investigating. At that time, however, I set all this aside to focus on the immediate concerns of being a new faculty member with courses to teach and a dissertation to finish! I didn't add sexuality to the course—but soon discovered it was already there.

Already (Hetero)sexed

Lee was gradually and carefully coming out as gay when he was in my first science teaching methods class. He and a girlfriend had recently split. Then during the semester, he told his parents and a few friends that he was gay. He looked forward to student teaching high school biology the next semester, and he was not particularly concerned about the gay thing. He intended to keep his private life out of the classroom. He wanted to be a science teacher, not a gay science teacher.

During my methods class, Lee didn't know I was gay, and I didn't know he was coming out. The natural moment for me to come out to the class that I was sure would arise never did. (No one returned from secondary school classroom observations with reports of homophobic language. That was supposed to be my cue!) And I was not Lee's university supervisor for student teaching the next semester. It was only near the end of Lee's student teaching that we unexpectedly met at a support group for gay students that I helped facilitate. ("What's *he* doing here?" we both wondered.) Lee told me that evening about some of his struggles in student teaching. We kept talking and later wrote together about the ironic symmetry of our experiences (Fifield & Johnson, 2000; Fifield & Swain, 2002): I wanted to add sexuality to science teacher education, only to find I couldn't imagine how to do it because my mind and science education were already overstuffed with encultured heteronormativity. Lee, on the other hand, simply wished to keep his sexuality out of the classroom but discovered it was already there in his colleagues and students, and in his own relationship with science.

Lee was cautiously coming out outside of school, and trying not to be out in school, so he was acutely attuned to language and risk.[11] He noticed when teachers and administrators (all men) who gathered at a teacher's house to watch college basketball jokingly called each other "cocksucker" and "faggot." He noticed when his students giggled about terms like *homozygous* and *phagocytosis*. He noticed that, as a gay white man, he called out students' racist and sexist language but didn't intervene when they casually said "fag" and "that's so gay," because he didn't want to raise suspicions about himself.

Students asked him where he was from, what he did for fun, what kind of car he drove, and if he had a girlfriend. He didn't pose as straight; he just wanted to avoid the issue. But heteronormativity doesn't appreciate ambiguity, even as it multiplies ambiguity. When he deflected questions, a class of 12th graders created a girlfriend for him, clipping her striking picture from magazines and leaving notes from her to him around the classroom. Lee interpreted this as playful affection that got out of hand. He told them to stop. In another class, several 10th-grade boys were less playful, scrawling "poop chute raider" on the board and desks in Lee's room, sometimes yelling it out or muttering other offensive comments he felt were aimed at him. He tried to tighten control by relying on worksheets and lectures more than class discussions. Lee endured the public violation of his private life mostly alone. His supervising teacher had been called to administrative duties for the term and was usually not in the classroom. Lee didn't report these experiences to his university supervisor. He did share stories with a few close friends.

Lee's accounts grounded my thinking about sexuality in the lived experiences of another and pushed me to look into my heteronormativity. I carried these into my first sustained readings of queer theory—catalyzed by Tierney (1997) and Pinar (1998)—that approached sexuality in education as a valence of knowing and being, rather than first and foremost as an equity concern. Around the time when Lee was telling me about the impossibility of keeping his private life out of the classroom, I was learning from Sumara and Davis (1998) that "queer theory asks not that pedagogy become sexed, but that it excavate and interpret the way it already *is* sexed and, further, that it begin to interpret the ways in which it is explicitly *heterosexed*" (p. 199, original italics). Lee's experiences helped me see the expansiveness of ambient heterosexuality (Murray, 1995) and heteronormativity, which are so ordinary and ubiquitous they typically pass without remark. Science classrooms, like most cultural spaces, are ambiently heterosexual: Heteronormativity shaped Lee's relationships with his students and influenced his instructional practices. Furthermore, heterosexuality necessarily exceeds the hetero. Reproducing "that unruly class 'heterosexual'" (Sumara & Davis, 1998, p. 199) relies on imagining kinds of people like poop chute raiders and hot girlfriends. Lee's stories helped me reimagine homosexuality as the "indispensable interior exclusion" (Fuss, 1991, p. 3) of heterosexuality, exemplifying the broader point that "the norm and its negated other are implicated and mutually constitutive of each other" (Luhmann, 1998, p. 151). I learned that heteronormativity is stranger than it thinks.[12]

As my first year at the University of Delaware ended, I sat with Will in a brew pub near campus where he proposed we think and write together about sexuality and science education, perhaps starting with an AERA conference paper. I thought that was a dubious and irresistible idea. We presented that AERA paper the following April (see Letts & Fifield, 2000).

Science and Self Entangled

Lee strongly identified with science—he was in science, science was in him. As his changing self-understandings as a gay man transformed his being, he no longer saw himself in the science he was teaching. He later wrote that he "felt like a hypocrite" teaching textbook genetics and reproduction because it "questioned, devalued, and denied the credibility of my sexual orientation" (Fifield & Swain, 2002, p. 185). Presenting Mendelian genetic crosses as archetypes of reproduction to 10th graders put what Lee knew about nature in conflict with what he was coming to know about himself.[13] Where was he (and others like him) in the natural order he was reproducing in his classroom? "The scientist in me and the 'natural laws' I had learned," he wrote, "conflicted with how I felt about myself" (p. 180). Lee's crisis as a knower-of and a known-about turned him from the flimsy objectivity of school science toward critical perspectives on scientific knowledge as cultured and partial.

Lee's sense of self and science was further at stake in his students' negotiations of self, science, and religion. Christianity, like science, was part of who he was. Lee's coming out was complicated by Bible passages he couldn't harmonize with his sexuality. He crafted a tenuous solution by holding science and religion apart and privileging genetics to secure his sexual orientation as inborn and natural. Using science this way helped justify his decision to embrace his sexuality instead of denying it or trying to change it.

Some of his students, though, were unconstrained by modernist niceties about science and religion as nonoverlapping magisteria.[14] During a class discussion about personality traits and the nature/nurture debate, some of Lee's students said that homosexuality was a choice, not a genetic trait. God, they were sure, would not create sinful genetic traits. Lee said these comments "stung" and tossed him again into uncertainty about scriptures he couldn't reconcile with his feelings of being a natural part of God's creation (p. 180). Was he both sinner *and* genetic reject? He wrote, "I wanted to tell my students the truth about me because I felt it would help them question and clarify their understandings, but I could not bring myself to do that" (p. 180).

Stranger Pedagogies

Lee wasn't ready to come out in the classroom. LGBTQ advocates would have wanted him to. Coming out of the closet was and remains central to LGBTQ identity politics in the United States. I had consumed a steady stream of books in the 1990s that detailed the struggles of closeted teachers and the virtues of coming out of the classroom closet.[15] By coming out, teachers could reclaim their voices, express their true identities, counter stereotypes, serve as role models, and enjoy more fulfilling lives. In *One Teacher in 10: Gay and Lesbian Educators Tell Their Stories*, Kevin Jennings (1994) described the stakes:

> A school with an openly lesbian or gay teacher is a better school. It is a school where truth prevails over lies; it is a school where isolated students at last have a place to turn to for support; it is a school where our nation's rhetoric about equality moves one step closer to becoming a reality. As gay and lesbian teachers win our freedom, we help to free our students, our colleagues, and our communities of the burden of bigotry. (p. 14)

Rhetoric like this replaced the hard stigma of being lesbian or gay with the soft stigma of enabling bigotry by being in the closet.

For many sexual minorities in the West, "the closet" and "coming out" have relatively recently become nearly obligatory (simultaneously, hetero- *and* homonormative) tropes that have helped many of us imagine and mark key passages in our lives.[16] But having been in the closet and come out (recognizing, of course, that coming out is never total and never ending), and then hearing Lee's stories, I noted how the dualism of "in/out" promises clarity that lived experiences of being in and/or out fail to deliver. I also witnessed how the limits on sense- and life-making afforded by the figures of the closet and coming out stifled alternatives for Lee and other students and early-career teachers I worked with. For instance, a first-year science teacher (who had been in my methods class after Lee) shared with me her disappointment at being out to only a few colleagues in her school after four years in college where she was out to everyone (Fifield, 2001a, 2002). In the same conversation, she described what I interpreted as her outspoken advocacy for diversity, including sexual and gender diversity, with students, teachers, and administrators in her school. Still, she felt the stigma of being closeted, of lying about herself. From inside the interpretive construct of the closet, she couldn't appreciate how much good she was doing in her school, nor how who she was in school was just a fraction of her many selves. The structure of being either "in" or "out" couldn't accommodate the reality of her nuanced and situated ways of being.

Coming out commits to revealing hidden truths and resolving ambiguity; it seeks to unite how one knows oneself with how one is known by others.[17] It's a pedagogy of transmitting proper knowledge about oneself and advancing "the dominant educational narrative in which one passes from ignorance to knowledge about both the 'self' and other" (Miller, 1998, p. 372). As authors of lives and texts, however, we know the fragility of transmitting knowledge *just so*. Our texts and lives are subject to myriad (mis)interpretations (including by ourselves) that are in large part beyond our control, and often beyond our awareness (see Evans, 2002). "Identity is never simply a question of self-authorship … [because it] is always already haunted by the other, by that which is not 'I'" (Sullivan, 2003, p. 149). As a temporary social (re)articulation of circulating norms, the "I" necessarily gives itself away to others (Butler, 2005; Ruffolo, 2007). Others' understandings of teachers who come out may not resemble the selves ("authentic" and otherwise) teachers intend to reveal (Talburt, 2000), which means that teachers who come out may also fail to fully recognize themselves in the result. This complicates the straightforward promises of coming out in the classroom.

Making my own sense of Lee's non-coming-out story, I was taken by the insistence of some students to place him as a sexual being. He thought his sexuality should have nothing to do with teaching but trying to keep his private life out of the classroom raised curiosities and suspicions. Lee's ambiguity piqued heteronormative imaginations that sought to resolve his strangeness into something familiar. Shane Phelan (2001) notes the paradoxical nature of the *stranger* as "neither us nor clearly them, not friend and not enemy, but a figure of ambivalence who troubles the border between us and them" (pp. 4–5). Phelan imagines gays, lesbians, and other sexual minorities in the United States as *sexual strangers*. Sexual strangers live and work among the majority but do not fully belong to it. They are troubling because they are close at hand and often hard to distinguish from nonstrangers. They "disrupt seemingly natural boundaries and borders" (p. 29), and "undermine and crack open from inside all polar categories of social order" (pp. 29–30). Still, "the position of the stranger is not only difficult, it is rewarding" (p. 8). Sexual strangers can act from positions of difference to challenge current social, political, and sexual relations, and to "offer one another and others new ways of questioning the current tight fabric of citizenship and national identity" (p. 8).

As I pondered Lee's and others' stories, Phelan's sexual strangers seemed to me like critical, queering pedagogues (Fifield, 2001b, 2004, 2012). In Luhmann's (1998) queering twist,

> pedagogy might be posed as a question (as opposed to the answer) of knowledge: What does being taught, what does knowledge do to students? … Pedagogy then begins to shift from transmission strategies to an inquiry into the conditions for understanding, or refusing, knowledge. ... Ignorance is not the opposite to knowledge but an opposition to knowing. (pp. 148–149)

What then are the (im)possibilities for a science pedagogy that takes up strangeness, being a sexual stranger, as a stance for inquiry into conditions for knowing and being? What are the (im)possibilities for a stranger science pedagogy that arcs from the familiar to the strange? I looked to Lee's and my own positions as sexual strangers for glimpses of stranger science pedagogies.

As Lee was learning to teach, he was also actively crafting, coming to know, and selectively (and with varying success) revealing and concealing the stranger within. He saw "seemingly natural boundaries and borders" fail as his sexuality refused to stay out of the classroom, the science he was teaching, and the understandings his students embraced and refused. As his self-understandings changed, so did the outbound perspectives he brought to bear on others, science, and the world. He came to better appreciate the social construction of science by witnessing the social construction of his self. He found himself anew in the biology he taught: a stranger missing from the "nature" school science delivers and yet present in his "nature" as he felt and understood it. As he reconstructed facts of life that acknowledged his life, he made new sense of the cultural life of scientific facts. Perhaps we catch sight here of a queering reconstitution of self-in-science/science-in-self in a move from the familiar to strange.

Of course, sexual minorities are not the only strangers in science classrooms. We are all in some ways strangers to ourselves, others, and the world because the indispensable interior exclusions of the norms we embody and/or refuse are indispensably interior to ourselves. We all transgress the tidy boundaries of kinds with which we (mis)identify. Looking within at exclusions we contain might take knowing and being beyond our delimited selves to enlightening entanglements with the others that make (our)selves (im)possible. Imagine balancing the lopsided arc of familiarizing pedagogies by joining them to committed practices of strangeness and making strange. A stranger science pedagogy might be one in which teachers and/as learners come out as strangers in defamiliarizing conversations with the selves, others, and worlds they think they know. Using science to make the strange(r) familiar can be powerful and rewarding—but this leaves us susceptible to comforting delusions of knowing. In stranger science pedagogies, we might enjoy the rewards of knowing while also remaining open to robust mysteries of self and science that rouse us from delusions.

Troubling Pedagogical Desires

Queer theorizing doesn't have to be about sexuality as normally conceived. For instance, queering may attend to how we enroll others in intimate labors of knowing and becoming. Pedagogical change in teacher education includes this "deeply social or dialogic situation of subject formation, the processes of how we make ourselves through and against others" (Luhmann, 1998, pp. 153–154). When others fail to fulfill roles in which our selves are invested, our selves may experience disruption and loss. Mourning such losses means "reorganizing one's working models of self and the world after the disorganization brought on by loss" (Davis, 2008, p. 309; Fifield, Grusenmeyer, & Ford, 2014). For emerging and experienced teachers alike, pedagogical change includes this play of self and other, risk and renewal, knowing and refusing. "Learning to teach … is a time when one's past, present, and future are set in dynamic tension" (Britzman, 2003, p. 31).

For 13 years at the University of Delaware, I was part of a team from biology, earth science, physics, and education that developed, taught, and studied an inquiry-based course block for elementary teacher education majors called the Science Semester (see Ford, Fifield, Madsen, & Qian, 2013). We were sympathetic with aims in the *National Science Education Standards* (National Research Council, 1996, 2000) and elsewhere (e.g., American Association for the Advancement of Science, 1989) to support meaningful understandings of science through inquiry pedagogies. We used problem-based learning (PBL; Duch, Groh, & Allen, 2001) to develop cross-disciplinary science content and practices through extended group investigations of science and inquiry pedagogies in thick and messy sociocultural contexts. This shifted responsibility for meaning-making toward students in collaborative groups and blurred disciplinary boundaries without discarding disciplinary perspectives. We *did* largely discard lecturing and moved knowledge making to group and individual research, laboratory and classroom activities, and discussions.

Our team investigated Science Semester students' changing understandings of inquiry pedagogies and of themselves as learners and future teachers (see Fifield et al., 2014; Fifield & Juck, 2006; Ford et al., 2013). We knew firsthand about elementary education majors' widely reported resistance to inquiry learning (e.g., Weld & Funk, 2005). Like reformers before us, we hoped that explaining the virtues of inquiry pedagogies and easing students' doubts about their abilities as science learners would help smooth the way. We weren't neophytes; we knew that inquiry pedagogies disrupt identities invested in the "teacher as knowledge-bearer" (Britzman, 2003, p. 79). Still, most of us carried the "unspoken assumption … that the 'better' the course, the smoother and less problematic the students' learning is likely to be" (Brown, 2006, p. 687). We wanted students to find pleasure in learning even as we embraced pedagogies that put their (and our own) identities at risk.

We taught mostly second-year students who conceived of themselves as part college student, part future teacher. Many told us that for as long as they could remember they had wanted to be elementary school teachers. They believed that good teachers are caring, knowledgeable, and enthusiastic. They told us that a caring teacher's enthusiasm makes students *want* to learn, and that good teachers express love for their students by sharing their knowledge and enthusiasm (Fifield et al., 2014). Like the future teachers in Goldstein and Lake (2000), ours spoke with ease about loving children. "I really just love kids. And I love the feeling of helping someone understand things," said one of our students. Another said, "I love seeing kids smile and knowing that I'm making their future brighter by teaching them and helping them understand how the world works." Behold entanglement: At stake in our students' desires for their learning in the Science Semester were the imagined children they enrolled in their nascent identities as loving teachers.

Science Semester students expected us, their professors, to provide the science content and hands-on experiences they needed to become good and loving elementary school teachers. They enrolled us in their "labor of identification" (Pinar, 1998, p. 33), just as we had enrolled them in our identities as change-making educators of future teachers (Fifield et al., 2014). Given these enrollments, it's not surprising that students were concerned about the "missing" lectures. What is a science course without lectures from professors? While friends in "normal" science courses took copious lecture notes, our "minilectures" were infrequent and brief. One of our students complained, "I'm teaching myself more than they're teaching me." Group investigations seemed to many of our students as burdensome and inefficient (while to others they were welcome relief from sitting through weeks of lectures). A student explained that she was used to professors who "go over … the important stuff they want you to know. ... You can just go to class and take notes, and just go over your notes [to study]." Many students' were also unsettled by their entangled responsibilities as learners and peer teachers in collaborative investigations. Halfway through one semester a student said, "I guess it just feels like it's a fuzzy line in between whether we are supposed to be acting as a teacher, or a student, or a future teacher. And it's very wishy-washy, nobody really knows."

Moreover, PBL seemed short on the fun and memorable experiences students associated with hands-on learning. On beginning-of-course surveys, nearly all our students believed that elementary teachers should use hands-on activities to make science fun (Ford et al., 2013). Hands-on activities were for our students more about making science *memorable* than *meaningful*. This notion of hands-on learning is a distant relative to inquiry learning based on sustained and purposeful investigation that may incorporate, but is not limited to, hands-on activities. An especially articulate student voiced, more vividly than most, concerns shared by others. Near the beginning of one semester she told us:

> Trying to pull research out of my butt, and like, collaborating with four other people in trying to come to some kind of conclusion doesn't work for me—at all. I am just not curious like a child. And exploration does nothing for me.

All these concerns eased as the course progressed, but never disappeared—even as students became better collaborators and developed more nuanced conceptions of inquiry learning (Ford et al., 2013). By the end of the semester, the student in the previous quote offered this reflection on her past, present, and future in dynamic tension:

> I want to be a good teacher. And I think that this experience has helped me a lot. It's just that it was very strange at first. ... This is going to sound really bad. But it's [i.e., inquiry learning] probably going to be easier to implement on my own students than it is to have it done to me. ... I still learn a certain way. Which happens to be memorization.

In the Science Semester we found ourselves with our students (and the fantasized children they already loved and would one day teach) in the sometimes unbearable paradox that "there can be no learning without conflict, but the conflict that animates learning threatens to derail the precarious efforts of trying to learn" (Britzman, 2003, p. 3). Just as some learning is a long-delayed response to teaching, certain pedagogical care-giving and receiving may for some, and for some time (and, who knows, forever?), feel like the withdrawal of care.[18] This queering pedagogy asks what we "can bear to know" (Luhmann, 1998, p. 150). Can we bear to know that "for the self to be more than a prisoner of its own narcissism, the self must bother itself" (Britzman, 1998, p. 32)? Can we bear our risky entanglements with others? Can (future) teachers bear strange ways of loving? And can science teacher educators bear the "awkwardness of learning" that marks "the emergence of an ambiguous, complex, irreducible and potentially problematic teacher self" (Brown, 2006, pp. 687, 677)?

Men in Stamen

As our work on *STEM of Desire* proceeds, I'm in my first year as a paraprofessional in special education. I enter an eighth-grade science class that's several days into a genetics unit. Dear old friend, school science genetics! You have borne such burdens in queering science education, yet you keep coming back for more.[19] I walk to the back of the classroom, passing slate-top tables. Solid. Heavy. All around is the familiar clutter of science stuff. And words, so many words! Posters on the wall speak of thinking scientifically. Genetics terms hang from a line strung across the foam panel ceiling. On a whiteboard is a labeled diagram of a flower. The teacher begins reviewing important terms. The pistil is the female part of the flower. It consists of the stigma, where pollen is deposited, and the ovary, which contains ovules with egg cells. The stamen is the male part of the flower. It consists of a slender stalk, the filament, tipped by the anther, which produces pollen grains that contain sperm cells. The teacher says you can remember the stamen is the male part of the flower because "men" is in it. This tip would later prove helpful on the quiz.

I start copying the flower diagram into a notebook for students who have difficulty keeping up and paying attention … and my mind wanders:

> How many times have I drawn this picture? A lot! Why do we use "female" and "male" for flower parts when these words are so loaded?[20] Do we just mean

> that egg cells are made in the ovule and sperm cells in the pollen? Biologists say that "females" produce the larger sex cells and "males" produce the smaller sex cells.[21] But that's not what "female" and "male" mean in everyday life. We press flowers to carry culture. What do eighth graders make of this? Why didn't I ask my undergraduates these questions? …

Lost in thought, I'm behind. The teacher moves on. I don't like my unfinished flower diagram. I take a photo of the teacher's drawing and recall an eighth-grade math teacher telling students that they will remember ideas better copying them by hand than taking a picture.

CONCLUSION

These reflections weren't written by the same people they're about. We're different people now in part because on-going conversations and debates by our past selves now pleasantly haunt our lives. As Britzman (1998) writes:

> Reading practices might be educated to attend to the proliferation of one's own identificatory possibilities and to make allowance for the unruly terms of undecidability and unknowability. One might think about the proliferation of identifications as a means to exceed—as opposed to return to—the self. (p. 85)

We've tried to exceed ourselves by "speaking and writing into existence denaturalized ways of being that are obscured or simply unthinkable when one centered, self-knowing story is substituted for another" (Miller, 1998, p. 368). But this was not easy. And maybe we didn't succeed. Formal education taught us to conjure the coherent narrative of development and progress. Now we're trying to write against coherence.

Simultaneously, we've tried to rise to the occasion as Bill Tierney (1997) formulates it (even as it sits in contradistinction to the visions of autobiography *just* advocated): "The challenge, then, is to map out the cartography of our lives that accounts for previous interpretations and creates the conditions for a reformulation of the academy" (p. 18). So our touchpoints are replete with modernist palimpsest while we actively try to write otherwise. Thus, our pursuit of queering science education traces normal and strange desires through the entangled (re)productions of subjects and objects, mapping as we go, understanding that both subjects and objects can desire and be desired, can know and be known about.

We hope these touchpoints have gestured to what emerged, unplanned, as we revisited and remained dissatisfied with what we understood and what we could accomplish with what we knew. We had broader, disparate interests than we could discipline, thus we were (and remain) quite undisciplined. But what we can see now in retrospect is an initial fascination with equity agendas in science education that, in fits and starts, pivoted towards tracing, troubling, and eliciting animating desires in/of science education. These desires include but exceed sexuality as normally conceived. Thus, queering science education is about sexuality and whole lot more, because sexuality exceeds itself.

Will was initially interested in a multicultural equity agenda for science education, with a specific focus on gender and sexuality. Steve was intrigued by the coconstitutive relations of selves (one's own with others') and science (education). We had a shared interest in mounting a critical response to heteronormativity. We came to see that "identity categories are never merely descriptive, but always normative, and as such, exclusionary" (Butler, 1992, p. 16). Our arc of interest started with the "queer" troubling of heteronormativity (and eventually its constitutive other, homonormativity); moved to realizing that sexuality is entangled with knowing and that it can be traced most everywhere; fully opened to the notion of queer as a verb, of queering as a practice of knowing, feeling, being; and drifted on to "sexuality" exceeding conventional conceptions via knowing and being as flows of desire. We came to understand that normalizing processes contain the conditions of their own undoing through unfaithful refusals, reinterpretations, and perverse repurposings. Reveal, disrupt, repurpose, repeat. Our touchpoints sketch (re)orientations toward the sketchy, shifting margins of science education.

We often wonder: What is there to love about science education when it denies its passions and ours, regularly fails others we care about, and is not what it claims to be? The perfect monster! Still, our affection endures because of, not despite, vexed relations with conventional science education and whatever technocratic reforms (the faithful companions of convention) currently hold sway. Sometimes we love what's bad for us; we want to look away but can't stop staring, touching. The normal and strange are intimates, though they usually deny it (just as we sometimes deny affection for conventional science education even as we arose/arise with it). From questionable subject positions around science education's interior and exterior margins, we seek ways to reveal, perturb, and reorient its and our own desires. Such seriously playful perversion of science education's (and our own and others') entangled desires is how we think about *queering* science education.

These are some insights into our vexed affair with science education. There are many loose ends—we hope generative, alluring, and compelling loose ends, rather than knotted ends, neatly tied off. Why do we fancy science education? Why do we care? What is our queer(ing) interest in it? While discontinuous and punctuated, our responses can be narrated in such a way to tell some stories about the intersections between queer theories and science education. But the object of our longing is even more distributed and bizarre than science (education). We're hooked on how that makes us feel. And what's not to love about that?

ACKNOWLEDGMENT

We thank Andy Gilbert for helpful feedback on a draft of this chapter.

NOTES

1 For example, see Graham (2014, chapter 5) and Olesen (2018).

2 See Barad (2007) for more on intra-action.

3 The other participants in this symposium were Leslie S. Jones, Jomo Mutegi, and Marisol Rodriguez-Muñoz.

4 Similarly, in Santavicca, Bazzul, and Witzig (this volume, chapter 15), Chase remarks that Bazzul and Sykes's (2011) queering critique of biology textbooks has been criticized by some for not being queer enough.

5 The NARST 2000 session, [Homo]Sexualities and Science Education: [Per]Forming a Research Agenda, included Steve Fifield and Barry Johnson (*Narratives of Silence: Confronting Queerness, Classroom Identities, and Disciplinary Knowledge in Science and Science Teacher Education*), John Mascazine (*Seasoned Science Teachers Tell Their Stories: Gay and Lesbian Issues and Concerns*), Connie Nobles and Will Letts (*Queering Science Education? You Don't Say*), Steven Rakow (*Improving the Climate for GLBT Students and Teachers: The Role of Professional Organizations*), and Anne-Marie Scholer (*Sexuality in the Classroom: One Teacher's Methods in a College Science Course*). Paired sessions at NARST 2002, Sexualities and Science Education I and II: Negotiating the Contested Intersections of Identity and Knowledge, included Steve Fifield (*What's in the Closet? Negotiating Identities in the First Years of Science Teaching*), Kathleen Greene (*Encountering and Combating Homophobia and Heterosexism in Science Education Methods Texts*), Will Letts (*What Is the 'Science'? and Who Are the 'All'?),* John Mascazine (an update of *Seasoned Science Teachers Tell Their Stories: Gay and Lesbian Issues and Concerns*), Connie Nobles (*Intersections of Sexuality, Science, and Attitudes*), and Vicky Snyder and Francis Broadway (*Queering High School Biology Textbooks and Pedagogy*).

6 Donna Haraway (1988) had already warned of these challenges. She wrote:

> There is a premium on establishing the capacity to see from the peripheries and the depths. But here there also lies a serious danger of romanticizing and/or appropriating the vision of the less powerful while claiming to see from their positions. To see from below is neither easily learned nor unproblematic, even if 'we' 'naturally' inhabit the great underground terrain of subjugated knowledges. ... *How* to see from below is a problem requiring at least as much skill with bodies and language, with the mediations of vision, as the 'highest' technoscientific visualizations. (pp. 583–584, original italics)

7 After René Magritte's *La Trahison des Images* (The Treachery of Images), 1928–29, which reads *Ceci n'est pas une pipe*.

8 Don't miss the homogenizing assumption in "LGBT perspectives" that such perspectives belong together. In addition to the problematic lumping of lesbian, gay, and bisexual, I was not at the time thinking carefully about the lived particularities and semantic problems of "transgender" and "cisgender." Dana Defosse, a graduate school colleague of mine, had introduced *cisgender* into broader discourse just four years earlier. For more on this, see Enke's (2012) critical reflection on cisgender. Also, I did not at the time include *queer, questioning, intersex*, and other categories that are now typically appended to LGBT.

9 Tierney (1997) and Pinar (1998) would soon open my mind to queer theoretical analyses of sexuality and desire that pivot from identity politics to the politics of identity. Further, critical responses to work on gender and science education (e.g., Nobles & Letts, 2000) would help me question "girls," "boys," "straight," "LGBT," etc., as markers of particular kinds of people and of work for and about those people. For example, in the late 1990s (and arguably still today) most gender equity initiatives in science education reduced gender to binary biological sex and treated "girls" and "boys" as stable and informative attributes of individuals in marginalized and dominant populations, respectively. This evacuated from *gender* its content as the embodied cultural meanings and practices of sexed bodies, and undermined possibilities for deconstructive critique. See Glasser and Smith's (2008) criticism of synonymizing sex and gender in education research and policy, including in STEM education.

10 Three years later I reflected on this:

> As I tried to contemplate what gender equity in science education might become, I remained powerfully influenced by heteronormative assumptions that exclude the diverse standpoints of sexual minorities from classroom culture and disciplinary knowledge in science and education. This is an example of how conceptions of sexual identity intersect with what counts as knowledge (Tierney, 1997). What does homosexuality have to do with science teaching? I looked through heteronormative lenses and did not see any compelling answers. As much as I had come out, I remained attuned to the epistemic and social privileges of the heteronormative closet (Sedgwick, 1990). Heteronormative ways of knowing bound my classroom identity and constrained the opportunities I offered my students to understand themselves, their future students, and their subject matters. (Fifield, 2001a, p. 10)

11 Being "out," like all being, is context-dependent, not all or nothing. See Mayo's (2007) reflections on some contextualities of outness.

12 "One of the central tenets of queer theory is that all sexualities, indeed all attempts at normativity, are bound to fail because achieving normal is an impossible task" (Mayo, 2007, p. 183).

13 You wouldn't know from the biology taught in schools that reproduction in most living things is not based on the pairing of individuals that are distinctly and stably either male or female (see Roughgarden, 2004).

14 "Nonoverlapping magisteria" is from Gould (1997).

15 For example, see Harbeck (1992), Jennings (1994), Kissen (1996), and McNaron (1997).

16 See Sedgwick (1990) for a foundational analysis of the closet and coming out in the West.

17 Coming out, before it was such a normal thing to do, was coming out as a stranger after passing (or trying to) as normal. But for some time now, the political imperative of coming out in LGBTQ identity politics has been to normalize the Other as pretty much like everyone else and so as entitled to equal rights. Instead, I wonder here about recouping coming out as revelation of the strange that does not resolve Otherness into familiarity. The social justice aspiration of coming out would then be elevating difference as a unifying feature of human being(s) rather than winning a tepid toleration of inconsequential difference that doesn't trouble the norm. Merely tolerating difference that is presented as inconsequential collapses equality into sameness under a modestly expanded notion of normality. See Duggan's (2002) discussion of the *new homonormativity*.

18 Noddings's (2005) elegant work, in contrast, seeks a tight linkage between teachers extending care and students feeling cared for. I see Noddings's desire for wholeness and completion in pedagogies of caring as a condition for boundless creative tensions, failures, and breakthroughs.

19 See, for example, Bazzul (2013), Bazzul and Sykes (2011), Fifield (2001a), Fifield and Swain (2002), Lemke (2011), and Snyder and Broadway (2004).

20 See Schiebinger (1993) for how pistils became female and stamens male.

21 "To a biologist, *'male' means making small gametes, and 'female' means making large gametes.* Period! By definition, the smaller of the two gametes is called a sperm, and the larger an egg. Beyond gamete size, biologists don't recognize any other universal difference between male and female" (Roughgarden, 2004, p. 23, original italics).

REFERENCES

Ahmed, S. (2006). *Queer phenomenology: Orientations, objects, others*. Durham, NC: Duke University Press.

American Association for the Advancement of Science. (1989). *Science for all Americans*. Washington, DC: Author.

Angelou, M. (1975). *Oh pray my wings are going to fit me well*. New York, NY: Random House.

Barad, K. (2007). *Meeting the universe halfway: Quantum physics and the entanglement of matter and meaning*. Durham, NC: Duke University Press.

Barad, K. (2012). On touching—the inhuman that therefore I am. *Differences: A Journal of Feminist Cultural Studies, 23*(3), 206–223.

Bazzul, J. (2013). *How discourses of biology textbooks work to constitute subjectivity: From the ethical to the colonial* (Doctoral dissertation). Retrieved from http://hdl.handle.net/1807/43477

Bazzul, J., & Sykes, H. (2011). The secret identity of a biology textbook: Straight and naturally sexed. *Cultural Studies of Science Education, 6*, 265–286.

Britzman, D. P. (1995). Is there a queer pedagogy? Or, stop reading straight. *Educational Theory, 45*(2), 151–165.

Britzman, D. P. (1998). *Lost subjects, contested objects: Toward a psychoanalytic inquiry of learning.* Albany, NY: State University of New York Press.

Britzman, D. P. (2003). *Practice makes practice: A critical study of learning to teach.* Albany, NY: State University of New York Press.

Brown, T. (2006). Negotiating psychological disturbance in pre-service teacher education. *Teaching and Teacher Education, 22*(6), 675–689.

Butler, J. (1990). *Gender trouble: Feminism and the subversion of identity.* New York, NY: Routledge.

Butler, J. (1992). Contingent foundations: Feminism and the question of "postmodernism." In J. Butler & J. W. Scott (Eds.), *Feminists theorize the political* (pp. 3–21). New York, NY: Routledge.

Butler, J. (2005). *Giving an account of oneself.* New York, NY: Fordham University Press.

Costa, V. B. (1995). When science is "another world": Relationships between worlds of family, friends, school, and science. *Science Education, 79*(3), 313–333.

Davis, B., & Sumara, D. (2000). Another queer theory: Reading complexity theory as a moral and ethical imperative. In S. Talburt & S. Steinberg (Eds.), *Thinking queer: Sexuality, culture and education* (pp. 106–129). New York, NY: Peter Lang.

Davis, C. G. (2008). Redefining goals and redefining self: A closer look at post-traumatic growth following loss. In M. S. Stroebe, R. O. Hansson, H. Schut, & W. Stroebe (Eds.), *Handbook of bereavement research and practice: Advances in theory and intervention* (pp. 309–325). Washington, DC: American Psychological Association.

Duch, B. J., Groh, S. E., & Allen, D. E. (2001). *The power of problem-based learning.* Sterling, VA: Stylus Publishing.

Duggan, L. (2002). The new homonormativity: The sexual politics of neoliberalism. In R. Castronovo & D. Nelson (Eds.), *Materializing democracy: Toward a revitalized cultural politics* (pp. 175–194). Durham, NC: Duke University Press.

Enke, A. F. (2012). The education of little cis: Cisgender and the discipline of opposing bodies. In A. Enke (Ed.), *Transfeminist perspectives in and beyond transgender and gender studies* (pp. 60–77, 210–217). Philadelphia, PA: Temple University Press.

Evans, K. (2002). *Negotiating the self: Identity, sexuality, and emotion in learning to teach.* New York, NY: RoutledgeFalmer.

Fifield, S. (2001a, January). *Identity, knowledge, and heteronormativity in science teacher education.* Paper presented at the Annual Meeting of the Association for the Education of Teachers in Science, Costa Mesa, CA. Retrieved from https://www.researchgate.net/publication/324263156

Fifield, S. (2001b, October). *Teaching on (the) edge: Negotiating the cultural intersections of "gay" and "science teacher."* Paper presented at the Center for the Study of the 21st Century, University of Wisconsin, Milwaukee, WI. Retrieved from https://www.researchgate.net/publication/324263343

Fifield, S. (2002, April). *What's in the closet? Negotiating identities in the first years of science teaching.* Paper presented at the Annual Meeting of the National Association for Research in Science Teaching, New Orleans, LA. Retrieved from https://www.researchgate.net/publication/324263371

Fifield, S. (2004, October). *Getting to know you? LGBT teachers and the epistemological prospects of coming out versus being a sexual stranger.* Paper presented at the Centre for Research into Professional Practice, Learning, and Education; Charles Sturt University, Bathurst, New South Wales, Australia. Retrieved from https://www.researchgate.net/publication/268260920

Fifield, S. (2012). *Teachers as sexual strangers* (Unpublished manuscript). Retrieved from https://www.researchgate.net/publication/324276952

Fifield, S., Grusenmeyer, L., & Ford, D. (2014). Pedagogical change and mourning in elementary teacher education. *Journal of Curriculum Theorizing, 30*(1), 75–86.

Fifield, S., & Johnson, B. (2000, April). *Narratives of silence: Confronting queerness, classroom identities, and disciplinary knowledge in science and science teacher education.* Paper presented at the Annual Conference of the National Association for Research in Science Teaching, New Orleans, LA.

Fifield, S., & Juck, M. (2006, April). *"You can't teach an old dog new tricks": Identity and resistance in an inquiry-based curriculum for preservice elementary teachers.* Paper presented at the Annual Meeting of the National Association for Research in Science Teaching, San Francisco, CA. Retrieved from https://www.researchgate.net/publication/324263324

Fifield, S., & Letts, W. (2014). [Re]considering queer theories and science education. *Cultural Studies of Science Education, 9*, 393–407.

Fifield, S., & Swain, H. (2002). Heteronormativity and common sense in science (teacher) education. In R. M. Kissen (Ed.), *Getting ready for Benjamin: Preparing teachers for sexual diversity in the classroom* (pp. 177–189). Lanham, MD: Rowman & Littlefield.

Ford, D. J., Fifield, S., Madsen, J., & Qian, X. (2013). The Science Semester: Cross-disciplinary inquiry for prospective elementary teachers. *Journal of Science Teacher Education, 24*(6), 1049–1072.

Fuss, D. (1991). Inside/out. In D. Fuss (Ed.), *InsideOut: Lesbian theories, gay theories* (pp. 1–13). New York, NY: Routledge.

Gieryn, T. F. (1999). *Cultural boundaries of science: Credibility on the line.* Chicago, IL: University of Chicago Press.

Glasser, H. M., & Smith III, J. P. (2008). On the vague meaning of "gender" in education research: The problem, its sources, and recommendations for practice. *Educational Researcher, 37*(6), 343–350.

Goldstein, L., & Lake, V. (2000). "Love, love, and more love for children": Exploring preservice teachers' understandings of caring. *Teaching and Teacher Education, 16*(8), 861–872.

Gould, S. J. (1997). Nonoverlapping magisterial. *Natural History, 106*(2), 16–22.

Graham, M. (2014). *Anthropological explorations in queer theory.* Farnham: Ashgate.

Halperin, D. (1995). *Saint Foucault: Towards a gay hagiography.* New York, NY: Oxford University Press.

Haraway, D. (1988). Situated knowledges: The science question in feminism and the privilege of partial perspective. *Feminist Studies, 14*(3), 575–599.

Harbeck, K. M. (1992). *Coming out of the classroom closet: Gay and lesbian students, teachers and curricula.* New York, NY: Harrington Park.

Harding, S. (1991). *Whose science? Whose knowledge? Thinking from women's lives.* Ithaca, NY: Cornell University Press.

Harding, S. (1998). *Is science multicultural? Postcolonialisms, feminisms, and epistemologies* Bloomington, IN: Indiana University Press.

Harding, S. (2004). Introduction: Standpoint theory as a site of political, philosophic, and scientific debate. In S. Harding (Ed.), *The feminist standpoint reader: Intellectual and political controversies* (pp. 1–15). New York, NY: Routledge.

Jagose, A. (1996). *Queer theory: An introduction.* New York, NY: New York University Press.

Jennings, K. (1994). *One teacher in 10: Gay and lesbian educators tells their stories.* Los Angeles, CA: Alyson.

Kissen, R. M. (1996). *The last closet: The real lives of lesbian and gay teachers.* Portsmouth, NH: Heinemann.

Leap, W. L. (2000). Book review of queering elementary education: Advancing the dialogue about sexualities and schooling. *Anthropology and Education Quarterly, 31*(4), 498–499.

Lemke, J. (2011). The secret identity of science education: Masculine and politically conservative? *Cultural Studies of Science Education, 6*, 287–292.

Letts, W. J. (1995, September). *Queering science: Using sexuality theory and research to critique the natural sciences.* Paper presented at Sexuality and Academia, Lavender Scholars Lecture Series, University of Delaware, Newark, DE. Retrieved from https://www.researchgate.net/publication/327515758

Letts, W. J. (1998, April). *But still, like dust, I rise: Reflections on sexual minority issues in science education.* Paper presented at the Annual Meeting of the National Association for Research in Science Teaching, San Diego, CA. Retrieved from https://www.researchgate.net/publication/327515736

Letts, W. J. (1999a, April). *The heteronormative nature of school science.* Paper presented at the Annual Meeting of the American Educational Research Association, Montreal, Canada.

Letts, W. J. (1999b). Making "boys" and "girls" in the classroom: The heteronormative nature of elementary school science. In W. J. Letts & J. T. Sears (Eds.), *Queering elementary education: Advancing the dialogue about sexualities and schooling* (pp. 97–110). Lanham, MD: Rowman & Littlefield.

Letts, W. J. (2001a). Boys will be boys (if they pay attention in science class). In W. Martino & R. J. Meyenn (Eds.), *What about the boys? Issues of masculinity in schools* (pp. 186–198). Buckingham: Open University Press.

Letts, W. J. (2001b). When science is strangely alluring: Interrogating the masculinist and heteronormative nature of primary school science. *Gender & Education, 13*(3), 261–274.

Letts, W. J. (2002a, April). *What is the 'science' and who are the 'all'? Queer renderings of science education.* Paper presented at the Annual Meeting of the National Association for Research in Science Teaching, New Orleans, LA.

Letts, W. J. (2002b). Re-visioning multiculturalism in teacher education: Isn't it queer? In R. Kissen (Ed.), *Getting ready for Benjamin: Preparing teachers for sexual diversity in the classroom* (pp. 119–131). Lanham, MD: Rowman & Littlefield.

Letts, W. J. (2004, January). *There is not stuff that I was necessarily hiding but ... I'm not going to advertise it out loud: Teachers negotiating gay, lesbian and bisexual identities in and out of school.* Paper presented at the Hawai'i International Conference on Education, Honolulu, HI.

Letts, W. J. (2005a, April). *The cultural plunge as queer identity exploration.* Paper presented at the annual meeting of the American Educational Research Association, Montreal, Canada. Retrieved from https://www.researchgate.net/publication/327515784

Letts, W. J. (2005b, September). *I don't go around saying I'm out at work, but there's no one who doesn't know: Negotiating identities by enacting strategies to resist homophobic labeling in schools.* Paper presented at the Schooling & Sex/ualities Conference, Melbourne, Australia.

Letts, W. J. (2006). I can't even think straight: Queer childhood and adolescence. *Journal of Gay & Lesbian Issues in Education, 3*(2–3), 163–166.

Letts, W. J., & Fifield, S. (2000, April). *Sexualities, silence, and science teacher education.* Paper presented at the Annual Meeting of the American Educational Research Association, New Orleans, LA. Retrieved from https://www.researchgate.net/publication/324263449

Letts, W. J., & Sears, J. T. (Eds.). (1999). *Queering elementary education: Advancing the dialogue about sexualities and schooling.* Lanham, MD: Rowman & Littlefield.

Letts, W. J., & Simpson, T. (2003, October). *How can there be so many truths in one story? Knowledge construction through 'othering' oneself.* Paper presented at the Annual Curriculum and Pedagogy Conference, Decatur, GA. Retrieved from https://www.researchgate.net/publication/327515863

Luhmann, S. (1998). Queering/querying pedagogy? Or, pedagogy is a pretty queer thing. In W. F. Pinar (Ed.), *Queer theory in education* (pp. 141–155). Mahwah, NJ: Lawrence Erlbaum Associates.

Lundin, M. (2014). Inviting queer ideas into the classroom: Studying sexual education from a queer perspective. *Cultural Studies of Science Education, 9*(2), 377–391.

Mayo, C. (2007). Queering foundations: Queer and lesbian, gay, bisexual, and transgender educational research. *Review of Research in Education, 31*, 78–94.

McNaron, T. A. H. (1997). *Poisoned ivy: Lesbian and gay academics confronting homophobia.* Philadelphia, PA: Temple University Press.

Miller, J. L. (1998). Autobiography as a queer curriculum practice. In W. F. Pinar (Ed.), *Queer theory in education* (pp. 365–373). Mahwah, NJ: Lawrence Erlbaum Associates.

Morris, M. (1998). Unresting the curriculum: Queer projects, queer imaginings. In W. F. Pinar (Ed.), *Queer theory in education* (pp. 275–286). Mahwah, NJ: Lawrence Erlbaum Associates.

Murray, A. (1995). Femme on the streets, butch in the sheets (a play on whores). In D. Bell & G. Valentine (Eds.), *Mapping desire: Geographies of sexualities* (pp. 66–74). London: Routledge.

National Research Council. (1996). *National science education standards.* Washington, DC: National Academies Press.

National Research Council. (2000). *Inquiry and the national science education standards: A guide for teaching and learning.* Washington, DC: National Academies Press.

Nobles, C., & Letts, W. (2000, April). *Queering science education: You don't say?* Paper presented at the Annual Meeting of the National Association for Research in Science Teaching, New Orleans, LA.

Noddings, N. (2005). *The challenge to care in schools: An alternative approach to education* (2nd ed.). New York, NY: Teachers College Press.
Olesen, V. (2018). Feminist qualitative research in the millennium's first decade: Developments, challenges, prospects. In N. Denzin & Y. Lincoln (Eds.), *The Sage handbook of qualitative research* (5th ed., pp. 151–194). Thousand Oaks, CA: Sage Publications.
Phelan, S. (2001). *Sexual strangers: Gays, lesbians, and dilemmas of citizenship*. Philadelphia, PA: Temple University Press.
Pinar, W. F. (Ed.). (1998). *Queer theory in education*. Mahwah, NJ: Lawrence Erlbaum Associates.
Roughgarden, J. (2004). *Evolution's rainbow: Diversity, gender, and sexuality in nature and people*. Berkeley, CA: University of California Press.
Ruffolo, D. V. (2007). Giving an account of queer: Why straight teachers can become queerly intelligible. In N. M. Rodriguez & W. F. Pinar (Eds.), *Queering straight teachers: Discourse and identity in education* (pp. 255–276). New York, NY: Peter Lang.
Schiebinger, L. (1993). *Nature's body: Gender in the making of modern science*. Boston, MA: Beacon Press.
Sedgwick, E. K. (1990). *Epistemology of the closet*. Berkeley, CA: University of California Press.
Snyder, V., & Broadway, F. (2004). Queering high school biology textbooks. *Journal of Research in Science Teaching, 41*, 617–636.
Sullivan, N. (2003). *A critical introduction to queer theory*. New York, NY: New York University Press.
Sumara, D., & Davis, B. (1998). Telling tales of surprise. In W. F. Pinar (Ed.), *Queer theory in education* (pp. 197–219). Mahwah, NJ: Lawrence Erlbaum Associates.
Talburt, S. (2000). *Subject to identity: Knowledge, sexuality, and academic practices in higher education*. Albany, NY: State University of New York Press.
Tierney, W. G. (1997). *Academic outlaws: Queer theory and cultural studies in the academy*. Thousand Oaks, CA: Sage Publications.
Weld, J., & Funk, L. (2005). "I'm not the science type": Effect of an inquiry biology content course on preservice elementary teachers' intentions about teaching science. *Journal of Science Teacher Education, 16*(3), 189–204.
Wylie, A. (2004). Why standpoint matters. In S. Harding (Ed.), *The feminist standpoint reader: Intellectual and political controversies* (pp. 339–351). New York, NY: Routledge.

ABOUT THE AUTHORS

Steve Fifield is an affiliated assistant professor in the School of Education at the University of Delaware (USA). He previously worked at the University of Delaware as a faculty and staff member in science teacher education and program evaluation. He has also been a grant writer at The Franklin Institute Science Museum in Philadelphia, Pennsylvania, and the Oregon Museum of Science and Industry in Portland, Oregon. He's currently a paraprofessional in special education services. Steve has a PhD in curriculum and instruction (with a minor in studies of science and technology) and an MS in ecology and evolutionary biology from the University of Minnesota–Twin Cities. He's interested in cultural studies of science and science education, including intersections with queer theoretical and Buddhist conceptions and practices of subjectivity and (non)self.

Will Letts is associate dean academic in the Faculty of Arts and Education at Charles Sturt University in Bathurst, NSW, Australia. His research interests include the cultural studies of science and science education, especially with respect to sex, gender, and sexuality; interrogating subjectivities in ECE/teacher education; and the enactment of pedagogical documentation in tertiary settings. He has loved learning about and doing science since he was very young. That's what led to him becoming a high school science teacher and a lecturer in science and technology. You can reach Will at wletts@csu.edu.au

BLAKE M. R. FLESSAS AND TIMOTHY D. ZIMMERMAN

3. BEYOND NATURE TALK

Transforming Environmental Education with Critical and Queer Theories

INTRODUCTION

As white, middle-class environmental practitioners/researchers—one queer and gender nonconforming, and one cis-identifying[1]—who have strong convictions about love of the environment *and* for all people, we see immense opportunity for environmental education to become a place where conversations about sexuality, race, power, desire, ability, and oppression intersect with deforestation, pollution, ocean acidification, loss of biodiversity, invasive species, fisheries decline, anthropogenic climate change, and all human-induced environmental degradation and interaction. Indeed, we see all environmental degradation as gendered, racist, classist, ableist, and oppressive. It then follows that our view of environmental education encompasses more than nature study or even complex human-nature interactions.

In this chapter we argue that environmental education must move beyond solely discussing nature while ignoring oppressive social structures to, instead, willingly engage, resist, and transform discourses of power and knowledge around forces including sexuality, race, gender, class, ability, and privilege as these circulate among, shape, and are shaped by the relations of humans and the environment. To be clear, we see this approach as an enhancing process, a need to acknowledge and address these forces whilst engaging in nature study and evoking wonder, awe, spirituality, and deep love for nature. To do this, we use critical and queer theories in tandem to reimagine environmental education. We then employ these theories to formulate recommendations for moving toward multiple epistemologies that subvert existing power structures. We refer to this praxis of reimagining environmental education as *queering environmental education.* We see *queering* as an action (a verb), the purposeful effort to deconstruct, destabilize, and problematize mainstream ideas of what is "normal" and "acceptable" and the hegemonic systems of society (e.g., patriarchy, institutionalized racism, heterosexism, etc.) that construct these ideas (Daring, Rogue, Volcano, & Shannon, 2012; Wilchins, 2004).

The idea of queering environmental education has been proposed before, in particular by Russell, Sarick, and Kennelly (2002) and more recently by Russell (2013). As we detail below, we find some recent efforts at queering environmental education to be focused on legitimizing environmental education so that it might be more readily assimilated into conventional educational systems. For example,

 | DOI:10.1163/9789004331068_005

Russell et al.'s (2002) vision of queering environmental education, which concludes that we must probe, unpack, and resist heteronormativity, still relies upon socially acceptable categories of inclusion, striving for an environmental education that recognizes and welcomes LGBT identities. This inclusion-as-assimilation approach echoes gay and lesbian activist groups' promotion of gay marriage as the primary goal of LGBT politics. Such an assimilationist approach, which we critique in detail below, plays directly into societal pressure on queers to fill the same roles as "normal" heterosexuals, thereby erasing difference and upholding the privileging of normalized identities. As Audre Lorde (1983) has taught us, though, the master's tools will never dismantle the master's house. Assimilating environmental education into mainstream education, where issues such as race, socioeconomic class, gender, and privilege are ignored, is not only detrimental to the goals of environmental education, but also serves to further alienate and marginalize those who are already oppressed by this system.

We frame these ideas by juxtaposing macro- and micro-level perspectives on, and manifestations of, discourses of power and knowledge. We employ critical theories and pedagogies at the macro end of the spectrum, which attempt to break down systems of oppression at the institutional level: capitalism, imperialism, and oppression writ large. On the micro end of the spectrum, we engage queer theories and pedagogies as perspectives for negotiating identity, subjectivity, and positionality at an individual and group level, and the ways they take up, negotiate, reproduce, resist, and reinvent broader sociocultural discourses. That said, we recognize the risks of oversimplification when applying a macro/micro binary. When employing queer modes of thought—a set of epistemic commitments to notions of intersectionality not assimilation, radical acceptance, and antihegemonic tendencies—we must resist simple binaries (e.g., male/female, heterosexual/homosexual, normal/abnormal) and remember that the micro and macro not only overlap, but also inform and even create each other.

To that end, we propose a critical queering of environmental education: an environmental education that does not hide from micro and macro forms of oppression but, instead, places all forms of oppression on equal footing with environmental degradation in a way that leads to liberation, reparation, and emancipation for all of nature, including all human beings. Our vision for queering environmental education incorporates critical, queer, *and* intersectional viewpoints to embody a liberatory pedagogy, while simultaneously moving away from the normalization of those excluded by science and science education (Fifield & Letts, 2014; Lemke, 2011) and, in our case, environmental education. We argue that the mechanism necessary to subvert the dominant educational narrative and push environmental education towards addressing natural complexity that fully integrates human complexity is to not only acknowledge oppression but draw it out, hold it in examination and contempt, and smash it. In so doing, we assert that rather than leaving the field of environmental education with useless smashed pieces, queering environmental education will allow us to reinvent those pieces, to jumble them up and create something new, to broaden the scope of the field and move it forward.

BACKGROUND

What Is Environmental Education?

Within a hegemonic society that bases conceptions of "normal" on white, heterosexual, male, upper middle-class, able-bodied experiences and perspectives, education systems serve to perpetuate the socially constructed hierarchies that marginalize and oppress those who do not fit the mold (Hursh & Martina, 2003), often through an assimilationist approach. This is true for environmental education as well.

Defining environmental education is difficult, especially since scholars have debated the goals, philosophical underpinnings, curricular approaches, even the definition, of environmental education for nearly 50 years (Reid & Payne, 2013; Stevenson, 2007; Wals, Brody, Dillon, & Stevenson, 2014). Several recent works detail the history and emergence of environmental education, from both U.S. and European perspectives (e.g., Carter & Simmons, 2010; Palmer, 1998). Here we provide a brief primer.

In the 18th and 19th centuries, philosophers such as Jean-Jacques Rousseau, and scientists and educators such as Louis Agassiz, advocated for an education focused on the environment and the study of nature. During this same time, transcendentalists like Ralph Waldo Emerson and Henry David Thoreau wrote of the spirituality and societal knowledge to be gained from nature, and conservationists such as John Muir spoke of the need to conserve nature in its pure form (Carter & Simmons, 2010). Contemporaneously with the early development of ecology as a formal discipline, the end of the Enlightenment, and the rise of modern science, the formation of nature studies and conservation education became the precursor to our modern notions of environmental education. In the 1960s, U.S. environmentalism and environmental education took hold, inspired by books such as *A Sand County Almanac* (Leopold, 1949), *Silent Spring* (Carson, 1962), and *The Quiet Crisis* (Udall, 1963), which brought environmental concerns into everyday conversations (Carter & Simmons, 2010), culminating in the creation of Earth Day and the Environmental Education Act in 1970. A number of international conferences and workshops on environmental education were held in the early 1970s. One in particular, the 1975 Belgrade International Workshop on Environmental Education, brought together representatives from 60 countries. The resulting Belgrade Charter identified the goal of environmental education as developing

> a world population that is aware of, and concerned about, the environment and its associated problems, and which has the knowledge, skills, attitudes, motivations and commitment to work individually and collectively toward solutions of current problems and the prevention of new ones. (UNESCO-UNEP, 1977, p. 15)

In the 1980s, however, environmental education began to fade away and became relegated to an extracurricular or supplemental component of ecology and science education. Practitioners have been fighting ever since to relegitimize environmental education's existence in mainstream education and in the eyes of society (Carter & Simmons, 2010; Gruenewald, 2004; Wals et al., 2014).

Despite the field's foundations in nature study and modern political activism, ongoing attempts to resurrect environmental education have moved it away from its political goals—especially environmental education designed for and taught in schools—and toward a science- or ecology-only stance, where educators teach almost exclusively the canonical, narrowly defined "science" of environmental issues and forego other factors. The field's commitment to action has often placed it at odds with this narrow view of science that focuses on rote memorization, narrowly applied theories, and professed impartiality (Stevenson, 2007), a view that drives historical, and often current, approaches to school science curriculum. With the move to solidify legitimacy, however, comes institutionalization, which we believe depoliticizes environmental education in a way that dangerously denies the social, political, cultural, and societal intersections of science, especially as they relate to environmental issues (Gruenewald, 2004; Hungerford, 2010; Martin, 1996).

Today, environmental education occurs in schools through outdoor programs such as nature walks, summer camps, after-school programs, and other extracurricular activities. Despite its roots in the radical politics and environmentalism of the 1960s, much of today's environmental education harkens back to late 19th- and early 20th-century ideas of nature study while ignoring the middle to late 20th-century foundations of the field. Though sometimes loosely tied to environmental issues such as climate change, contemporary environmental education is more often than not taught in school classrooms in the context of environmental science, ecology, or earth sciences. Measuring water quality indicators such as pH, temperature, and nitrogen and phosphorous levels often leads to classroom discussions of anthropogenic causes and impacts, such as run-off from golf courses. Discussions about who has access to golf courses, who decides which fertilizers to use and how much, the transnational agrochemical corporations those fertilizers come from, and who is ultimately impacted by the fertilizer use are strikingly absent.

Programs for school-aged people at outdoor nature camps are similarly narrow. These programs, which often focus on particular species or groups of species (e.g., monarch butterflies, birds) or certain ecosystems (e.g., salt marshes, peat bogs), place most of the attention on the science of nature and rarely mention micro-level positionalities—where one stands in relation to others, the power or lack thereof that one's identity may hold in a social situation—or macro-level social structures that greatly impact these natural systems. Descriptions of these programs often tout the science education standards that will be covered during the program, inherently privileging Western scientific knowledge (Mueller & Bentley, 2009), but rarely address micro- or macro-level constructs that perpetuate systems of oppression of nature and people. Nature center programs for adults are even less explicit about the science content and just as silent about the relationship between environmental degradation and oppression. Ecotourism—where people pay large sums of money to travel to "exotic" places with nature guides to see rarely encountered, charismatic megafauna such as whales or gorillas—often invokes notions of conservation (Small, 2011). However, those who design and lead ecotourism experiences steer clear of uncomfortable topics such as white privilege, sexuality, and gender at the macro or micro levels. This is depoliticized environmental education that is silent about marginalization of the poor; people of color; trans and gender nonconforming people; and people whose sex, sexuality, or identities fall outside the expectations that larger systems of power impose upon us all.

Repoliticizing Environmental Education: Environmental Degradation Is Political

We believe environmental education is necessarily political because environmental degradation is driven by the power structures put in place by social and cultural norms at both the micro and macro levels of systems of oppression. For example, we see environmental degradation as

- *Racist.* Many of the people negatively impacted by environmental degradation are people oppressed based upon their (real or perceived) race or ethnicity. A growing body of research demonstrates the health benefits of intact natural landscapes. However, U.S. neighborhoods that house primarily people of color are often concentrated, through histories of systemic oppression, in areas with high levels of air, water, and soil pollution and with highly fragmented, degraded, or de-natured natural spaces (Bullard, 2000; Taylor, 2002, 2011). The recent exposure of Flint, Michigan, residents, the majority of whom are black and brown people, to water laden with lead and other poisonous substances is only the latest example of environmental racism.
- *Heterosexist.* As "progressive" as environmentalism seems to be, systemic micro-level exclusion of LGBTQ people from the environmental arena is still prevalent. For example, Jourdan Imani Keith (2014) describes being forced off public transit in Seattle, Washington, by people "hurl[ing] insults … attacking gay people" (para. 4). As a result, Keith and her spouse "protected [themselves] by driving [their car] while wanting to protect our environment [by taking mass transit instead]" (para. 5).
- *Classist.* Often, high-quality natural areas and "pro-environmental behaviors" are only accessible to those with some amount of wealth. For example, here in the United States, the movement toward buying local, organic, higher-quality, and nutrient-rich food as an environmentally sustainable action often means paying more. These higher prices limit access to high-quality, sustainably produced food to those with higher incomes. Other examples are, not surprisingly, similar to environmental racism, since classism and racism often co-occur (e.g., asthma rates are highest in poor neighborhoods).
- *Geopolitical.* There are many nations in the global south that will be disproportionately impacted by sea level rise as a result of climate change. Sea level rise is linked to high carbon output by nations in the global north, many of which, including the United States, have failed to sign international treaties

> or take serious action to reduce their impact on these less powerful nations. The more environmentalism is tied to capitalism and consumerism, the more it continues dominant hierarchies and narratives of the globalized world.

Furthermore, because of environmental education's roots in politics, environmental educators are in a unique position to *acknowledge* the politicality of education more easily than one might be able to within, say, a mathematics classroom. The mathematics classroom is, of course, political (see our section below on critical pedagogy), but there exists a long and intensive history of this type of "fact-based" education being considered apolitical and unbiased, in contrast with environmental education (see Stevenson, 2007). In a field cultivated in the politically tumultuous 1960s, we as contemporary educators must not forget our political roots while working in the context of a multiplicity of oppressions. Such an approach must move beyond the dualistic notion that educators are either "biased" or "unbiased," since in the construction of identity, one develops their own lens through which they see the world and themselves, driven by power relations that shape our values through shared cultural discourses that normalize certain perspectives. This lens is shaped by one's particularly situated cultural, socioeconomic, political, familial, environmental, and societal experiences. We argue that preparing contemporary environmental educators to move beyond the dualities of biased/unbiased, subjective/objective, opinions/facts, etc., requires that environmental education deeply engage with critical pedagogy, which we see as a first step to establishing reflexive praxis.

CRITICAL PEDAGOGY, ECOPEDAGOGY, AND QUEERING

Critical pedagogy is an engaged and liberatory pedagogy that cultivates active learners who construct their knowledge through personal agency, critique, and a reconsideration of the status quo (Freire, 1970/1993; hooks, 1994). With roots in critical theory of the neo-Marxist Frankfurt School, and first described by Paulo Freire in his book *Pedagogy of the Oppressed*, critical pedagogy seeks to problematize and disrupt power structures within neoliberal education frameworks. The idea that education is inherently political is foundational to critical pedagogy. The distribution of power within the educational system, from the macro (federal and district level, or even societal level) to the micro (within a classroom or between two students), constantly demonstrates this politicality. Kincheloe (2008) provides one example of the reification of existing oppressive paradigms:

> A district supervisor who writes a curriculum in social studies, for example, that demands the simple transference of a body of established facts about the great *men* and great events of American history is also teaching a political lesson that upholds the status quo. (p. 9, original italics)

Critical pedagogy, in contrast, is dedicated to liberation from oppression by centering the student within the discourse of what is to be learned, to what end, and by/for whom. By developing a critical consciousness, students are encouraged not only to survive, question, and critique oppression but to *actively combat* systemic oppression.

A form of critical pedagogy specific to environmental education is the idea of an *ecopedagogy*. Richard Kahn (2010) asserts that while an uncritical sustainable-development rhetoric is being used to legitimize globalization and imperialism and to obscure the deeper origins of the ecological crisis, critical pedagogy lacks an essential ecological dimension. In an effort to unite ecologically focused education and critical pedagogy, Kahn proposes an ecopedagogy rooted in the critique of domination (Kellner, 2010) and adapted from a South American movement by the same name (Kahn, 2010). The underlying idea of ecopedagogy is that in order to effectively challenge the global ecological crisis, we must address its cultural roots and question the anthropocentric worldview that portrays nature as the proper subject of human domination (Bowers, 2004; Gruenewald, 2004; Kahn, 2010; Kellner, 2010). These ideas strike at macro-level power dynamics as root causes of environmental degradation. Environmental education and critical pedagogy both benefit from the ecopedagogy fusion, Kahn asserts, and we need these frameworks to build a sustainable future (Kahn, 2010; Kellner, 2010).

Even as ecopedagogy strives to be a liberatory departure from environmental education, it is distinctly lacking in queer modes of thought. The ecopedagogy literature employs a binary voice of good versus evil, mainstream versus radical, oppressive versus liberatory. As Bowers (2004) argues, ecopedagogy is founded on a Western imperialist worldview. As such, ecopedagogy focuses on a macro-level critique of worldviews and global perspectives, without addressing micro positionalities that queer theories help us visualize. We argue that to move towards a liberating environmental education, we must apply an actively queer*ing* lens to all that we do. We apply this conceptualization of the act of queering to critical pedagogy and its synthesis with environmental education.

"Queer" as Action

Daring et al. (2012) lay out the use of the word *queer* in all of its versatility: as a noun, adjective, and verb. Historically, the word queer has meant strange or odd—something outside the norm. In the 20th century, it began to be used as a slur referring to homosexuals, specifically gay men. More recently queer has been reclaimed as an umbrella term for the many identities referenced in various versions of the LGBTQQIAP[2] acronym and as an identity of its own (Daring et al., 2012). Queer identity brings us back to that original meaning of the word, back to being outside the norm and beyond: to being at odds with it. "Queer is by definition *whatever* is at odds with the normal, the legitimate, the dominant. … It is an identity without an essence," a positionality (Halperin, 1995, p. 62). The common words *gay* and *lesbian* have come to signify a desire to fit in and to normalize the identities to which they refer, whereas queer represents a celebration of the abnormal, against assimilation (Luhmann, 1998; Pinar, 1998). Considering queer as an adjective, then, we delve even deeper into this meaning. We can see social power dynamics at play by examining the binary of normal versus abnormal. "Whatever behaviors (or desires, thoughts, etc.) fall into the category labeled 'normal' are dominant, intelligible, visible, and in many cases, powerful. Other behaviors will fall into the 'abnormal' category and become subordinate, unintelligible, invisibilized, and su-, re-, and o-ppressed" (Daring et al., 2012, p. 12). When queer becomes a verb, then, it signifies that queer intention—deconstructing, destabilizing, and problematizing mainstream ideas of what is "normal" or "acceptable" at the micro level of individuals and small groups (Krywanczyk, 2007, p. 2) and in macro-level hegemonic social systems (e.g., patriarchy, institutionalized racism, heterosexism) (Daring et al., 2012). Queer "has a deliberately disruptive, political component that is inextricable from yet not reducible to sexuality" (Whitlock, 2010, p. 82). Queer*ing* moves beyond the inclusion of queer identities and towards queer modes of thought that problematize and pull apart normative and normatized[3] conceptions of identity and ideology, opening space to nurture new ways of thinking, feeling, doing, being.

Pedagogy? How Queer!

"To queer [pedagogy] is to venture into controversial, intellectually complicated, nuanced terrain with students" (Krywanczyk, 2007, p. 5). Whereas a queer pedagogy can refer to a number of things, our definition of/intention in queering pedagogy is to focus less on the inclusion of LGBTQQIAP etc., identities and more on the method of the pedagogy itself: encouraging critical thought and the willingness to delve into complexity; investigating, questioning, and re-evaluating assumptions; and interrogating processes of normalization, especially around sex, gender, "natural" desires, and the relations with other dimensions of cultural discourse and personal experience. Queering pedagogy must engage with intersectionality and share aspects of critical pedagogy, queer theory, cultural studies, and ecofeminism (Gough & Gough, 2003; Pinar, 1998). *Intersectionality* is a term originally coined by Kimberlé Crenshaw (1989) to refer to interactions of various power structures and systems of oppression, specifically in the lives of marginalized people. In its original use, intersectionality refers to the unique discrimination black women face and the failure of mainstream feminism and antidiscrimination laws to recognize their intersecting identities. The concept can be translated to any situation in which power structures are at play (Adewunmi, 2014).

Perhaps the first published attempt to tease out the meaning of a queer pedagogy was a 1993 paper by Mary Bryson and Suzanne de Castell, "Queer Pedagogy: Praxis Makes Im/perfect." In their effort to bridge the distance between queer theory and queer pedagogy, the authors propose several possible meanings of the latter and eventually settle on a direction in line with our own queering of pedagogy over 20 years later:

> It seems that a worthwhile avenue for the elucidation of a queer praxis might be to consider the value of an actively *queerying* pedagogy—of queering its technics and scribbling graffiti over its texts, of colouring outside the lines so as to deliberately take the wrong route on the way to school—going in an altogether different direction than that specified by a monologic destination. This seems a promising approach indeed for refashioning pedagogy in the face of the myriad institutionally sanctioned "diversity management" (Mohanty, 1990) programs that, today, threaten to crowd out and silence most opportunities for radical emancipatory praxis (de Castell & Bryson, 1992). (Bryson & de Castell, 1993, p. 299, original italics)

The authors queerly refuse to propose a direction to move in, but challenge others who "still believe" in a queer pedagogy to "tell it like it is, or, at least, how it might be" (p. 301). "Queer pedagogy displaces and decenters—queer curriculum is noncanonical" (Pinar, 1998, p. 3). It has no absolute rules, no clear-cut definition. As Pinar (1998) highlights, queer theorists do not focus on specific institutional issues, but rather emphasize the big picture, the "structures that make oppressive institutional practices possible" (p. 14). Therefore, queering pedagogy can be a way to use those specific institutional issues as a springboard for educating about macro systems, troubling our earlier macro/micro dichotomy.

QUEERING ENVIRONMENTAL EDUCATION

Responding to Tactics of Inclusion

Mueller and Tippins (2012) warn that "if not careful, teachers may implicitly teach students that their local knowledge, face-to-face conversations, narratives, cultural traditions and ceremonies are not as important as the standardized curriculum" (p. 7). In the same fashion, normative environmental education that outwardly adheres to modes of thought driven by dominant groups, such as straight, or white, or male, or middle-class students or some combination thereof, implicitly ignores nondominant modes of thought. This perpetuates notions and enactments of environmental education that are exclusive of many nondominant constituencies, thereby discouraging participation by those to whom normative environmental education is not relevant, and are, therefore, directly responsible for environmental education's own exclusivity.

Lee (2012) argues that connecting learning to its sociocultural context will help transform learning environments into places where youth are excited to invest their time and energy in science. In recent years, liberal environmental education settings have been focused on developing a rhetoric of diversity and inclusion, seeking ways to bring underrepresented people into the field. This has manifested in the creation of financial aid programs for outdoor education settings and other special programs but has strategically ignored the voices of the actual people these settings seek to include. The goals of making environmental education a field in which students of varying genders, sexualities, racial and economic backgrounds, and identities are encouraged and inspired are achievable not by promoting a rhetoric of diversity and inclusion, but by shifting our paradigmatic understandings of what environmental education *is*. When we realize the field is privileging specific identities and excluding others, the question should not be how to include those who are excluded, as is so often the case in well-meaning liberal circles, because this only reifies normatization. Indeed, the very question acknowledges the lack of inclusivity and thus the *existence of* exclusivity deeply engrained in the current model of environmental education. Therefore, we argue that the goal should instead become a deep dismantling and reworking of the way we conceptualize the purpose and practice of environmental education.

Heteronormativity and Queer Desire

One of the issues that must be explored and problematized when beginning to queer environmental education is the patriarchal heteronormativity that is prevalent in the field, especially taking form in outdoor, experiential, and adventure education (Warren, 1996). For example, Russell et al. (2002) provide an excellent treatment of the male domination of adventure education. This stems from (and contributes to) the traditional representation of nature in Western art and research as gendered, sexualized, and racialized. In the colonial imagination of the men who contributed to the foundations of environmental education (e.g., Rousseau, Emerson, Muir), nature is envisioned either as a pure, virginal, feminine (white, heterosexual) woman, or as something dark, mysterious, and wild, waiting to be conquered (Evans, 2002; Gough & Gough, 2003). When nature is viewed in this way, it perpetuates a raced, classed, hetero-patriarchy and creates a heterosexist atmosphere (McClintock, 1996; Russell et al., 2002). As a result, the status quo of oppressive society is reproduced within the contexts of outdoor, experiential, and adventure education, perpetuating the same global capitalist systems that arguably created the environmental crises environmental education works to combat.[4]

Queer Politicality

In order to further queer environmental education, we must consider the political tone of that which is queer. As we mention above, environmental education is (should be) radically political.[5] Yet the assimilationist environmental education that seeks assimilation into mainstream educational discourse is constantly at odds with this notion (Gruenewald, 2004). Queering resists assimilation and can help us recenter politicality within environmental education. Mical Boles-Friscia's (2010) thesis, *Writing to Transgress*, discusses the conflict around politics in an outdoor school program. The program is designed to host public school students in an educational science-camp setting, and the desire to cater to a wide variety of students is often cited as the root of a perceived need to remain apolitical. When volunteering at the program, Boles-Friscia was asked to remove a T-shirt that brought up complex issues of gender, race, and class privilege through a sardonic song lyric, because the topics were deemed inappropriate for the setting.[6] Discussions of politics in some contexts are, indeed, alienating at times, especially for students who are not accustomed to them. "Unfortunately, when it comes to systemic oppression, the absence of dialogue only supports the status quo" (Boles-Friscia, 2010, para. 8). By insisting, "we don't talk about that here," the program unintentionally upholds normatization of white, middle-class, cisgender straight, male identities and thus the oppression of those who do not fall into these categories.

It's Supposed to be Confusing? Addressing Complexity

The authors of this piece recognize, and have experienced, the inaccessibility of queer theory to those who are not versed in the particular theoretical language of this work. We also realize, though, that an author may make their piece as accessible or inaccessible as it needs to be in order to achieve its goal. One goal of this paper, for example, is to be "intelligibly queer" and "queerly intelligible" (Ruffolo, 2007, p. 255; Whitlock, 2010): to give an account of queer that can be accessed by anyone moderately versed in academic language. Complexity and nuance are inherent to a queer analysis. In investigating that which is queer in the world, we construct an understanding of the complex web of positionalities of all beings. We must examine, then, not only *what* it is that confuses us (about queer theory, about the positionalities of diverse beings in the natural and social environment) but *why* it confuses us: the internalized heterosexist repulsion to that which we do not understand—we must queer our own perspectives. If we merely tolerate confusing and inaccessible contributions to queer theory, we unintentionally perpetuate the oppressive tolerance rhetoric that is exactly what queer pedagogy is fighting against (Whitlock, 2010). Tolerance does not complicate power—the power remains in the hands of the majoritized and relies on their generosity to tolerate the minoritized (Ruffolo, 2007; Whitlock, 2010). But *queer* is a radical politic that is complex and disruptive, and *queering* must complicate and disrupt; if we can rely on that complexity and disruption in queer theory, and thus queer pedagogy theory, then we can begin to wield it.

Towards a Queer(ing of) Critical Environmental Pedagogy

Integrating critical pedagogy into environmental education is crucial to creating critically engaged students who are conscious of and value the environment. In environmental education we focus almost solely on the natural environment as something outside of us, outside of society, removed from our immediate lives and surroundings. We "go out" into the "great outdoors" and forget that not only is the natural environment ("nature") literally all around us and present in every aspect of our lives, it is with a critical eye directed toward society—the social environment—that we can come to truly see our connection as humans to the natural environment.

While Kahn's ecopedagogy focuses on the hegemonic systems of capitalism and imperialism, our approach complicates these ideas further by including queer(ing) analyses of discursive power and identity. Postmodernism tells us that the types of power we can readily see as oppressive are not the only types of power that subjugate. The idea of a normative/nonnormative binary is produced through discourse—it is not (only) imposed upon us by the state. The cops are not the sole entity that polices the reproduction of a heterosexual-cissexual norm, rather society polices itself: A father does not let his son wear a dress, a teacher organizes a girls versus boys spelling competition, etc. This social dialogue creates the set of standards to which we then hold each other; discourse creates the norm and it creates the queer (Wilchins, 2004). The ideas of discursive power help us reconnect the micro and macro in critical and ecopedagogy as micro discourses inform and create macro systems. This can propel us toward a synthesis of the critical and the queer as we have imagined them.

Russell (2013) speaks of the perceived naturalness of certain types of queer, both within human and nonhuman species, in his (re)orientation of a queer ecopedagogy. He asserts that in validating queer identities it is not enough for environmental educators to assert that homosexuality is natural by teaching about gay animals, like the penguins at the Toronto Zoo, but that a queer ecopedagogy would "further [inquire] into queer desires," narratives, and utopias (p. 23). As he explains,

> the larger story is perhaps more queer than the simplistic notion of "gay penguins," considering that members of an endangered species—likely because of anthropogenic pressures—were forcibly separated from their objects of affection in order to propagate their species by the very animals that threatened them in the first place! (p. 23)

Here queering the discourse around endangered species engages the politicality of environmental education while revealing the multidimensionality of power structures. Russell's queer ecopedagogy "promotes embodied attentiveness and reflection on being or feeling queer in the world" (p. 13), leading to an engagement with otherness and reflection on *all* oppressed beings, including but not limited to queer-identifying humans.

Taking Russell's proposal of a queer ecopedagogy and putting it into conversation with Freire's critical pedagogy of the oppressed, we begin to see the formulation of a critical queer environmental pedagogy. Leveraging hooks's (1994) ideas of engaged pedagogy, we argue that queering environmental education requires an engagement with critical practice. hooks's (1994) holistic engaged pedagogy "is more demanding than conventional critical or feminist pedagogy" (p. 15) because of its emphasis on joy in learning, self-care, and the "self-actualization" of the students *and* the educators. This dual process of self-actualization requires that we, as educators, present ourselves as authentic people to our learners and to ourselves;

this authenticity that can be achieved through a deep understanding of our own identity, experience, socialization, and knowledge. Only in doing so can we fully support the self-actualization of learners and do justice to the environmental content we engage. Queering self-actualization challenges claims to a stable, actualizable self, and twists this idea into a continual, open-ended process of becoming, queering, reflecting, critiquing, searching, becoming.

TOWARD QUEER, CRITICAL, ENVIRONMENTAL PEDAGOGICAL PRACTICE

Things Practitioners Can Do

As implied by Fifield and Letts (2014) and Lemke (2011) in their critiques of science education, when environmental educators are exposing systemic oppression to/for their learners, they must simultaneously and continuously interrogate their own internalized, implicit biases. As educational researchers and practitioners, we, the authors, engage in this hard work in many ways and call upon our colleagues to do the same. To guide our own interrogations, we often share our personal experiences and challenging situations with friends and colleagues within and across affinity groups. We ask others to review our curriculums for majoritarian problems (i.e., worldviews, stories, assignments, etc., that utilize and reify dominant paradigms; see Chapman, 2007), and we reflect on our curriculums, noting problems with or absences of intersectionality. We suggest environmental education practitioners use these techniques, among others. In addition, we recommend that environmental educators engage in both pedagogical and curriculum design changes that move toward addressing intersectional oppression. Below we provide examples of techniques we find useful for moving environmental education practice beyond nature talk and toward queering of environmental education.

- *Name oppression.* We advocate bringing oppression to discussions in environmental education contexts, whether on the trail, in the classroom, or at the visitors center, by naming forms of oppression as teachable moments. Examples from ecojustice can prove useful. For example, if learners mention pollution—a common topic that many people associate with environmental education—help them recognize that the most polluted environments are predominantly those in which poor and/or racially and ethnically marginalized groups live. Classism and racism concentrate pollution where people are already struggling and have fewer resources to fight back.
- *Design antioppression.* At Tamarack Farm, a farm and wilderness summer camp for teens emphasizing community building, simple living, and social justice, we deliberately allocate time for programming that prompts campers to interrogate racism, sexism, classism, Western-centric knowledge, gendered norms, and more within the microcosm of camp and in the broader world. This empowers participants to build, critique, and problem-solve the failures of a camp community that reflects their desires for a liberatory society. In the classroom,

we provide space and support to critique and explore what might otherwise be overlooked. For instance, teachers might start by addressing normatized notions of sex and sexual binaries. Instead of solely using monogamous, heterosexual pairings in stories about ecology (e.g., the mommy bunny and the daddy bunny), discuss examples of a plurality of desires (e.g., nonhuman beings [whales, fish, deer, flowers, etc.] changing sex, acting in nonheterosexual ways). A teacher can begin with typical nature-show examples of male-female pairings, followed by examples of sequential hermaphroditism in marine wrasses and gastropods. Then push further by presenting and questioning the naturalization, normalization, and essentialization of both hetero and queer desires through examples of nature (Mortimer-Sandilands & Erickson, 2010; Russell, 2013).

- *Identify macro and micro.* Help learners look beyond critiques of "unsustainable" environmental practices and behaviors (e.g., high carbon-footprint activities like driving, not turning off the lights, not recycling) as harmful to the environment to develop big-picture perspectives that include political, economic, social, and cultural worldviews. For example, at Tamarack Farm we mapped out what a teenager ate for breakfast, tracing not only where the food came from but what processes it went through, how it travelled, who was involved, how they were affected, and anything else campers felt was connected. This map could contain anything from the interactions between a migrant farm worker and their employer to food deserts to capitalism.
- *Question "naturalness."* "Human nature" is often used to justify oppressive structures. The idea that humans are inherently greedy and self-serving, for example, apparently means capitalism is our best bet. But who decides what is natural and how is it justified? By examining histories of the naturalization of sexed bodies, social gender, sexualities, and broader desires, we uncover the formation of dominant narratives based on narrow views of the perceived naturalness of nonhuman beings. And while gay penguins may help us rewrite this script, what happens when that desire becomes normatized?
- *Address complexity.* Simplifying issues to make them easier to teach silences perspectives that do not fit neatly into simple explanations. By leaning into complexity and investigating nuance in subject matter, we can push learners toward much deeper understandings of natural and social environments and, in turn, encourage critical thought in other facets of their lives and education. For example, at an outdoor science program for sixth graders, we avoided using a "good versus evil" binary (not polluting is good/polluting is bad), which collapses complexity and constricts possibilities for difficult discussions and creative solutions. We chose to engage students in a discussion about convenience and need, exploring the things they and their families rely on to survive that may have negative environmental impacts, challenging them to develop nonbinary solutions to nonbinary problems.
- *Acknowledge privilege.* Backpacking and camping gear is expensive, as is taking an ecotourism trip. If a child's parent(s) or guardian(s) are working one or more

low-wage jobs, there may be no opportunity for vacation. Growing up in a poor, rural village with no plumbing, one's "free time" may be spent retrieving water, not traveling to national parks. Privileging of certain knowledge—specifically Western scientific knowledge—dominates the discourse in most environmental education programs in the "developed world," especially the United States. For example, Western science might claim to identify a "new" species when, in reality, that species has been known by local, indigenous peoples since the time the people and the species began to coexist.

- *Examine assumptions*. Underlying assumptions that are left unexamined perpetuate both macro- and micro-level injustices. On the macro level they support systemic oppression, while at the micro level they contribute to microaggressions.[7] Queering calls upon us to interrupt these cycles. For example, the general idea that people don't engage in proenvironmental behavior because they "don't know" or "don't care" is frequently inaccurate (see Kollmus & Agyeman, 2002). These behaviors may be financially inaccessible. Or one may feel unwelcome in some settings, such as farmers markets, where the clientele is perceived as specific types of people unlike oneself. Assumptions about bodies are prevalent in environmental education. Many environmental education programs are focused on "*seeing* the beauty of nature" or getting out into nature in ways that do not accommodate diverse physical, sensory, and cognitive abilities.

CONCLUSION/CONTINUATION

The critical queer environmental pedagogy we argue for here regards contingency, positionality, and politicality as essential elements within any environmental education setting. In the moment of engagement between the educator and learner, the educator must hold attention to their own positionality, the positionality of the others, and the layers of power dynamics and privilege at play and how they intersect, intertwine, and interact with the environmental content at hand. These moments necessarily shift and are contingent upon the particular individuals involved, the space of their interaction, the environmental issues (e.g., deforestation, species decline) under examination, and myriad other dynamics. Such an approach takes training, attention, and awareness that, as an approach, it is always under examination and reconstruction as macro- and micro-level oppressions change, as environmental issues change, and as awareness and critiques grow and shift. This is the praxis of queer critical environmental education.

For 11 years Russell et al.'s (2002) call to environmental educators to queer their perspectives and pedagogies was (willfully?) ignored until Russell (2013) took it up. Today we revive and further these calls as an urgent and necessary step towards a liberatory environmental education. Critical thinking is promoted in contemporary, forward-thinking environmental education settings, but it is often reserved for considering environmental harm and falls short of addressing complexity and nuance, power dynamics, and systems of oppression that are interwoven with environmental harm. "Critical thinking" has become a meaningless buzzword, promoted as a nebulous goal towards which learners should strive without even identifying the reason. Actual critical thinking will not be promoted as a benchmark by the government, because it is the state and other macro-level constructed systems of hegemonic oppression that stand to suffer. It is time to not only acknowledge the micro- and macro-level systems of oppression that exist within our own educational communities, but to create a new environmental education that pointedly deconstructs them. By ignoring them we reinforce the hegemonic norm. But by queering our pedagogies we can combat complacency, otherness, and even boredom, and disrupt the ideas that are holding us back as we seek more just relationships with and within the world.

NOTES

1. *Cisgender* is an adjective referring to people whose gender identity corresponds to their assigned sex at birth (e.g. somebody who was assigned male at birth and identifies as a man) (SMYRC, 2013). See Enke (2012) for a cultural analysis of *cis* as an identifier.
2. Lesbian, gay, bisexual, transgender, queer, questioning, intersex, asexual, pansexual, etc.
3. While the term *normalized* conjures images of postconflict foreign relations, *normatized*, on the other hand, refers to making something *normative* rather than *normal* by, for example, taking something that was previously *othered* and creating a new norm based upon it.
4. Kahn (2010) talks at length about connections between environmental crises, capitalism, and neoliberalism. We do not have space to go deeply into it here, but it is well worth following up.
5. *Radical* is a botanical term for root. We use it here to evoke "radical" political potential by acknowledging political (grass) roots.
6. The lyric was from a Ben Folds Five song: "Y'all don't know what it's like/Bein' male, middle class and white."
7. Microaggressions are small acts of often unintentional insult or offense that communicate racism and other oppressive messages to marginalized people and that accumulate daily, causing real damage to individuals (Microaggressions Project, 2010).

REFERENCES

Adewunmi, B. (2014, April 2). Kimberlé Crenshaw on intersectionality. *New Statesman.* Retrieved from http://www.newstatesman.com/lifestyle/2014/04/kimberl-crenshaw-intersectionality-i-wanted-come-everyday-metaphor-anyone-could

Boles-Friscia, M. (2010). We don't talk about that here. *Writing to transgress* (Unpublished division III thesis). Hampshire College, Amherst, MA.

Bowers, C. A. (Ed.). (2004). *Re-thinking Freire: Globalization and the environmental crisis.* New York, NY: Routledge.

Bryson, M., & de Castell, S. (1993). Queer pedagogy: Praxis makes im/perfect. *Canadian Journal of Education, 18*(3), 285–305.

Bullard, R. D. (2000). *Dumping in Dixie: Race, class, and environmental quality* (3rd ed.). Tustin, CA: Westview Press.

Carson, R. (1962). *Silent spring.* Boston, MA: Houghton Mifflin.

Carter, R. L., & Simmons, B. (2010). The history and philosophy of environmental education. In A. M. Bodzin, B. S. Klein, & S. Weaver (Eds.), *The inclusion of environmental education in science teacher education* (pp. 3–16). New York, NY: Springer.

Chapman. T. K. (2007). Interrogating classroom relationships and events: Using portraiture and critical race theory in education research. *Educational Researcher, 36*(3), 156–162.
Crenshaw, K. (1989). Demarginalizing the intersection of race and sex: A Black feminist critique of antidiscrimination doctrine, feminist theory and antiracist politics. *The University of Chicago Legal Forum, 1989*(1), 139–167.
Daring, C. B., Rogue, J., Volcano, A., & Shannon, D. (2012). Queer meet anarchism, anarchism meet queer. In C. B. Daring, J. Rogue, A. Volcano, & D. Shannon (Eds.), *Queering anarchism* (pp. 5–17). Oakland, CA: AK Press.
Enke, A. F. (2012). The education of little cis: Cisgender and the discipline of opposing bodies. In A. Enke (Ed.), *Transfeminist perspectives: In and beyond transgender and gender studies* (pp. 60–77). Philadelphia, PA: Temple University Press.
Evans, M. M. (2002). Nature and environmental justice. In J. Adamson, M. M. Evans, & R. Stein (Eds.), *The environmental justice reader: Politics, poetics, & pedagogy* (pp. 181–193). Tucson, AZ: University of Arizona Press.
Fifield, S., & Letts, W. (2014). [Re]considering queer theories and science education. *Cultural Studies of Science Education, 9*(2), 393–407.
Freire, P. (1970/1993). *Pedagogy of the oppressed* (Rev. ed.). New York, NY: Continuum.
Gough, N., & Gough, A. (2003). Tales from Camp Wilde: Queer(y)ing environmental education research. *Canadian Journal of Environmental Education, 8*(1), 44–66.
Gruenewald, D. A. (2004). A Foucauldian analysis of environmental education: Toward the socioecological challenge of the earth charter. *Curriculum Inquiry, 34*(1), 71–107.
Halperin, D. M. (1995). *Saint Foucault: Towards a gay hagiography*. Oxford: Oxford University Press.
hooks, b. (1994). *Teaching to transgress*. New York, NY: Routledge.
Hungerford, H. R. (2010). Environmental education (EE) for the 21st century: Where have we been? Where are we now? Where are we headed? *The Journal of Environmental Education, 41*(1), 1–6.
Hursh, D., & Martina, C. A. (2003). Neoliberalism and schooling in the US: How state and federal government education policies perpetuate inequality. *Journal for Critical Education Policy Studies, 1*(2), 1–13.
Kahn, R. V. (2010). *Critical pedagogy, ecoliteracy, & planetary crisis: The ecopedagogy movement*. New York, NY: Peter Lang.
Keith, J. I. (2014). Homophobia's hidden carbon count. *Orion, 33*(3). Retrieved from https://orionmagazine.org/article/homophobias-hidden-carbon-count/
Kellner, D. (2010). Afterword: Mediating critical pedagogy and critical theory: Richard Kahn's ecopedagogy. In R. Kahn (Ed.), *Critical pedagogy, ecoliteracy, & planetary crisis: The ecopedagogy movement* (pp. 151–155). New York, NY: Peter Lang.
Kincheloe, J. L. (2008). *Critical pedagogy primer*. New York, NY: Peter Lang.
Kollmuss, A., & Agyeman, J. (2002). Mind the gap: Why do people act environmentally and what are the barriers to pro-environmental behavior? *Environmental Education Research, 8*(3), 239–260.
Krywanczyk, L. (2007). Queering public school pedagogy as a first-year teacher. *The Radical Teacher, 79*, 27–34.
Lee, Y.-J. (2012). Identity-based research in science education. In B. J. Fraser, K. Tobin, & C. J. McRobbie (Eds.), *Second international handbook of science education* (pp. 35–45). Houten: Springer.
Lemke, J. (2011). The secret identity of science education: Masculine and politically conservative? *Cultural Studies of Science Education, 6*(2), 287–292.
Leopold, A. (1949). *A Sand County almanac and sketches here and there*. Oxford: Oxford University Press.
Lorde, A. (1983). The master's tools will never dismantle the master's house. In C. Morage & G. Anzaldua (Eds.), *This bridge called my back: Writings by radical women of color* (pp. 98–101). New York, NY: Kitchen Table/Women of Color Press.
Luhmann, S. (1998). Queering/querying pedagogy? Or, pedagogy is a pretty queer thing. In W. F. Pinar (Ed.), *Queer theory in education* (pp. 141–155). Mahwah, NJ: Lawrence Erlbaum Associates.
Martin, P. (1996). A WWF view of education and the role of NGOs. In J. Huckle & S. Sterling, (Eds.), *Education for sustainability* (pp. 40–55). London: Earthscan.

McClintock, M. (1996). Lesbian baiting hurts all women. In K. Warren (Ed.), *Women's voices in experiential education* (pp. 241–250). Dubuque, IA: Kendall Hunt.
Microaggressions Project. (2010). *Microaggressions: Power, privilege, and everyday life* [website]. Retrieved from http://www.microaggressions.com/about
Mortimer-Sandilands, C., & Erickson, B. (Eds.). (2010). *Queer ecologies: Sex, nature, politics & desire.* Bloomington, IN: Indiana University Press.
Mueller, M. P., & Bentley, M. L. (2009). Environmental and science education in developing nations: A Ghanian approach to renewing and revitalizing the local community and ecosystem. *Journal of Environmental Education, 40*(4), 53–63.
Mueller, M. P., & Tippins, D. J. (2012). Citizen science, ecojustice, and science education: Rethinking an education from nowhere. In B. Fraser, K. Tobin, & C. McRobbie (Eds.), *Second international handbook of science education* (pp. 865–882). Dordrecht: Springer.
Palmer, J. (1998). *Environmental education in the 21st century: Theory, practice, progress and promise.* London: Routledge.
Pinar, W. F. (1998). Introduction. In W. F. Pinar (Ed.), *Queer theory in education* (pp. 1–47). Mahwah, NJ: Lawrence Erlbaum Associates.
Reid, A., & Payne, P. (2013). Handbooks of environmental education research: For further reading and writing. In R. B. Stevenson, M. Brody, J. Dillon, & A. E. J. Wals (Eds.), *International handbook on environmental education* (pp. 529–541). New York, NY: Routledge.
Ruffolo, D. V. (2007). Mapping a post-queer terrain: Radicalizing poststructural bodies (dissertation abstract). *Higher Education Perspectives, 3*(2), 9.
Russell, C. L., Sarick, T., & Kennelly, J. (2002). Queering environmental education. *Canadian Journal of Environmental Education, 7*, 54–66.
Russell, J. (2013). Whose better? (re)orientating a queer ecopedagogy. *Canadian Journal of Environmental Education, 18*, 11–26.
Small, E. (2011). The new Noah's Ark: Beautiful and useful species only. Biodiversity conservation issues and priorities. *Biodiversity, 12*(4), 232–247.
SMYRC (Sexual and Gender Minority Youth Resource Center). (2013). *Definitions and vocabulary* [pamphlet]. Portland, OR: Author.
Stevenson, R. B. (2007). Schooling and environmental education: Contradictions in purpose and practice. *Environmental Education Research, 13*(2), 139–153.
Taylor, D. (2002). *Race, class, gender, and American environmentalism* (U.S. Forest Service, Pacific Northwest Research Station, General Technical Report PNW-GTR-534). Portland, OR: U.S. Department of Agriculture. Retrieved from https://www.fs.fed.us/pnw/pubs/gtr534.pdf
Taylor, D. (2011). The evolution of environmental justice activism, research, and scholarship. *Environmental Practice, 13*(4), 280–301.
Udall, S. L. (1963). *The quiet crisis.* Layton, UT: Gibbs Smith.
UNESCO-UNEP. (1977). *Final report: The International Workshop on Environmental Education.* Paris: UNESCO. Retrieved from http://unesdoc.unesco.org/images/0002/000276/027608eb.pdf
Wals, A. E. J., Brody, M., Dillon, J., & Stevenson, R. B. (2014). Convergence between science and environmental education. *Science, 344*(6184), 583–584.
Warren, K. (1996). *Women's voices in experiential education.* Dubuque, IA: Kendall/Hunt Publishing Co.
Whitlock, R. U. (2010). Getting queer: Teacher education, gender studies, and the cross-disciplinary quest for queer pedagogies. *Issues in Teacher Education, 19*(2), 81–104.
Wilchins, R. A. (2004). *Queer theory, gender theory: An instant primer.* Los Angeles, CA: Alyson.

ABOUT THE AUTHORS

blake m. r. flessas is an outdoor environmental educator on Turtle Island (North America). They have lived and worked on both the west and east coasts. Their passions are fostering racial justice through education, manifesting queer space in society and nature, and empowering youth to build the world in which they want to exist. blake believes science and magic have a lot in common and outdoor education is a great place for them to meet. blake began writing about queering environmental education while at Hampshire College, and they strive to connect theory and practice while running around with kids in the woods.

Timothy D. Zimmerman is Jonathan Lash Chair and assistant professor of environmental education at Hampshire College in Amherst, Massachusetts (USA). With degrees in ecology, marine science, and education, Tim strives to improve environmental conservation and protection through education. As a learning scientist, his research and teaching focus on designing, implementing, and studying curriculum materials that foster environmentally sound decision-making. During his nearly 30 years of teaching, he's taught many topics including marine ecology, oceanography, educational theory, curriculum design, museum learning, research methodology, and environmental education. Presently, his teaching focuses on preparing the next generation of environmental educators.

ANDREW GILBERT AND EMILY M. GRAY

4. WONDER IN THE SCIENCE CLASSROOM

PROLOGUE

Emily is a sociologist. She hated science at school. She was one of those people to whom science was a boring world of right or wrong answers, of doing experiments correctly. She was growing up queer in the 90s, the age of Madonna's *Erotica*, a time when not fitting in was cool, was the thing to be, as hard as it was. Emily got low scores in science and spent her late teens working in gay clubs, returning to education at 23 to study sociology—finally finding a way to make sense of the world. Queer theory gave her a framework for understanding herself: where gender is performative (Butler, 1990), and sexuality is too. Emily now understands that the social world is not necessarily "real," and that we all have a hand in (re)producing it all the time through the ways we speak about and act in it.

Andrew is a science educator. He was raised in suburban Washington, DC, by a single mother, where money was always tight. The experiences in this working-class environment shaped his understanding of social justice, particularly in terms of gender and class. In ninth grade, he moved in with his father to attend high school in an upper middle-class community. He didn't fit easily into this new environment. He rebelled against the structure and banality of this suburban existence through punk and metal music and shunned anything remotely associated as "popular." However, despite his angst toward this John Hughes-esque high school experience, life in suburbia as a straight, newly middle-class, white male presented few challenges. He attended university right out of high school and finished four years later with a degree in geology and a license to teach earth/physical science.

QUEERING SCIENCE EDUCATION THROUGH WONDER

In an effort to imagine a science education where passion and interest are central to the pursuit of knowledge, we question the commonly normalized notions of science in favor of a vision of inclusive, joyful, and queer possibilities. For us, conceptualizing possibilities for queering science education through wonder is about contesting binaries, exploring margins, and countering essentialist thinking that characterizes intertwined patriarchal and neoliberal agendas that dominate how the sciences are approached in schools. Schools often are overly focused on cognition, but wonder intentionally engages both the cognitive and emotive as part of the learning process. Consequently, the focus of this chapter is squarely on K–12 settings, which we term

 | DOI:10.1163/9789004331068_006

school-based science education (SBSE). SBSE is generally expected to reify the rules, activities, and practices of science itself, where the enactment of science in schools is often governed by hypermasculine patriarchal structures that position it as a producer of absolute truths (Bazzul & Sykes, 2011; Snyder & Broadway, 2004; Tuana, 1989). Science, nature of science, and scientific discourse are often policed in particular and predictable ways.

Furthermore, SBSE is freighted with neoliberal agendas that frame success in science as mainly an engine for prosperity (Gilbert, 2011, 2013a; Rudolph, 2002). When we speak about neoliberalism here, we follow the work of Wendy Brown (2015) who describes neoliberalism as "a peculiar form of reason that configures all aspects of existence in economic terms" (p. 17). Because neoliberalism as a political ideal and technique of the self has taken hold in so many national contexts, it is perhaps not surprising that success within the discipline of school science is framed as a vehicle for prosperity. These neoliberal agendas directly impact SBSE contexts by stripping resources from schools, pitting schools to compete against one another, and heaping market-driven pressures on children as a means for profit by private industry (Kovacs, 2011). Gilbert (2013a) contended that in neoliberalism, science and STEM education are wielded as tools for the nation-state to outcompete rival nations. The technocratic mission to frame scientific understanding as a library of knowledge to master in order to harness its economic potential has consistently failed to pique the interest of children and future teachers in learning science in school (Gilbert, 2013b; Milne, 2010). Tytler (2007) termed this phenomenon a *crisis of interest* in science education. This so-called crisis results from experiences with school science that often misrepresent the nature and processes of science through pedagogy steeped in memorization and regurgitation of disconnected science facts (Bybee, 1994; Kenny et al., 2014). We question whether this is truly a crisis of interest or, rather, a result of years of uninspired pedagogy that not only misrepresents science but also actively constrains the thinking of children. Ultimately, by trying to domesticate children's science thinking, we fail to feed their need for feral wandering by reducing science to binary notions of intellectual gain or loss. This pedagogy slowly starves both interest in and meaningful engagement with science, which survive only if children cultivate them by other means.

Today's adults have often been taught by a system of neoliberal schooling that views wondering and freedom of thought as dangerous—or at least superfluous—distractions that must be tamped down in favor of a real-world approach that supposedly demands certainty, as well as simplistic binary notions of correctness, in order to maximize performance measured solely by standardized test scores. In this system, wondering is an inefficiency that impedes what schools have already determined children need to know. Ironically, it is the real world where uncertainty, doubt, and anxiety dominate our lives (Cvetkovich, 2012). Schooling therefore leaves many children and adults unprepared to skillfully navigate human realities of life. But what if, as Cvetkovich (2012) so convincingly argues, uncertainty, doubt, and anxiety are reconfigured to be creative spaces, as spaces of wonder in themselves?

Queer theory's take on the affective turn seeks to depathologize shame (Probyn, 2005), failure (Halberstam, 2011), and uncertainty (Cvetkovich, 2012). Within such a paradigm, the academic success of students in SBSE need not be our ultimate goal (albeit a desirable one); rather, we embrace failure and ambivalence as part of the welcoming space that is opened up through reconfiguring the science classroom.

Here we follow the work of Halberstam (2011), who views failure as wondrous, joyous, and pedagogical:

> We will wander, improvise, fall short, and move in circles. We will lose our way, our cars, our agenda, and possibly our minds, but in losing we will find another way of making meaning in which … no one gets left behind. (p. 25)

Halberstam's statement acts as a guide for us through this chapter—a way of thinking about wonder, queerness, and SBSE as well as how we want to work with these ideas. We want to engage with the risky business of learning and with the possibility of failure in our endeavor to reconfigure science education.

For many teachers, opening up unwieldy spaces for children to ask questions and ponder the unknowable can be a frightening aspect of the job—particularly with younger children (Gilbert, 2009) within the confines of a neoliberal milieu that stifles experimentation in favor of standardization. A queering of SBSE allows us to embrace this pedagogic vulnerability as opposed to being shamed by it. A queer lens suggests we need to own this space to try on our vulnerabilities and to see where they lead.

In this chapter we argue that to open SBSE to its fullest human potential, educators must embrace the queer fullness of science content and processes to consider desire, love, awe, fear, and an entire range of human emotion, including our relationships and interactions within the Universe. We argue for the conceptualization of a generative theory of *wonder* to realize some of the possibilities of queering SBSE.

THEORETICAL FOUNDATIONS

In some ways it is easier to begin a discussion of wonder by delineating what wonder is not. Hadzigeorgiou (2012) compellingly argued for the distinct differences between wonder and curiosity, and he worked to untangle these concepts:

> Yet wonder is not the same as curiosity. What deserves to be stressed here is a delicate distinction between curiosity and wonder: Curiosity is actually the drive to investigate or study something while wonder is a state of mind or feeling. Moreover, wonder has an aesthetic dimension, which can be totally absent from curiosity. (p. 987)

These emotive notions are clearly expressed when scientists describe the role of wonder in their work and the contemplative states that wandering, questioning, and appreciating bring. It is this connection to visceral contemplation that distances wonder from curiosity. The beauty and power of wonder lie in this direct connection to aesthetics and emotion (Stolberg, 2008; Zembylas, 2007), while curiosity is framed within the cognitive realm. Hadzigeorgiou (2012) further argued that building pedagogical approaches steeped in wonder will facilitate student awareness of

- understanding as incomplete, where there is always more to learn;
- the multiple connections across and between scientific concepts to be considered; and
- the beauty of the natural world and scientific phenomena.

It is this awareness of beauty and admiration, which are steeped in the emotive realm, that ultimately provides the separation between wonder and curiosity.

Including aesthetics in the pursuit of scientific understanding fosters higher levels of engagement with science (Hadzigeorgiou, 2005; Wickman, 2006). The connection between the emotive and scientific understanding nurtures students' intense desire to understand the world (Gilbert, 2011; Leafgren, 2009). The emotive exists, despite usually being repressed and ignored in SBSE, and is ever-present in scientific understanding. These emotional cues serve as important common ground for students and teachers to connect science content to everyday experiences (Stolberg, 2008). Consequently, this repressed emotional engagement with science must be acknowledged, excavated, lifted up, and embraced in classrooms to drive science pedagogy (Hadzigeorgiou, 2012; Zembylas, 2007).

A small number of researchers in science education have investigated wonder as a solution to students' persistent disinterest in the sciences (Hadzigeorgiou, 2001, 2005). Working with school children in New Zealand, Milne (2010) argued,

> It is about children's science; children personalising their science activity, leading to their development of creative explanations of natural phenomena. It requires the children to be involved in exploration, inquiry, explanation and making connections and is often, can be, should be, based around or ignited by aesthetic experiences that promote affective and often emotional responses associated with the dispositions like fascination, anticipation and engagement and awe, wonder & interest that spark curiosity and can lead to the use of scientific inquiry to develop explanations of natural phenomena. (p. 110)

Milne moves beyond the standard approach to inquiry by directly centering the notions of "awe, wonder & interest" (p. 110). These ideas represent an important contribution concerning the use of wonder and how we might operationalize teaching science with primary children. Gilbert (2013b) also articulated the importance of wonder-based approaches to provide benefits for adult students, particularly those who carry negative associations with science.

These voices stand in defiance to science as a tool for the nation-state and are clearly focused on individuals and communities of practice. Current discussions regarding STEM and science education more broadly are typically operating on *presumed interest* of the student, meaning if the content is well-aligned, accurate, and focused on real-word issues, the students' interest will follow. We and others

(Girod & Wong, 2002; Hadzigeorgiou, 2005, 2012; Stolberg, 2008; Wickman, 2006) argue that efforts to engage with science education that are bereft of connections to emotions and the essential role of the aesthetic in science contexts will continue to struggle to engage and inspire students.

We wish to present a queer version of science education here, one that firmly positions science as wondrous. Such a positioning of wonder at the heart of science acknowledges and celebrates the uncertainty, beauty, and joy that encompass scientific activity. Lastly, the spirit of wonder is not about some touchy-feely, Pollyanna notion of the precocious child. Rather we conceptualize wonder as a theoretical and embodied positioning that pushes boundaries in an effort to imagine the possibilities. Thus, we envision it as a subversion as called for by Butler (1993)—a particular effort to subversively disrupt discursive narratives/practices/identities associated with SBSE and to challenge science to enact the ideals it claims to promote regarding wonder.

ENACTING PURPORTED IDEALS OF SCIENCE

Wonder itself is often described by scientists as a desire that burns within and fuels them to pursue answers to the unknown. These ideas have been articulated by several scientists over the years, such as Albert Einstein (1931), Rachel Carson (1965), and E. O. Wilson (2013). These pronouncements often delineate the special role that wonder has played in driving science forward, while simultaneously celebrating that our understandings of natural phenomena are necessarily incomplete. Wonder allows the scientist to bask in the beauty of the unknown, of future possibilities, and of untamed thinking to generate unlikely hypotheses that lead to new discoveries. This position is exemplified in the following excerpt from physicist and BBC television personality Brian Cox (2011):

> The collapsing remnants of the most luminous stars represent the edge of our understanding of the laws of physics and therefore the edge of our understanding of the wonders of the Universe. This is exactly where every scientist wants to be. Science is a word that has many meanings; some might say science is the sum total of our knowledge of the Universe, the great library of the known, but the practice of science happens at the border between the known and unknown. (p. 5)

Cox sets up a vision for ambiguity as a wondrous place. He also clearly articulates the important difference between the products and processes of science.

These messages have been repeated over and over by scientists both famous and obscure. In her final book, *The Sense of Wonder*, Rachel Carson (1965) beautifully illustrates similar notions regarding the beauty and importance of engaging children with the wondrous spaces between the known and unknown. Prized naturalist and award-winning science writer E. O. Wilson once summed up this commitment to looking at science with a sense of wonder in his Pulitzer Prize-winning book *Letters to a Young Scientist*. Wilson (2013) quipped, "The ideal scientist thinks like a poet and only later works like a bookkeeper" (p. 74). It may be naive to consider this position as commonly accepted across the sciences, and it may be that these highly successful scientists draw upon notions of wonder because their success is driven by the connection between the aesthetic and scientific phenomena. Perhaps it is this that drives their success in understanding the world around them.

Despite these proclamations regarding the potential of wonder, this method of connecting the range of human emotion with scientific observations of natural phenomena remain undertheorized and rarely enacted in SBSE. In turn, SBSE typically avoids notions of complexity and distances children from these so-called fanciful notions of wonder and imagination (Leafgren, 2009). In essence, queering SBSE (as well as science) through wonder opens it to the ideals that science espouses.

WONDERING AND WANDERING

The goal of this section is to wander through the wonderings of Emily and Andy. We also wish to engage in our own untamed thinking by moving away from the artificial singular voice that can mask individual desire, experience, emotion, and the self in academic writing.

Andy

Over my 20 years working as a scientist, teacher, and general science devotee, I've found that scientists typically approach new dilemmas and data with unabashed enthusiasm, and this same energy is also exhibited by amateur science lovers. This excitement regarding science phenomena is often used by others to derisively poke fun at the "nerdy" scientist. Children who overtly proclaim strong science identities are not immune from similar derision from classmates and the occasional teacher. This becomes a sort of science shaming that acts to homogenize thought and bound thinking, domesticating the thoughts of the child scientist. Similarly, children whose thinking challenges others (particularly adults/teachers) represent wildness, the frightening feral-thinking child. Public intellectuals in science (e.g., Brian Cox, Rachel Carson) call for exactly this type of untamed thinking, which they argue is an important tool for inspiring learning and triggering intrinsic desires toward science learning. Teachers and schools view the untamed child as an object to be controlled at all costs, and they are often subjected to draconian measures, such as losing recess, isolation from peers, being medicated, and other means to tame their wildness. The daydreaming, wondering, and imaginative child is also not safe from these methods of domestication. These contexts frame the untamed child as an object of ridicule or one that needs fixing that, in turn, isolates the child and constrains the emotive joy of new ideas and possibilities.

My current research into wonder and science teaching stems from my own desire to *untame* my thinking, to take some risks and postulate openly about the challenges science educators at all levels face. Wondering as generative theory represents rethinking and challenging deeply held assumptions about science and science learning to envision new possibilities. Wonder demands that we ask tough questions of what we know and desire within SBSE. For instance, wondering provides a framework to imagine science as committed to justice through the joyful pursuit of understanding the world and our responsibilities in it as human beings. Instead, for three centuries science as a gender(ing) practice has been used to support male-dominated moral and political choices that, among other things, have further subjugated women (Harding, 1986). Feminist scholars have challenged patriarchy in the sciences and the notions that science is about serving human interests by dominating nature. This highlights how science itself is used by and for male interests (rather, straight male interests) and has created constructs that depict women as weak, emotional, and lacking rational thought. These depictions are also supposedly "value-neutral," and thus represent "truth" (Keller, 1996), and consequently create gendered scientific identities that distort what it means to be successful in science. In a longitudinal study of three women of color who successfully navigated the opportunity gap to obtain access to scientific careers, Johnson, Brown, Carlone, and Cuevas (2011) wrote:

> Identity must be understood not to imply that all individuals have full choice in the identities they can author, but as something that operates within the matrix of oppression. Not all identities are available to all individuals. When the women were in their most constrained positions in the matrix of power, there were some identities that were simply closed to them, simply not an option. (p. 359)

This study deeply problematized the notion that entry into science was a simple task of creating more opportunities for women. Rather, becoming oneself in relation to science involves systemic issues of identity and agency. We now know that crafting selves at the intersection of "racial," gender, and scientific identities is no simple matter. This is an example of how I share with future teachers my wonderings about what it might entail to create science free from oppression, so I can then invite them into their own wonderings. Together we create avenues of engagement that help us better understand the context, history, and possible future of our science practices.

Emily

Queer theories enable us to trouble normalizing discourses, including science, and to challenge the assumptions of rational, value-free (ergo masculine and heteronormative) domains. A queer reading of science may reorient it to become a semiformal, accessible, welcoming, and comfortable space (Youdell, 2011). As we illustrated earlier, public intellectuals like Brian Cox, Rachel Carson, and E. O. Wilson have depicted ways of being within science that frame passions, careers, and lifelong interests in the natural world. Halberstam (2011) reminds us that science is full of stories of genius and kooky losers. In popular culture, the science nerd has some kudos—think here of Sheldon in *The Big Bang Theory*, Willow in *Buffy the Vampire Slayer*, and Benedict Cumberbatch's recent portrayal of the Enigma-code-breaker Alan Turing in *The Imitation Game*. All of these characters represent the unorthodox, the sometimes brilliant, and the failures and meanderings that may lead to eventual success within scientific research.

For Halberstam (2011), the goal of thought and experimentation *should* be to get lost:

> Terms like *serious* and *rigorous* tend to be code words, in academia as well as other contexts, for disciplinary correctness. They signal a form of training and learning that confirms what is already known according to approved methods of knowing, but they do not allow for visionary insights or flights of fancy. (p. 6, original italics)

Enabling students to wander through science—to traipse, to meander—allows us to envision more powerful, just, and joyful alternative learning spaces within the sciences (Halberstam, 2011), and to push for such inclusive learning spaces by wrestling with uncertainty in order to create welcoming spaces for all young people (Gilbert, 2014). By deploying the notion of wonder we argue that inviting students to feel, desire, explore, and discover offers them opportunities to engage in more expansive ways with science. Science, ultimately, is a longing, a desire to know the unknown—the many kooky characters that populate scientific research are testament to such desires, and maybe it is wonder itself that allows for such kookiness. By queering science, we might go some way to answering Britzman's (1995) provocative question: What "if gay and lesbian theories were understood as offering a way to rethink the very grounds of knowledge and pedagogy in education?" (p. 151). If we use queer theory as a tool with which to challenge both authority and the ever-present force of normalization, we can question the existence of certainty and reorient the science classroom as a space that facilitates such questioning.

Emily and Andy

Queer theory encourages us to excavate the facades surrounding the so-called truths of our everyday existence. Queer theory reveals the arbitrariness of social structures and makes us see the performative nature of our behaviors (Butler, 1990) relating to identity that are (re)produced to enact those arbitrary constructs. Queer theories, then, are liberatory in that they free our thinking and open everything to question.

In the spirit of Broadway (2011), who called for broader ways of conceptualizing queer curricular and pedagogical ideas to transform science education, our vision is that wonder can be a vehicle to deliver on calls for bringing queer theory and pedagogy to bear in SBSE, and science education more broadly. The goal is to imagine, dare we say wonder, what science education could look like if it was designed more in step with the ways both children and adults think—a science less fettered by rules and normalizing constrictions toward one of desire, interest, and possibility. In essence, using wonder as a lens to queer blurs the accepted boundaries of SBSE. To queer the science classroom—to encourage students to get lost, to wander without aim, and to wonder within science education—gives us powerful tools with which to rethink the learning and teaching of science in schools. These goals are not an effort to help students feel comfortable; rather, these activities are meant to intentionally trigger discomfort, vulnerability, and joy as means to enact more realistic engagements with science. Scientists purport an ideal of science as a welcoming space where many things are unknown and where they take joy in the discomforts of the unknown. We think SBSE can be such a space too.

QUEERING WONDER WITH QUEER THEORY

We also acknowledge, however, that wonder is itself a problematic framework for thinking about education, learning, and teaching. Firstly, the connection of wonder to religious faith and an unquestioned acceptance of scientific phenomena being explained solely as the will of God can be dangerous territory for a teacher of science. This danger results from the utilization of wonder alone as a tool for understanding within the science classroom. This strips away a desire to know in which explanatory evidence is subsumed by belief (Silverman, 1989; Weinstein, 1998). Although we may agree with the premise of these critiques, we also recognize that they are set squarely in the dominant masculinist culture of science that regulates how we are supposed to relate to scientific ideas. Einstein (1931) problematized simplistic notions of God and belief, and he pointed toward an idea of religiousness where scientific discovery was ultimately a similar emotional experience:

> The most beautiful thing we can experience is the mysterious. It is the source of all true art and science. He to whom this emotion is a stranger, who can no longer pause to wonder and stand rapt in awe, is as good as dead: his eyes are closed. This insight into the mystery of life, coupled though it be with fear, has also given rise to religion. To know that what is impenetrable to us really exists, manifesting itself as the highest wisdom and the most radiant beauty which our dull faculties can comprehend only in their most primitive forms—this knowledge, this feeling, is at the center of true religiousness. In this sense, and in this sense only, I belong in the ranks of devoutly religious men. (p. 6)

Einstein deftly describes that beautiful condition of being human, where seeking knowledge and knowing is enlightenment, albeit one wrapped in human frailty and our limited capacity to understand the complexities of the surrounding universe and the nature of our own existence. Wondering queerly helps us to keep our minds open to the possibilities, much like the mind-bending concepts involved in quantum mechanics, the immensity of the universe, and geologic time. In this manner, queer perspectives push the boundaries surrounding science and religion, and tease out how each can be prone to authoritarian narratives that do not provide space for multiple ways of knowing. The key for us is that when science cannot provide ready-made answers for the mysteries in our world, uncritical surrender to the supernatural need not be the next step after wonder, as some of the critiques of wonder may suggest.

We also wish to explore how children and young people might *feel* about science, and to forge new ways of thinking about how these feelings might be harnessed by science educators. We understand feelings here to be both embodied sensations and cognitive experiences (Cvetkovich, 2012). Here we draw upon writers from queer theory's affective turn who argue that through the exploration of feelings we, as researchers and educators, can cultivate a more embodied understanding of the messiness of inquiry and its open-endedness—knowing it in your bones and guts as well as your brain (Ahmed, 2004; Cvetkovich, 2012). We argue that engaging students in science classrooms with their feelings, and specifically the sensation of wonder, may allow us to break away from binaries— namely, simplistic right-and-wrong depictions of previously discovered science "facts"—toward an emotionally embodied doing of science.

We do, however, acknowledge Ahmed's (2004) cautionary note about the ways in which we experience the classroom as educators vis-à-vis the ways in which students experience our classrooms. Ahmed argues that

> to receive an impression is to make an impression. So we may walk into the room and "feel the atmosphere," but what we may feel depends upon the angle of our arrival, … [and] the pedagogic encounter is full of angles. How many times have I read students as interested or bored, such that the atmosphere seemed one of interest or boredom, … only to find that students recall the event quite differently! (p. 41)

We may *feel* a sense of wonder, excitement, or boredom emanating from our students, but they may be feeling something entirely different within the same moment. And so it is from within this space that queering the notion of wonder allows us to lose our way; to get lost; to try, fail, and succeed and not necessarily find the answers. We all represent the fuzzy edges at the frontiers of science, although those edges differ for each of us based on our own degree of science content understanding. The coming to know new ideas feeds, rather than satiates, our need to know, and consequently triggers new wonders.

Wonder is also often considered the sole domain of children. This is typically seen in people's commonsense vision of children, as well as in science education literature, such as Milne's (2010) notion of children's science that clearly delineates wonder as the property of the young. Queering wonder challenges this position and argues that wonder itself is a human trait and not one that fades as a natural part of the aging process (Gilbert, 2013b). Further, queering accepts wondering with all its associated classroom "risks." Queering imagines a process of wonder that is less romantically childlike as it opens people's curiosities to more feral spaces than are usually accommodated in science classrooms. In Jen Gilbert's (2014) book,

Sexuality in School: The Limits of Education, the author argues that a sexual charge runs through school:

> Sexuality animates education; teaching and learning are invested with an erotic frisson that propels and sabotages the practices of education. … [L]ove, curiosity and aggression fuel our engagements with knowledge. And yet education—its practices, procedures, rules, structures, and relations—can be undone by the wildness of sexuality. (p. x)

Gilbert addresses the fear with which children are often viewed because they represent wildness and uncontrolled sexuality and the uncertainty of the adults they will become. In this respect, queer theories inform wonder by inviting teachers and other anxious parties to embrace, find joy in, and be inspired by the wondrous wildness of children and their unfettered imaginative possibilities.

CONCLUDING THOUGHTS

In our view, the potential of children and young people within the sciences is nearly boundless when we consider alternative ways of conceptualizing the collective possibilities of a scientifically inspired populous. We wish to move the discussion beyond mere literacy in science toward a notion of nurturing a fascination for the world around us. The goal becomes building positive experiences of and attitudes towards science by constructing a hearty sense of wonder. To this end, this chapter has wandered, traipsed, and illustrated how approaching science with wonder represents an intentional effort to bring emotion and joy into science classrooms to imagine new possibilities. Ultimately, we consider this work to be a joyous journey. Consequently, we describe these processes with words like emotion, aesthetic, need, desire, beauty, connection, and joy. In the spirit of reparative reading (Sedgwick, 2003), we think not only of queering science through wonder but also about how wonder offers important possibilities and approaches to queer theory. For the "reparatively positioned reader, it can seem realistic and necessary to experience surprise. Because there can be terrible surprises, however, there can also be good ones" (p. 22). Queer theory, then, encourages us to approach surprise with wonder because of the possibilities for understanding the complexities of the world with joy and an opening up of our human capacities for thinking about our everyday experiences.

The experience of writing this chapter has been illustrative of such a process. Sara Ahmed's (2004) work on unhappy queers shaped Andrew's reparative reading and openness with a surprise. Her work clarified that his interest in wonder was born of his own unhappiness with the state of science education, a place where he saw himself outside of the view of many in the field. He found this to be simultaneously troubling and joyous.

> Objects that promise happiness are passed around, accumulating positive affective value as social goods. When we pass happy objects around, it is not

> necessarily the feeling that passes. To share in such objects, or have a share in such objects, means simply that you *share an orientation toward those objects as being good.* (p. 4, original italics)

More problematically, Andrew's discomfort with wanting to infuse joy into the work of queer theory arose from his own privilege of not having to fear that his sexual identity, race, or even social status made him a subject for ridicule or violence. He knew that many who do not enjoy that privilege often remain "unhappy with this world" (Ahmed, 2004, p. 9), but perhaps willing to challenge it. For Emily, queer theory is already joyous *because* it allows her to be unhappy, to remain confused, to sit in the mess of the world and to find comfort there.

In many ways, the thinking of scientists and queer theorists is very much aligned through the notion of uncertainty. Lemke (2011) argued that depictions of science and science content in SBSE are often oversimplified and that choices to simplify represent value judgements. Those judgments intentionally strip complexity from the sciences in an effort to diminish uncertainty. Conversely, those scientists wishing to adhere to a more honest scientific complexity must be willing to embrace uncertainty, yet, paradoxically, their own subjectivities do not always allow this to happen. In our wanderings, while considering both what makes a theory queer and what it means to queer science, we are struck by the similarities between the goals of utilizing wonder in science and what Fifield and Letts (2014) hoped to achieve through conceptualizing science education with a queer lens:

> By delving deeply into queer theory and other critical interpretive frameworks we can create conditions for more intentional experiences of and skillful inquiries into being(s) in the science curriculum. This puts love of learning, and learning to love, in a new light if learning is putting the self at risk by mindfully excavating and engaging the contingent relations that lie behind the illusion of an essential, autonomous self, and love is opening in awareness to the complicities of being(s). (p. 399)

Wonder is one of many possible critical interpretive frameworks that may help to facilitate a science education that is aligned more closely with what science claims to be: an uncertain space that is both unknown and pleasurable. We envision such a framework as enabling the science education classroom to become more expansive; allowing, encouraging, and experiencing wonder with students of science may provide a platform for a way of thinking that is more aligned with the ideals that science claims at its core. Successful scientists describe vulnerability as part and parcel of the desire to learn more. This is evidenced by Cox (2011) when he described the true wonder of science as the space between the known and the unknown. In SBSE (and science teacher education, for that matter), we want to help students see that vulnerability is something to embrace.

Queering SBSE through wonder helps us better understand vulnerability as a positive and necessary source and outcome of untamed thinking. Such a stance accepts uncertainty and, in fact, finds that space pleasurable. A theory of wonder embraces the wisdom of queerness: that everything is arbitrary and open to interrogation. Queerness helps us face and begin to transfigure the terrifying and uncertain. Vulnerability becomes a way of thinking, living, and being as opposed to something we must fear, hide, or flee. Queer theory tells us that the things we fear are often the result of arbitrary social conditions and that this arbitrariness runs parallel within SBSE. A theoretical positioning of wonder provides a philosophical lens in the face of the enduring mysteries and grandeur of nature; science constructs understandings that ultimately enrich and further ennoble nature's mysteries rather than resolving, dissolving, or demystifying them.

We find an interesting tension when aligning queerness and wonder with the way science icons and public intellectuals champion the ideals of science. We assume these scientists would not consider their positions on wonder queer visions of science, but, whether intended or not, those scientists (e.g., Einstein, Wilson, Cox, and others) clearly elevate the role of unfettered thinking in science born from wonder. For us, queering reconnects wonder and wholeness to bring science to its fullest human potential as a practice filled with both cognition and emotion. Therefore, we see holding science to its ideals as a queer project that embraces subjectivities and multiple ways of knowing and thinking. The ideals of science fly in the face of many common scientific practices; wonder as generative theory pushes back and demands that we keep these ideals in the foreground.

Queer frameworks provide guidance for unearthing and disrupting the oft unquestioned discourses surrounding how science operates by demonstrating the distance between practices and ideals. Queer theory provides wonder with a fearless and unapologetic approach to incorporating emotion back into the everyday work of SBSE. Making a space for queer wonderings in the science classroom allows us to improvise, fall short, and move in circles (Halberstam, 2011). It allows us to get lost, to try, to fail, to succeed, and to try again. It allows us to celebrate these failures and successes, to sit in the mess, and to be happy about it.

REFERENCES

Ahmed, S. (2004). *The cultural politics of emotion.* New York, NY: Routledge.

Bazzul, J., & Sykes, H. (2011). The secret identity of a biology textbook: Straight and naturally sexed. *Cultural Studies of Science Education, 6*(2), 265–286.

Britzman, D. (1995). Is there a queer pedagogy? Or, stop reading straight. *Educational Theory, 45*(2), 151–165.

Broadway, F. (2011). Queer (v.) queer (v.): Biology as curriculum, pedagogy, and being albeit queer (v.). *Cultural Studies of Science Education, 6*(2), 293–304.

Brown, W. (2015). *Undoing the demos: Neoliberalism's stealth revolution.* New York, NY: Vanderbilt Books.

Butler, J. (1990). *Gender trouble: Feminism and the subversion of identity.* New York, NY: Routledge.

Butler, J. (1993). *Bodies that matter: On the discursive limits of sex.* New York, NY: Routledge.

Bybee, R. (1994). *Reforming science education: Social perspectives and personal reflections.* New York, NY: Teachers College Press.

Carson, R. (1965). *A sense of wonder*. New York, NY: HarperCollins.

Cox, B. (2011). *Wonders of the universe*. London: Harper Collins.

Cvetkovich, A. (2012). *Depression: A public feeling*. Durham, NC: Duke University Press.

Einstein, A. (1931). *Living philosophies: A series of intimate credos* (pp. 3–7). New York, NY: Simon and Schuster.

Fifield, S., & Letts, W. (2014). [Re]considering queer theories and science education. *Cultural Studies of Science Education, 9*(2), 393–407.

Gilbert, A. (2009). Utilizing science philosophy statements to facilitate K–3 teacher candidate's development of inquiry-based science practice. *Early Childhood Education Journal, 36*(5), 431–438.

Gilbert, A. (2011). Visions of hope and despair: Investigating the potential of critical science education. In C. Malott & B. Porfilio (Eds.), *Critical pedagogy in the 21st century: A new generation of scholars* (pp. 401–418). Charlotte, NC: Information Age Publishing.

Gilbert, A. (2013a). *Courageous pedagogy: Enacting critical science education*. Charlotte, NC: Information Age Publishing.

Gilbert, A. (2013b). Using the notion of 'wonder' to develop positive conceptions of science with future primary teachers. *Science Education International, 24*, 6–32.

Gilbert, J. (2014). *Sexuality in school: The limits of education*. Minneapolis, MN: University of Minnesota Press.

Girod, M., & Wong, D. (2002). An aesthetic (Deweyan) perspective on science learning: Case studies of three fourth graders. *The Elementary School Journal, 102*(3), 199–226.

Hadzigeorgiou, Y. (2001). The role of wonder and 'romance' in early childhood science education. *International Journal of Early Years Education, 9*(1), 63–69.

Hadzigeorgiou, Y. (2005). Science, personal relevance and social responsibility. Integrating the liberal and the humanistic traditions of science education. *Educational Practice and Theory, 27*, 82–93.

Hadzigeorgiou, Y. (2012). Fostering a sense of wonder in the science classroom. *Research in Science Education, 42*, 985–1005. doi:10.1007/s11165-011-9225-6

Halberstam, J. (2011). *The queer art of failure*. Durham, NC: Duke University Press.

Harding, S. (1986). *The science question in feminism*. Ithaca, NY: Cornell University Press.

Johnson, A., Brown, J., Carlone, H., & Cuevas, A. (2011). Authoring identity amidst the treacherous terrain of science: A multiracial feminist examination of the journeys of three women of color in science. *Journal of Research in Science Teaching, 48*(4), 339–366.

Keller, E. (1996). *Reflections on gender and science*. New Haven, CT: Yale University Press.

Kenny, J. D., Hobbs, L., Herbert, S., Jones, M., Chittleborough, G., Campbell, C., Gilbert, A., & Redman, C. (2014). Science Teacher Education Partnerships with Schools (STEPS): Partnerships in science teacher education. *Australian Journal of Teacher Education, 39*(12), 43–65.

Kovacs, P. E. (2011). *The Gates Foundation and the future of U.S. "public" schools*. New York, NY: Routledge.

Leafgren, S. (2009). *Reuben's fall: A rhizomatic analysis of disobedience in kindergarten*. Walnut Creek, CA: Left Coast Press.

Lemke, J. (2011). The secret identity of science education: Masculine and politically conservative? *Cultural Studies of Science Education, 6*(2), 287–292.

Milne, I. (2010). A sense of wonder, arising from aesthetic experiences, should be the starting point for inquiry in primary science. *Science Education International, 21*, 102–115.

Probyn, E. (2005). *Blush: Faces of shame*. Minneapolis, MN: University of Minnesota Press.

Rudolph, J. (2002). *Scientists in the classroom: The cold war reconstruction of American science education*. New York, NY: Palgrave.

Sedgwick, E. (2003). *Touching feeling: Affect, pedagogy, performativity*. Durham, NC: Duke University Press.

Silverman, M. (1989). Two sides of wonder: Philosophical keys to the motivation of science learning. *Synthese, 80*, 43–61.

Snyder, V., & Broadway F. (2004). Queering high school biology textbooks. *Journal of Research in Science Teaching, 41*, 617–636.

Stolberg, T. (2008). W(h)ither the sense of wonder of pre-service primary teachers' when teaching science? A preliminary study of their personal experiences. *Teaching and Teacher Education, 24*, 1958–1964.
Tuana, N. (1989). *Feminism and science.* Indianapolis, IN: Indiana University Press.
Tytler, R. (2007). *Re-imagining science education engaging students in science for Australia's future.* Camberwell, VIC: Australian Council for Educational Research.
Weinstein, M. (1998). *Robot world.* New York, NY: Peter Lang.
Wickman, P. (2006). *Aesthetic experience in science education: Learning and meaning making as situated talk and action.* Mahwah, NJ: Lawrence Erlbaum Associates.
Wilson, E. O. (2013). *Letters to a young scientist.* New York, NY: Liveright.
Youdell, D. (2011). *School trouble: Identity, power and politics in education.* London: Routledge.
Zembylas, M. (2007). Emotional ecology: The intersection of emotional knowledge and pedagogical content knowledge in teaching. *Teaching and Teacher Education, 23*, 355–367.

ABOUT THE AUTHORS

Andrew Gilbert is an associate professor of science education in the College of Education and Human Development at George Mason University in Fairfax, Virginia (USA). He is currently associate director for the Center for Social Equity through Science Education. His main research interests center on inquiry-based instruction, social justice, and the exploration of wonder and emotion in science learning. He has published his work in a range of outlets and is also the author of the book *Courageous Pedagogy: Enacting Critical Science Education.* He can be reached at agilbe14@gmu.edu

Emily M. Gray's interests within both research and teaching are interdisciplinary and include sociology, cultural studies, and education. Her key research interests lie with questions related to gender, social justice, and student and teacher identity work within educational policy, and with wider social justice issues within educational discourse and practice. She is also concerned with popular culture, public pedagogies, and audience studies, particularly with online "fandom" and with media and popular culture as pedagogical tools. Feminist and queer theories of identity shape Emily's research and how she uses theory to examine, understand, and analyze the complexities of teacher identities within a range of educational settings.

HELENE GÖTSCHEL

5. TEACHING QUEERING PHYSICS

An Agenda for Research and Practice

Being a professor of gender in engineering and information technology at a university of applied sciences, I am always pondering the meanings of gender and diversity in science, technology, engineering, and mathematics (STEM). For instance, I challenge the predominant binary social attributions of technological competence and gender, and I integrate the findings into my teaching of basic physics. In addition, I study processes of naturalization and normalization in physics education. To me it is obvious to scrutinize engineering from the perspectives of gender studies and queer theory.[1] For example, even today, in mechanical and electrical engineering, women and young people with nonbinary gender identities are still not accepted as "proper" members of the scientific culture. Homosexuality, too, seems to threaten the masculinity presumably inherent in legitimate technical actors. Normative ideas about users are inscribed in the artefacts of engineering science. Study materials use masculine language and orient their examples almost exclusively towards "normal" white male biographies, and thus address only one segment of the engineering student population.

At the same time, at our university and, as far as I know, at other universities in Germany, hardly any of the current discussions and considerations about the relevance of gender exceed equity initiatives for women in STEM, which mostly reinscribe conventional norms of knowing and being. However, in my research project, *teaching queering physics*, I am studying methods and techniques for teaching both physical knowledge and competencies of questioning representations and norms of physical talent, technological competence, heroes of the history of physics, presumptions of heterosexual normativity, hegemonic masculinity, and much more. This project brings to the fore norms and normative processes, as well as the resulting multiple exclusions of some people from STEM subjects, where these insights can be integrated into recommendations on teaching. In this chapter I wish to outline an agenda for research and practice of teaching queering physics, and to inspire other scholars and teachers to fruitfully bridge the gap between queer and deconstructive theories on one side and traditional positivist sciences on the other side.

 | DOI:10.1163/9789004331068_007

TEACHING QUEERING PHYSICS WHILE TEACHING PHYSICS

As a former professor for gender and interdisciplinary studies in pedagogy, I used to teach theories and research results from gender studies and queer theory in courses for students of science and science education at Darmstadt Technical University. Now, as a physics professor at Hannover University of Applied Sciences and Arts, I teach students of electrical, mechanical, and industrial engineering. My students are mostly male with diverse ages, education and/or work experiences, first languages, nationalities, and ethnic and social backgrounds. They might differ in their sexual orientation and gender identity, too, but they keep these secret, because LGBT+[2] issues remain a great taboo in engineering in Germany.

Teaching queering physics in (under)graduate science and science education is not limited to adding knowledge about queer theory to the curriculum of future engineers and science teachers. Teaching queering physics is an act of explicitly or implicitly translating complex theories and deconstructive methods to students of science education and science. This will enable students to reflect on their understandings of physics and physical knowledge and on their participation in the (re)production of "(hetero)normative facts," and by doing so to enact and engage in critical science literacy. The first and foremost aim of my physics course is to offer an inviting, open, and encouraging lecture format. Therefore, I ask myself the challenging question: How can I teach my first-semester engineering students queering physics while I teach them experimental physics in a four-hour lecture? How can they be enabled to critically question and transcend norms around gender, sexuality, desire, narratives, cultures, facts, and other powerful social categories that shape hegemonic presentations of physics and physicists? I treat these critical abilities as *competencies of queering* that are needed by all science and engineering students, and certainly those in my own discipline of physics.

In this chapter I offer some examples from my current teaching practice and exploratory research in a field I term "applied gender research in physics" or, more specifically, "teaching queering physics." The queering strategies I have been using so far could be described as scrutinizing what is supposed to be normal and what is invisible or silenced, making the familiar strange, reflecting on the discursive production of physical knowledge, and revealing narratives of physics as hegemonic presentations. At best these strategies will turn my physics course now and then into a space welcoming queer perspectives and identities, as well as into a training environment for queering thinking. Bridging the gap between traditional positivist sciences and queer and deconstructive theories poses several questions. First, can deconstructive approaches that perceive gender and sexuality as performative and material-discursive entities be applied to physics? Second, can we translate these reflections into empirical research? Third, can the resulting findings inspire a constructive treatment of gender and sexual heterogeneity in the training of science and engineering students, as well as of future science teachers? And finally, can these considerations be translated into teaching practice? My answer to all four questions is "yes," on the following grounds.

WELCOMING QUEER PERSPECTIVES AND IDENTITIES

German discussions of best practices of "gender and diversity in academic teaching" in STEM fields are unconvincing (e.g., Leicht-Scholten & Schroeder, 2014). The discussions do not relate to a gender-informed pedagogy: They do not understand gender in a deconstructive way, nor are they up-to-date with queer theory and other intersecting theoretical discourses. They take knowledge of physical science rather half-heartedly into consideration and do not challenge the two-gender hegemony. Discussions on queering science and science education hardly exist (Balzter, Klenk, & Zitzelsberger, 2017). One rarely finds detailed suggestions for physics that exceed a deficit-model perspective that advocates for progressive, student-centered teaching and the addition of historical and contemporary female role models. Therefore, my first steps to examine the intersections of physics education and queer theory were in a 2013 workshop designed for future science teachers, entitled Queer Physics, which I taught as a visiting professor at the Darmstadt Technical University, in Germany.[3]

The Queer Physics workshop is a case study for curriculum development in teaching *queer* physics, where I for the first time included lesbian, gay, bisexual, trans*, queer, questioning, and intersex identities and perspectives in a course on gender and physics (Götschel, 2015). Showing that deconstructive and queer analytical perspectives can indeed be applied to physics and physics teaching methods was the concern of the workshop. My introductory presentation explicated the theoretical foundations of deconstructive gender studies, queer theory, and new materialism theory. I familiarized the participants with a new systematization of gender research on physics consisting of people, cultures, and the knowledge of physics. The students in science teacher training then learned about emerging research results and international activities from LGBT+ points of view. Afterwards, the participants worked in small groups with selected materials such as journal articles and an American Physical Society (APS) report on changing the workplace and education climate in physics for LGBT+ people (Simmons & Barthelemy, 2013). Having concerned themselves particularly with the APS report, students discussed the political implications of workplace culture in physics.

Spontaneous feedback offered by the participants and a formal seminar evaluation underlined the importance of integrating queer perspectives into physics. The students even requested a follow-up seminar to study Barad's (2012) queer performativity of nature in more detail. This clearly demonstrates how looking into queer knowledge can motivate students to learn (more) about physics, in this case quantum physics, and current theories of gender studies. To apply deconstructive gender studies, queer theory, and new materialism theory to the philosophy of quantum mechanics requires quite a bit of background knowledge that cannot possibly be dealt with entirely in such a short workshop. Yet, these considerations are interrelated with two issues that need more systematic coverage in teacher training: presenting challenging views on (a) physics as a field of human endeavor which produces sedimented knowledge, and (b) the general idea of heteronormativity.

Teaching *queering* physics, however, is "more fundamental and foundational than the laudable, but limited, goal of including [LGBT+] … students/identities/subjectivities/perspectives" (Fifield & Letts, 2014, p. 399). It is to move beyond the inclusion of diversity and toward the analysis of how physics contributes to the production of hierarchical differences. Teaching queering physics is about the question of how future teachers and scientists can confront themselves with deconstructive approaches and empirical research that perceive gender, sexuality, desire, and identity as performative and discursive entities. Moreover, it is the act of teaching the critical strategy of scrutinizing the normative; the cultural narratives of physics, such as its power over nature or its fundamental usefulness; and stories of the outstanding physicist and the proper engineer.

TEACHING QUEERING: A THEORETICAL FRAMEWORK

Before we can go into the arguments put forward for queering physics, we need to systemize the areas of physics that can be scrutinized from queering perspectives and to develop a theoretical background in deconstructive gender studies, queer theory, and new materialism. At a glance, physics and queer theory do not strike one as having much of anything in common. In the hegemonic understanding of physics, no gender identity, sexual orientation, or desire is associated with either the objects of investigation in physical science or with their descriptions via mathematical formulas and principles (Götschel, 2010). It is obvious that this opinion is not true. For 40 years, interdisciplinary studies have been looking at the interplay between gender and physics. And yet, this point has to be scrutinized more systematically. So far, hardly any studies include queer theory and only a few studies include deconstructive theories (e.g., Hasse, 2000; Pettersson, 2013). Since physicists have gender and sexual identities, it does make sense to inquire about the democratic principles underlying equal opportunities for people in physics from a queering point of view. Furthermore, in (Western) society, physical scientists encounter a bias towards a two-gender hegemony (Kessler & McKenna, 1978), heteronormativity (Warner, 1991) and mononormativity[4] (Pieper & Bauer, 2005).

In addition to the people in physics, we can examine its nonhuman actors, cultures, narratives, analogies, terms, and facts from queering perspectives. Physics is a community of practitioners with a hegemonic masculinist and heteronormative image dominated by heroes rather than heroines. Therefore, more and more empirical studies analyze the culture of physics and investigate scientific institutions such as laboratories, research groups, subsections of physics, and academic societies and organizations (e.g., Hasse & Trentemöller, 2008; Traweek, 1992); and the way physics is perceived by the public, presented in the media, and narrated at conferences and in textbooks (e.g., Erlemann, 2015; van der Veen & Cook-Gumperz, 2010). In (under)graduate teaching too, as this chapter will show, questions about binary thinking and heteronormativity in the culture of physics are relevant. Moreover, since the knowledge of physics is developed by human actors examining the material world, social standards, relations, and conceptions of normality enter into physical knowledge production (Götschel, 2011). In this context, we can ask how binary conceptions of diverse sexes, sexualities, genders, and desires influence the way we construct, experience, and understand the world. Additionally, quantum physicists discovered the indefiniteness and (as Barad, 2007, would call it) *queerness*[5] of matter, showing that nature, and with it the object of physical observation itself, refuses separateness and, therefore, reacts fundamentally queerly.

Within the context of physics, deconstructive gender studies examine the role gender plays in terms of people, culture, and knowledge production (Götschel, 2011). Gender is no longer seen as a binary, stable biological or cultural entity. Rather, it is a culturally constructed event created via repeated, citational, performative actions (Butler, 1990). This view (a) questions the tradition in gender studies assuming a differentiation between the biological sex and social gender, and (b) makes material-physical gender also available for a discourse-theoretical analysis. Taking a "linguistic turn"—in other words, taking the efficacy of language and discourse seriously—deconstructive gender researchers examine, on the one hand, the production of the binary gender identities "male"/"female" in the process of performativity and, on the other hand, how the materiality of sex is sedimented by performative, citational repetitions and solidified in its meaning.

Reflecting on queering physics also means considering the social significance of what Butler (1990) termed the *heterosexual matrix*. Western society maintains a two-gender system with unambiguous and unchangeable identities: One is either male or female. Socially and culturally speaking, (anatomical) sex, (social) gender, and sexual desire are each seen as binary entities and mutually related to each other (Butler, 1993). Within this ordered system, sexual identity places people in socially unequal positions, and sexuality, as well as mononormative, heteronormative erotic desire, decides whether an individual is socially accepted or stigmatized. Sex, gender, sexuality, and desire are critically investigated in queer theory (Krass, 2003). Here deconstructive critiques of the traditional understanding of "gender as a part of the essential self" are combined with LGBT+ studies, which conceive of sexual behavior and sexual identities as social constructs. Sexuality and desire are no longer viewed as a matter of one's personal interest but as powerful political/ social categories regulating and channeling people, giving or denying them rights and privileges (Hark, 2010). Queer theory questions normality, normativity, and deviation in relation to sex, gender, sexuality, and desire and includes the queer reading of "texts" and discourses, as well as a theoretical reflection on queerness itself.

Both gender researchers and physicists look at bodies: The former study human bodies, the later are predominantly concerned with physical, nonliving, material bodies. Consequently, the restrictive binary system of human identity in a broader sense must be taken into account. Following Czollek, Perko, and Weinbach (2009) and Degele (2008), I view queer theory as a cognitive movement that challenges the normative concepts of identity, takes a closer look at heterosexuality and heteronormativity, includes further regulative social constructs, and advances an open view on identity. From this pluralistic-queer standpoint, the identity of physical bodies can be seen as undetermined and inconclusive, a view which questions normative processes and standards.

Furthermore, we need to understand that nonhuman, material bodies develop in processes of materialization. New materialism theory (Alaimo & Hekman, 2008) stimulates these considerations by challenging the separation between nature and culture in Western societies. To my mind, new materialism links up to the deconstructive concepts of sedimentation and naturalization of sex and extends this performative-theoretical notion of matter with considerations about the event-like character and materialization of physical matter in quantum mechanics (Barad, 2007).[6] Materialization therefore is a particularly adaptable aspect of gender studies in physics. In recent years, this approach also brought about a "material turn" in the social and cultural sciences, that is, researchers attending to agencies of material objects in their studies.[7]

New materialism theory conceives of phenomena or matter as active entities capable of action and change, not preceding their interactions but emerging through them. In this context Barad (2007) speaks of *agential realism*. Her theory covers the "epistem-onto-logical" interrelationship between knowing and being, as well as political and ethical issues.[8] She follows Niels Bohr (1931) in his physical-philosophical version of quantum mechanics.[9] In Bohr's so-called Copenhagen Interpretation, the observable, unusual behavior of elementary particles in comparison to macrophysics only comes into being while being observed. Barad takes the matter one step further, describing the *intra-active* process which creates the materiality of both the scientist and the physical object during a physical experiment. Similar to discursive practices in an intersubjective process, material practices, too, bring about materiality in intra-active processes. According to Barad, such material intra-actions need to be included in a queer physical theory along with the creation of materiality through discursive performativity (Barad, 2007). In addition, her deliberations exemplify the effect that insights gained via gender studies in physics have on transdisciplinary gender studies.

QUEERING PHYSICS NARRATIVES

In order to relate physics, gender, desire, and social norms to each other it makes sense to examine the lack of diversity among the scientists who are portrayed in stories of the advancement of physics. Only a few studies have addressed this issue, and those have focused primarily on the role of women and ethnic minorities in physics. Apart from a few Asian physicists, it is predominantly white, straight, middle-class male researchers whose contributions have been recognized in the annals of history, despite the image of physics as objective and universal. For example, material bodies, their physical characteristics, and associated natural laws have been named almost exclusively after white male actors. One of the few exceptions are bosons. These elementary particles were named after the (British) Indian physicist Satyendra Nath Bose to honor his contributions to quantum physics (James, 2004). Another example is the so-called Randall-Sundrum models (also known as the five-dimensional warped geometry theory on the fundamental forces of the universe) developed by the U.S. physicists Lisa Randall and Raman Sundrum. Donna Strickland is only the third female to receive the renowned Nobel Prize (2018) for her achievements in physics, after Marie Curie (1903) and Maria Goeppert-Mayer (1963). Michael Faraday is another rare exception. His working-class background played an important role in his innovative contribution to electromagnetism (Whitten, 2001).

These historical accounts tell much about the concept of the successful predominantly white, straight, middle-class male physicist. On one hand it reflects that women, working-class minorities, and people of color hardly had access to physics education for centuries—a tradition that left its mark until today (Seymour & Hewitt, 1994). On the other hand, these accounts ignore the work of women and minorities in the field who have made important contributions to physics (Byers & Williams, 2006; Yount, 1998). I know of no records on gender and sexual minorities, namely lesbian, gay, trans*, or intersex researchers, who have earned a place in the history of physics. Similarly, I have found no sources that document any out LGBT+ researchers in university physics departments in Germany. In the United States, the case is different: APS, the largest worldwide organization of physicists, takes the issue of gender and sexual minorities increasingly seriously. In addition to programs for women and minorities, the society recently started a program for LGBT physicists to work toward "a more inclusive physics community" (APS, 2017, para. 2).

Having excluded these lives for a long time, the heteronormative historical perspective on physics is coming under fire. Only after her death did it become known that Sally Ride, a popular astrophysicist and astronaut, had spent 27 years of her private and working life with educator Tam Elizabeth O'Shaughnessy. Astronomer and science journalist Lisa Grossmann (2012) contemplates the significance of Sally Ride's sexual orientation:

> Ride was one of my childhood heroes. I dressed as her for Halloween when I was aged eight and my lifelong passion for space was first budding. Ride's legacy is mostly one of inclusion: bringing more women into science, encouraging girls to think they can do anything. … And I felt that same sense of recognition when my girlfriend read me this line from Ride's obituary, that she was "survived by her partner of 27 years, Tam O'Shaughnessy." … She was neither out nor deeply in the closet, it seems. Ride doesn't appear to have kept her partnership with O'Shaughnessy particularly secret. They worked together and wrote books together. Would they have legally married if they could? We may never know. But Ride's sex was visible in a way that her sexuality wasn't. (paras. 3, 5, 9)

Other physical scientists are working to change the conditions that constrain possibilities to a posthumous coming-out like Sally Ride's. Instead, they actively campaign for LGBT+ rights and visibility in the sciences. For example, Professor Nergis Mavalvala is a Pakistani-born American astrophysicist at the renowned Massachusetts Institute of Technology. She lives in Boston with her female partner and her young son (Sohrabji, 2010). She views herself as an "out, queer person of color" (Venkatraman, 2012). In 2012, she delivered a keynote address at Out to Innovate: STEM with Pride, the conference of the National Organization of Gay and Lesbian Scientists and Technical Professionals (NOGLSTP), where she argued strongly for more diversity in physical science (NOGLSTP, 2017).

I teach physics in a way that challenges these normative narratives about the kinds of people who can advance science and engineering. I try to sensitize my students to the narratives told in schools, public media, textbooks, and lecture rooms about the kinds of people who will successfully master physics examinations and become proper engineers (i.e., people who look, think, and behave as engineers should; see Slaton, Cech, & Riley, this volume, chapter 17). I want my students to be aware of the supposed limited diversity of people in physics today and in the past, and to reflect on the discouraging messages this limited diversity gives to them. Whenever possible I include information about the contexts in which particular physical science concepts were developed. I mention briefly that people from different professional fields, such as medical doctors, beer brewers, clerics, and researchers and inventors without academic degrees, have successfully concerned themselves with classical mechanics. I discuss who had access to formal education and financial research support in 18th-century Europe. I explain how observed physical phenomena could only become established knowledge after long-lasting negotiation processes. For instance, Robert Mayer, a German ship's physician, was the first to state the principle of energy conservation. However, it took contemporary physicists a long time to acknowledge Mayer's discovery.

Until the 19th century, only a few women could access physics. Whenever I can, I mention female physicists and female academics working in physics, such as Èmilie du Châtelet, who translated Newton's works and made them known in France (Zinsser, 2007). In contrast to all the sportsmen and racing cars that fill physics textbooks, I complement these traditionally gendered objects with personal, female, nonwhite, or nonbinary historical and contemporary examples. From time to time I include short instructional films in my lecture in which, for example, professors of color show their expertise by demonstrating and explaining physical experiments. In general terms, I make visible a greater diversity of past and present physicists with different gender identities and expressions; sexualities and desires; class affiliations; and ethnic, cultural, national, or religious backgrounds to broaden the picture of the proper physicist and to show role models to all my students. I want my students to understand that they should not accept without question the normative message that they would need to be as brilliant as Newton or Einstein to understand physics. The famous physicists of the 18th century were not geniuses from the very beginning (Fara, 2004; Terrall, 2011); neither do my students need to be exceptional to understand basic physics.

Telling these biographies and narratives of those underrepresented or silenced in the history of physics is a conventional technique to include some outsiders in the picture. Using these stories to reflect on the processes and mechanisms of othering in the history of physics supports my students to question their understanding of physics in a subversive way. Therefore, I understand my approach as teaching queering physics narratives. Yet, a lot more research in both the history of physics and physics education needs to be done to provide material for teaching physics as a cultural activity of a huge variety of people connected via various networks. In the case of classical mechanics, I wish I could refer to more biographies, read "against the grain," of people who financially, academically, technically, or practically supported the development of this subfield. In addition, I need more case studies of people who were excluded from participating in the development of physics in the 18th century or who are overlooked in narratives of 21st-century physics. Most of all, to better reveal the discursive production and canonization of classical mechanics, I need comprehensible representations of research that scrutinize the idea of physics as being developed by genial individuals in a cumulative progress of scientific knowledge.

QUEERING PHYSICS CULTURE

The culture of physics is as masculine, white, and heteronormative as the culture of engineering (Gonsalves, Danielsson, & Pettersson, 2016; Lord et al., 2009). Teaching physics in electrical, mechanical, and industrial engineering therefore reinforces the exclusiveness of STEM fields. Deconstructive gender studies, queer theory, and new materialism theory provide a framework to analyze the narratives, images, and cultures of physics and to examine their impact on inclusions and exclusions in STEM. For a long time, German-speaking gender researchers[10] focused exclusively on the situation of white middle- and upper-class women in physics, and mainly on the professional careers of women from historical or social science perspectives. Women's experiences as outsiders within tell much about the silenced, the other, and the normal of physics culture (Hasse & Trentemöller, 2008). In contrast, Anglo-American studies expanded their analysis towards ethnic minorities as well as first-generation academics. Recently, researchers have developed an interest in the situation of sexual and gender minorities.

Professional organizations reflect efforts to counter the exclusion and isolation of gender and sexual minorities in STEM. For example, NOGLSTP was established in 1983 to fight homophobia in the academic workplace.[11] This spawned a growing number of LGBT+ networks in academic fields, such as mathematics, astronomy, chemistry, and information technology. In 2012, the APS hosted the first session at a major physics conference on sexual and gender diversity issues (Cofield, 2017). One of the presenters, Elena Long, had recently founded LGBT+Physicists, a resource network on sexual and gender diversity in physics.[12] The experience of the physicists organized in these groups can tell us about the inclusions and exclusions of the physics community and how these have changed over time.

U.S. trans* activist Professor Savannah Garmon was one of the contributors to 2012 APS session. Having conducted research in Paris, Copenhagen, Toronto, and Tokyo, she now lectures at Osaka Prefecture University in Japan and studies quantum optics and quantum solid-state physics. On her website, *Leftytgirl*, Garmon (2012) blogs about a variety of feminist, trans*, and social justice issues, including her experiences with her male-to-female transition and how she first feared for her career in physical science. She is one of seven scientists who talk in an interview with *Physics Today* about their coming out in physics, how their sexual or gender identity overlaps with their careers, and their suggestions to make STEM fields more welcoming (Feder, 2015). Asked about her experiences of bias from other physicists because of being trans*, Garmon says:

> I have been in situations [as a postdoc] where people thought they were saying things amongst themselves, and they thought that had no impact, but actually it did, it definitely did. It went deeper than feeling shut out. It felt like I didn't belong, like I shouldn't be there. I felt I didn't have anyone to talk to about the work environment. And maybe that also played into me being hesitant to engage in physics conversations with some of the students. Sadly, that could mean a missed opportunity for collaborations, and maybe that's an opportunity for good science that gets passed up. It also hurt on an emotional level. There were days when I wanted to engage more strongly with my colleagues, but the environment as a whole just felt so deteriorated that more so I just wanted to leave. Those are missed opportunities too. It's unpleasant to think back on. (paras. 24–26)

An APS study of the climate for LGBT+ people in physics sheds light on the community and culture of physics and its dominant norms (Atherton et al., 2016). Survey and interview respondents (n = 324) said that in many physics environments social norms established expectations of closeted behavior. They reported exclusion, marginalization, and trouble identifying allies to help mitigate isolation. Moreover, transgender and gender-nonconforming physicists encountered the most hostile environments, which put many of them at risk for leaving their workplace or school. The report recommends developing a training program on inclusive workplaces and mentorship practices that incorporates the needs of LGBT+ physicists and their allies. It also recommends establishing an APS Forum on Diversity and Inclusion "that works toward a more inclusive, diverse, and equitable society for all physicists, including those who identify as LGBT, women, racial and ethnic minorities, persons with disabilities, and others" (p. 9).

Compared to the working conditions in Germany, where there is no LGBT+ network in physics or engineering, these activities by APS are really marvelous. The world's largest professional organization for physicists promotes an active, engaged, and diverse membership that includes women, ethnic minorities, and LGBT+ persons. Yet, APS narrates LGBT+ physicists as vulnerable victims rather than as an enriching source of diversity for scientific development. The "LGBT Physicists" webpage describes harassment as a serious issue in academic science "that negatively impacts climate, retention, and productivity" (APS, 2017, para. 1). The main argument put forward seems to be the (neoliberal) idea that people who could openly show their authentic selves in professional environments may work much more efficiently than those using up their energies to conceal their real identity.

We should examine whether the APS LGBT+ network has an enduring impact. Will LGBT+ people gain better access to physics and will more of them enter the field? Will the inclusion of physicists "who identify as LGBT, women, racial and ethnic minorities, persons with disabilities, and others" (Atherton et al., 2016, p. 9) have an impact on research and knowledge production in physics? Will their visible presence influence the society's majority members—white, male, healthy, heterosexual physicists—and change the culture of the American physics community in academia, national labs, and industry? What about negative side effects, such as a backlash in the APS community against equal opportunities? These and further questions need to be studied.

I am convinced that it is important to bring the competencies of queering physics culture into the physics lecture theater. Whenever possible I mention LGBT+ physicists, and I try to challenge the image of physics as "hard and difficult" as much as its masculinity, heteronormativity, and naturalization. I use innovative methods such as discussions, poster presentations, and student participation in experiments, and I teach from time to time with a cuddly toy. Some of these activities—as the evaluation results of my course show—do not fit the picture some of my students have of physics and leaves them unsettled and confused. Others, mostly female and/or minority students, appreciate my provocative teaching. They experience physics as more welcoming and less difficult than they expected.

Masculinity is a salient feature in German physics textbooks because of the predominant use of masculine nouns and illustrations of male people. Role models for female and genderqueer students, by contrast, are absent in these textbooks. Additionally, most physics textbooks use stereotypical examples and exercises. The lecture notes used in my department (see Haussmann, 2010) show many examples of male physical scientists, sportsmen, and other hegemonic males, in addition to racing cars, weapons, and machines, all of which symbolize manliness and reiterate the masculinization of physics. Therefore, critically examining physics textbooks is another way to understand and scrutinize the unspoken norms and values of the culture of physics. Some of the more progressive books add sportswomen and animals to the pictures. But this does not always succeed in undermining cultural stereotypes, as I will show in two examples.

To open a space for reflection on normativity, I challenge masculinity without further comment by choosing "atypical" sports or genders for learning tasks. Physics in textbooks is quite often explained with the help of sportsmen's activities that can be characterized as rough, tough, and competitive. However, when it comes to aesthetics, sportswomen are represented in the textbooks. In one of my tasks on calculating the conservation of angular momentum, I replaced the usual female ice skater with Evgeni Plushenko, the Russian three-time world champion in men's figure skating. I let my students watch his quadruple and triple jumps and excellent pirouettes for five minutes. Many of the male mechanical engineering students were visibly uneasy studying Plushenko's "feminine" body movements, and started giggling. I read this as a disturbance of the robust associative ties between physics, the appropriate presence and body language of engineers, and a certain kind of masculinity. By choosing the unexpected and atypical, I teach physics in a queering way.

In a task on kinematics, taken from my university's set of exercises, a lion is chasing an antelope for a short time, and the students are asked to calculate if the antelope will escape or be caught by its hunter. I use this exercise in a queering way to explain to my students the thinking patterns of physics. I present the task "lion hunts antelope" and then show a short film about lionesses teaming up to hunt a zebra. The idea here is to enable the participants to see that natural movement patterns such as acceleration and braking, change of direction, and the necessary teamwork of a pride hunting together are left out of conventional analyses and end up as rectilinear and regular movements in the task. Moreover, the students see that female lions hunt as a team while male lions do not. In this way, the students can question the still prevailing cultural scripts of the lone hero and of active masculinity and passive femininity as they play out in physics (Terrall, 2011).

Exploring these topics with my students in the physics course can be understood as teaching queering physics culture by making the familiar strange. We need more research on masculinity in physics culture(s) to explore, for example, cultural variation among engineering, physics, and other disciplinary and departmental settings. Reviews of research on international, national, and local cultures in physics would be helpful to develop further strategies for queering physics culture.

QUEERING PHYSICS KNOWLEDGE

By investigating the production of academic knowledge, science and technology studies (a transdisciplinary field combining philosophy, history, and sociology) has changed the image of physics from a field of indestructible truth and substantial knowledge into a field of human endeavor and processes that gradually stabilize physical knowledge. We may safely assume that binary heteronormative gender systems are part of the social construction of natural science knowledge, and therefore cultural beliefs about gender, sexuality, identity, and desire, too, are in the equation when it comes to concepts, research questions, models, and natural laws in physics.

This does not necessarily mean, though, that gender, sexuality, and desire can be easily assigned to physical objects and their mathematical descriptions. Rather, it is about how cultural notions associated with gender, sexuality, and desire (Lucht, 1997) shape and inspire physicists' experiences and materialize, as it were, in physical models and theories. As an example, binary gender symbolism is inscribed in the metaphors of thermodynamics. The dichotomy of male activity and usefulness versus female passivity and uselessness produces normalized and naturalized binary scientific categories such as work and heat, energy and entropy, control and disorder, culture and nature (Kovács, 2013).

Binary heteronormative gender systems also show up in particle physics. According to the standard model, all matter consists of quarks and leptons. Here we could ask why hundreds of particles are classified in a model with several generations or families which, in turn, are arranged in dyadic pairs. A queering point of view highlights the fact that heteronormative and mononormative conceptions of pair, family, and generation are conceptual schemes we use and rely on in physics long before we have adequate evidence for doing so. Moreover, those schemes shape the very efforts to justify them and, because they are presumed, they become part of our understanding and storying of the world (Pickering, 1984). Choosing exactly this heteronormative and mononormative model to describe the "nature" of particles, physicists make subtle statements about the nature and normality of a two-gender system, heterosexual desire, and monogamy (Götschel, 2006). As far as I know, a systematic analysis of heteronormative conceptions in physics knowledge has not yet been carried out, except for studies of the (hetero)sexually loaded language of high-energy physics (Traweek, 1988, 1992).

In "Nature's Queer Performativity," Barad (2012) takes the matter one step further. She holds that such diverse entities in the sciences as colonial amoeba, neural receptors in stingrays, lightning, and dinoflagellate microorganisms are queer creatures. In her extensive philosophical reflections on quantum mechanics, she plausibly argues that atoms are "ultraqueer … [entities because of] their radically deconstructive ways of being" (Barad, 2012, p. 25). She uses quantum leaps and quantum erasers to illustrate her point. According to the atomic model, electrons revolve around the nucleus of an atom on stable paths, or orbits. Between these orbits, they make discontinuous movements, so-called quantum leaps. Being at no time between the paths, the electrons nevertheless change from one orbit to another. A far more complex case is that of quantum erasers. Experiments in quantum mechanics on the wave-particle duality revealed that starting an experiment by asking, Are you a particle? generates a particle, while the experimentally realized question, Are you a wave? actually creates a wave. For many years, this phenomenon had been read as the interplay between measuring tools and matter. Conversely, more recent experiments showed that the "disruption" created on the very moment of measuring can be deleted or "erased" afterwards and with it the manifestation of matter as a particle or a wave. From this Barad infers that quantum leaps and quantum erasers confirm that an atom is a queer entity, since its original "identity" is ambiguous and thus exists in a radically deconstructive way (Barad, 2012).

However, as I am not teaching quantum mechanics, but classical mechanics, my students do not get to know the particle-wave duality. Therefore, I do not see a possibility to *explicitly* discuss the entanglement of matter and meaning (Barad, 2007) in my basic physics course. I am interested to see if and how scholars apply Barad's understanding of materialization to other areas of physics beyond quantum mechanics. Anyhow, I *implicitly* use Barad's concept in my teaching when I keep in mind that processes of selection, experiment, observation, and labeling create at the same time physical matter on a macroscopic scale *and* future engineers with physical knowledge. Besides the example I discuss next, I do not know of any queer or queering research on classical mechanics concepts, their conditions of development, or the historical actors who did the work.

An exception could be Scheich's (1985) feminist analysis of the theory of impetus, a medieval precursor of the theory of conservation of momentum. Scheich followed up historical findings that the theory of impetus was a physical as well as an economic theory, combining concepts of motion, work, and property. This understanding of force in which the producer transfers a force to the product, Scheich argues, only takes into account the experience of skilled (male) workers in craft, trade, and commerce in early capitalism. Women at that time were increasingly restricted to their negligible "natural place" of biological reproduction work, in which they were seen as unable to produce added value. Production by women was, therefore, neglected in the economic theory applied to the new mechanical philosophy of nature. Moreover, biological reproduction work remained invisible when Newton's laws of motion laid the foundation of classical mechanics. In the same way as women's reproductive work was necessary for the continued existence of society but not visible in its economic concepts, physical objects will continue moving (uniformly forward in a straight line) forever without any application of a force that puts and keeps the object in motion. Therefore, I critically introduce this new theory of motion in my course. I also deconstruct the vague concept of "force" as a trick to shift the description of movements from a qualitative understanding of motion to its quantitative calculation. With these sparse but complex examples, I try to show my students that the physics knowledge they learn is not just "out there," objective and universal, but negotiated and fabricated.

A vivid example of the naturalization of physics knowledge is the standardization of the measure of length during the French Revolution. The French Committee for Weights and Measures dedicated its efforts during the French Revolution towards replacing the anatomically motivated measures of the cubit (stretching from the middle finger to the elbow), foot, and yard with a naturally derived linear measure. Aiming to find a new standard close to the Parisian cubit, the scientists looked for a partial meridian that connects the North Pole and the equator, stretches over 45°N latitude, leads through a well measured region, and starts as well as ends at sea level. "By chance" just one meridian fit this description: the distance between Duenkirchen via Paris to Barcelona (Jedrzejewski, 2002). Thus, based on the said conditions, the Paris authorities described a "natural measure of length" which came close to the Parisian cubit and could only be defined by a longitude that crossed through the French capital. This example illustrates the discursive production of science, because supposedly objective, naturally derived units and concepts are frequently based on political and social requirements as well as selection processes which are naturalized in physics.

Objectivity, mathematization, and popularization of knowledge (Daston & Galison, 2007) are other issues I discuss in my physics course. But teaching queering physics knowledge is very time consuming. Teaching queering physics will remain challenging as long as understanding physics as a human endeavor is *not* one of the objectives of an applied physics course, while becoming a "proper" engineer is. Moving this project forward calls for more educational research on teaching queering physics knowledge and, most of all, the production of user-friendly, university-level course materials that support the queering of physical science knowledge, rather than perpetuating and glorifying its objectivity and universality.

QUEERING PHYSICS REPRESENTATIONS

I do not (yet) see a possibility to teach new materialism in my mechanics course. Nevertheless, new materialism supports me to reflect on the agency of physical matter and the processes of its materialization which normally are taken for granted. Physics is not just a 300-year-old collection of knowledge that is sedimented in textbooks. Physics is coming into being every moment my students and I teach, learn, talk, think, demonstrate, proof, or calculate in the lecture theatre. Therefore, there is a need to unfold physics in a queering way, or at least to reflect the performance of physics from a queering perspective.

As an example, I try to include the students in the actual experiments as much as possible. To physically experience Newton's third law of motion, also called the law of interaction, two participants face each other standing on skateboards, each of them holding one end of a rope. It does not matter if one student or the other or both students pull at the rope, the skateboards move towards one another as formulated by the law of interaction. Whenever I perform this experiment with my first-semester students in electrical, mechanical, and industrial engineering, patterns of gender inequality, heteronormativity, and desire are (re)produced. When the experiment was repeated in mechanical engineering, by chance a slim female participant had replaced a stout male student. As before, the two skateboards moved towards each other, no matter who pulled at the rope. However, since the slim woman was lighter than her corpulent predecessor, her skateboard accelerated faster than in the previous experiment. Observing this, a male student shouted, "They are in love with each other!" and all the spectators in the lecture theatre burst out laughing. Two skateboards, a rope, two students, the audience and cultural narratives of physics, gender, and desire played out a choreography that eroticized and sexually charged the space in a specific way, produced a binary understanding of gender, defined heterosexuality as the normal sexual orientation for physicists, put the female student in place "as the other of physics," and materialized physical teaching aids—by following physical laws—as powerful objects for demonstrating the straightness of physics (Götschel, 2017).

My (male) teaching assistant and I were very upset about the multiple hierarchies that were created during this experimentation in the lecture theatre. I had been working hard to unobtrusively create a climate of confidence, and it was the first time in this course that a female student participated actively in an experiment in front of the male-dominated auditorium. We understood that her participation was not experienced as comfortable but provocative, creating uneasiness, laughter, and unprofessional behavior expressed via a joking atmosphere. Assuming that this incident would make it harder for the few female students in class to participate in future experiments, I addressed the issue in the following lecture on angular acceleration. I started the lesson with a short film entitled *Angular Acceleration*, which had been produced in 2014 for a German short-film festival and had, contrary to its title, little to do with physics—it told a lesbian love story. Afterwards, I briefly mentioned the events of the previous lesson. I pointed out that every student should enjoy participating in the experiments and that no one should mar the experience with gendering or racializing remarks and jokes. Moreover, since we do not know anything about the participants' sexual orientation, I clearly condemned any form of comments imputing heterosexuality to other participants. By condemning the heterosexual presumption, I was calling attention to and criticizing the underlying heteronormativity that was so dominantly represented in the lecture theatre. I call this strategy queering physics representation.

The study of materiality and material practice is an up-and-coming field of research in science education (Milne & Scantlebury, forthcoming). There are also interesting studies of material practices in the physics lab (Gonsalves et al., 2016; Danielsson, 2007). Now we need research to understand these normative processes in the lecture theatre. For instance, a research participant-observer in my lecture could critically analyze my own entanglement with the masculinization, normalization, and materialization of physics. Moreover, the impacts of my innovative teaching could be studied in greater detail than is provided by a standardized questionnaire for the annual evaluation.

CONCLUSION

With this chapter, first and foremost, I wanted to provide examples of teaching queering physics. I have shown how approaches which perceive gender, sexuality, desire, and matter as performative and material-discursive entities can be applied to physics in meaningful ways. Critical theories such as the performativity of gender (Butler, 1993), queer theory (Krass, 2003), and new materialism (Barad, 2007) act as a theoretical framework to scrutinize how physics contributes to the production of hierarchical differences. As teaching queering physics is a transdisciplinary endeavor, these reflections must draw from different fields of research. Historical approaches can help to challenge normative narratives about the progress of physics and the kinds of people who successfully contribute to it. Cultural studies, social science, pedagogy, and gender studies offer methods to analyze masculinity and heteronormativity in physics. A cross-national review of these diverse analyses could provide some orientation in this highly complex field and begin to trace gaps in the research. Higher education research needs to elaborate in a systematic way how the resulting findings inspire a constructive treatment of gender and sexual heterogeneity in the training of science and engineering students, as well as of future science teachers. Educational research can inform the production of higher education course materials to help instructors and students rethink physics as a cultural activity. In addition, scholars in higher education could support the curriculum development needed in STEM fields to open space for these endeavors. Although materiality and material practice are rising areas of interest in science education, further research needs to be done to understand material-discursive dynamics in the lecture theatre and teaching laboratory.

In my project of teaching queering physics, I face a complex problem of concurrently teaching and learning multiple things: imparting physics knowledge to engineering students, applying deconstructive and queer theories to my teaching of physics, offering new perceptions of physics, teaching competencies of queering to my students, and reflecting on my teaching approaches. Ruth Hubbard aptly described the situation as "like trying to look out of the rear window to watch oneself push the bus in which one rides" (Hubbard, 1983, p. 67). As the task is challenging on all these different levels, a collaborative research team would be my first choice. But I am still seeking funding for team research on understanding and challenging physics and gender in material-performative interactions. Drawing inspiration from deconstructive gender studies, queer theory, and new materialism, my approach so far has been practice-oriented and highly eclectic. Nevertheless, as I have shown in this chapter, teaching queering physics to engineering students can bring critical attention to normative narratives of physics, masculinity and heteronormativity, the negotiation and fabrication of physics, and physics as a performative act.

ACKNOWLEDGMENTS

I would like to thank my valued colleagues Chris Baudy, Robin Bauer, Rylee Hühne, and Florian C. Klenk for stimulating discussions about queering and teaching. I am deeply grateful to Steve Fifield and Will Letts who gave me marvelous encouragement and extremely helpful feedback that strengthened this chapter.

NOTES

[1] In German-speaking countries, queer theory has been increasingly discussed within the subject of gender studies. Far from being mainstream, most of the discourses around queer theory take place in social sciences and cultural studies.

2 In recent years, the acronym LGBT (lesbian, gay, bisexual, trans*) has been extended in political and academic papers to LGBTIQQ to include intersexuality, queer, and questioning or other forms of sexual and gender identity. Often, the longer version is replaced by LGBT+. While trans (without the asterisk) is used to describe trans men and trans women, trans* (with the asterisk) includes all transgender identities such as two-spirit, nonbinary, genderqueer, gender-fluid, genderless, and other gender nonconforming identities.

3 My workshop was part of the Queer Week Series, which included 15 events, such as presentations by guest speakers, open seminars, political discussions, poster displays, and workshops. It was conducted in December 2013 by my colleague Florian C. Klenk and me in cooperation with the queer division of the General Students' Committee at Darmstadt Technical University.

4 Mononormativity scrutinizes the production of knowledge, technologies of power, and praxes that create the dyadic heterosexual couple as the "natural" form of human relationship and that devalue other forms of living together as pathological.

5 Queerness, synonymous for peculiarity or quirkiness, describes the condition of being queer. Straightness would be the opposite, the condition of being "normal."

6 Simply put, a performative-theoretical notion of matter means that we cannot assume existing objects. They materialize—in other words, they are created in the act of speaking and acquire their meaning during discourse.

7 New materialism theorists criticize social and cultural sciences for their "oblivion to material objects" in feminist theories, science and technology studies, and the philosophy of natural sciences. This criticism does not—as the term may suggest—refer to dialectic or historical materialism in Marxist sociology.

8 With her neologism epistemontology, also called onto-epistemology, Barad emphasizes the necessity to overcome the Western tradition of segregating ontology (what is in the world) and epistemology (knowing what is in the world).

9 In the 1920s, quantum mechanics, or quantum physics, had been established as a branch of physics that looks at the unusual behavior of matter at the atomic and subatomic level, (re)actions which at that time had not been observed with macro forms. Such processes allow for mathematical predictions, but the philosophical meanings of these processes are not easily discerned.

10 In German-speaking countries, "gender studies," literally called "gender research" (Geschlechterforschung), is a program of study at some universities but "queer studies" is not. Therefore, I speak of gender (studies) researchers rather than of queer studies researchers in German-speaking countries.

11 See http://www.noglstp.org/about/a-brief-history/

12 For materials from this session and related resources, see http://lgbtphysicists.org/media.html

REFERENCES

Alaimo, S., & Hekman, S. (2008). *Material feminisms*. Bloomington, IN: Indiana University Press.

APS (American Physical Society). (2017). *LGBT+ physicists*. Retrieved from http://www.aps.org/programs/lgbt/index.cfm

Atherton, T. J., Barthelemy, R. S., Deconinck, W., Falk, M. L., Garmon, S., Long, E., & Reeves, K. (2016). *LGBT climate in physics: Building an inclusive community*. College Park, MD: American Physical Society. Retrieved from https://www.aps.org/programs/lgbt/upload/LGBTClimateinPhysicsReport.pdf

Balzter, N., Klenk, F. C., & Zitzelsberger, O. (Eds.). (2017). *Queering MINT. Impulse für eine dekonstruktive Lehrer_innenbildung*. Opladen: Barbara Budrich.

Barad, K. (2007). *Meeting the universe halfway: Quantum physics and the entanglement of matter and meaning*. Durham, NC: Duke University Press.

Barad, K. (2012). Nature's queer performativity. *Kvinder, Kön & Forskning, 21*(1–2), 25–53.

Bohr, N. (1931). *Atomtheorie und Naturbeschreibung*. Berlin: Springer.

Butler, J. (1990). *Gender trouble: Feminism and the subversion of identity*. New York, NY: Routledge.

Butler, J. (1993). *Bodies that matter*. New York, NY: Routledge.

Byers, N., & Williams, G. (2006). *Out of the shadows: Contributions of twentieth-century women to physics*. Cambridge: Cambridge University Press.

Cofield, C. (2017). March meeting session highlights LGBT+ issues for physicists. *APS News*. Retrieved from https://www.aps.org/publications/apsnews/201204/mrchmtghighlite.cfm

Czollek, L. C., Perko, G., & Weinbach, H. (2009). *Lehrbuch Gender und Queer. Grundlagen, Methoden und Praxisfelder*. Weinheim & Munich: Juventa Verlag.

Danielsson, A. (2007). *The gendered doing of physics: A conceptual framework and its application for exploring undergraduate physics students' identity formation in relation to laboratory work*. (Unpublished licentiate dissertation). Uppsala University, Uppsala, Sweden.

Daston, L. J., & Galison, P. (2007). *Objectivity*. Cambridge, MA: The MIT Press.

Degele, N. (2008). *Gender/queer studies. Eine Einführung*. Paderborn: Fink.

Erlemann, M. (2015). Geschlecht in physikalischen Fachkulturen. Ethnographische Sondierungen. In T. Paulitz, B. Hey, S. Kink, & B. Prietl (Eds.), *Akademische Wissenskulturen und soziale Praxis. Geschlechterforschung zu natur-, technik- und geisteswissenschaftlichen Fächern im Vergleich* (pp. 156–174). Münster: Westfälisches Dampfboot.

Fara, P. (2004). *Newton: The making of genius*. New York, NY: Columbia University Press.

Feder, T. (2015, February 27). LGBT physicists: The interviews. *Physics Today*. Retrieved from http://physicstoday.scitation.org/do/10.1063/PT.5.9034/full/

Fifield, S., & Letts, W. (2014). [Re]considering queer theories and science education. *Cultural Studies of Science Education, 9*, 393–407.

Garmon, S. (2012). *Sexual and gender diversity in physics* [blog post]. Retrieved from https://leftytgirl.wordpress.com/2012/12/19/sexual-and-gender-diversity-in-physics/

Gonsalves, A. J., Danielsson, A., & Pettersson, H. (2016). Masculinities and experimental practices in physics: The view from three case studies. *Physical Review Physics Education Research, 12*(2). Retrieved from https://journals.aps.org/prper/pdf/10.1103/PhysRevPhysEducRes.12.020120

Götschel, H. (2006). Die Welt der Elementarteilchen – Genderforschung in der Physik. In S. Ebeling & S. Schmitz (Eds.), *Geschlechterforschung und Naturwissenschaften – Einführung in ein komplexes Wechselspiel* (pp. 161–187). Wiesbaden: VS.

Götschel, H. (2010). Gender goes Physical – Geschlechterverhältnisse, Geschlechtervorstellungen und die Erscheinungen der unbelebten Natur. In R. Becker & B. Kortendiek (Eds.), *Handbuch Frauen- und Geschlechterforschung* (pp. 842–850). Wiesbaden: VS.

Götschel, H. (2011). The entanglement of gender and physics: Human actors, workplace cultures, and knowledge production. *Science & Technology Studies, 24*(1), 66–80.

Götschel, H. (2015). Queere Physik?! In S. Huch & M. Lücke (Eds.), *Sexuelle Vielfalt im Handlungsfeld Schule. Konzepte aus Erziehungswissenschaft und Fachdidaktik* (pp. 207–230). Bielefeld: Transcript Verlag.

Götschel, H. (2017). Drehmomente fallender Pinguine. Queer-dekonstruktive Perspektiven in der Physik. In N. Balzter, F. C. Klenk, & O. Zitzelsberger (Eds.), *Queering MINT. Impulse für eine dekonstruktive Lehrer_innenbildung* (pp. 125–147). Leverkusen-Opladen: Verlag Barbara Budrich.

Grossman, L. (2012, July 27). Why Sally Ride's sexuality really matters. *New Scientist* [online]. Retrieved from http://www.newscientist.com/article/dn22115-why-sally-rides-sexuality-really-matters.html#.VRh5LeE71wg

Hark, S. (2010). Lesbenforschung und Queer Theorie. Theoretische Konzepte, Entwicklungen und Korrespondenzen. In R. Becker & B. Kortendiek (Eds.), *Handbuch Frauen- und Geschlechterforschung* (pp. 108–115). Wiesbaden: VS.

Hasse, C. (2000). *Kraftfeltet, kulturelle læreprocesser i det fysiske rum*. (Unpublished doctoral dissertation). Copenhagen University, Copenhagen.

Hasse, C., & Trentemöller, S. (2008). *Break the pattern! A critical enquiry into three scientific workplace cultures: Hercules, caretakers and worker bees*. Tartu: Tartu University Press.

Haussmann, G. (2010). *Vorlesungsskript zur Vorlesung Experimentalphysik 1* (Unpublished lecture notes). Hochschule Hannover, Hannover.

Hubbard, R. (1983). Have only men evolved? In S. Harding & M. B. Hintikka (Eds.), *Discovering reality: Feminist perspectives on epistemology, metaphysics, methodology, and philosophy of science* (pp. 45–69). Dordrecht: D. Reidel.

James, I. (2004). *Remarkable physicists: From Galileo to Yukawa*. Cambridge: Cambridge University Press.

Jedrzejewski, F. (2002). *Histoire universelle de la mesure*. Paris: Ellipses.

Kessler, S., & McKenna, W. (1978). *Gender: An ethnomethodological approach*. New York, NY: Wiley.

Kovács, Á. (2013). Gender in chemical thermodynamics: Metaphors and worldviews in Atkins' physical chemistry. In H. Götschel (Ed.), *Transforming substance: Gender in material sciences* (pp. 147–181). Uppsala: Skrifter från Centrum för genusvetenskap.

Krass, A. (2003). *Queer denken. Gegen die Ordnung der Sexualität (Queer Studies)*. Frankfurt am Main: Suhrkamp.

Leicht-Scholten, C., & Schroeder, U. (Eds.). (2014). *Informatikkultur neu denken–Konzepte für Studium und Lehre: Integration von Gender und Diversity in MINT-Studiengängen*. Wiesbaden: Springer Vieweg.

Lord, S. M., Camacho, M. M., Layton, R. A., Long, R. A., Ohland, M. W., & Wasburn, M. H. (2009). Who's persisting in engineering? A comparative analysis of female and male Asian, Black, Hispanic, Native American, and White students. *Journal of Women and Minorities in Science and Engineering, 15*, 167–190.

Lucht, P. (1997). Frauen- und Geschlechterforschung für die Physik. *Koryphäe – Medium für feministische Naturwissenschaft und Technik, 23*(4), 28–32.

Milne, C., & Scantlebury, K. (Eds.). (forthcoming). *Material practice and materiality: A re-introduction into science education*. New York, NY: Springer Publishing.

NOGLSTP (National Organization of Gay and Lesbian Scientists and Technical Professionals). (2017). *Out to innovate 2012 keynote address video*. Retrieved from http://www.noglstp.net/outtoinnovate/past-events/out-to-innovate-2012/2012-out-to-innovate-keynote-address-video

Pettersson, H. (2013). Boundaries of masculinity and plasma physics: Toward an understanding of the making of masculinities in experimental physics. In H. Götschel (Ed.), *Transforming substance. Gender in material sciences* (pp. 57–81). Uppsala: Skrifter från Centrum för genusvetenskap.

Pickering, A. (1984). *Constructing quarks: A sociological history of particle physics*. Edinburgh: Edinburgh University Press.

Pieper, M., & Bauer, R. (2005). Polyamory und Mono-Normativität. Ergebnisse einer empirischen Studie über nicht-monogame Lebensformen. In L. Méritt, T. Bührmann, & N. B. Schefzigm (Eds.), *Mehr als eine Liebe. Polyamouröse Beziehungen* (pp. 59–69). Berlin: Orlanda Frauenverlag.

Scheich, E. (1985). Was hält die Welt in Schwung? Feministische Ergänzungen zur Geschichte der Impetustheorie. *Feministische Studien, 4*(1), 10–32.

Seymour, E., & Hewitt, N. M. (1994). *Talking about leaving: Why undergraduates leave the sciences*. Boulder, CO: Westview Press.

Simmons, E. H., & Barthelemy, R. S. (2013, June 21). Climate change. *Inside Higher Ed*. Retrieved from https://www.insidehighered.com/advice/2013/06/21/making-academic-departments-welcoming-lgbt-staff-and-students-essay

Sohrabji, S. (2010, October 4). MIT astrophysicist Nergis Mavalvala wins MacArthur grant. *Parsi Khabar* [Blog post]. Retrieved from http://parsikhabar.net/individuals/mit-astrophysicist-nergis-mavalvala-wins-macarthur-grant/2708/

Terrall, M. (2011). Heroic narratives of quest and discovery. In S. Harding (Ed.), *The postcolonial science and technology studies reader* (pp. 84–102). Durham, NC: Duke University Press.

Traweek, S. (1988). *Beamtimes and lifetimes: The world of high energy physics*. Cambridge, MA: Harvard University Press.

Traweek, S. (1992). Border crossing: Narrative strategies in science studies and among physicists in Tsukuba Science City, Japan. In A. Pickering (Ed.), *Science as practice and culture* (pp. 429–465) Chicago, IL: University of Chicago Press.

Van der Veen, J., & Cook-Gumperz, J. (2010). *The role of narratives in the co-construction of community identity in physics* [Preprint]. Retrieved from http://web.physics.ucsb.edu/~jatila/papers/physics-narratives_2010.pdf

Venkatraman, V. (2012, June 1). Gravitational wave researcher succeeds by being herself. *Science*. Retrieved from http://www.sciencemag.org/careers/2012/06/just-herself

Warner, M. (1991). Introduction: Fear of a queer planet. *Social Text, 13*(29), 3–17.
Whitten, B. L. (2001). Standpoint epistemology in the physical sciences: The case of Michael Faraday. In N. Tuana & S. Morgen (Eds.), *Engendering rationalities* (pp. 361–379). Albany, NY: State University of New York Press.
Yount, L. (1998). *Asian-American scientists*. New York, NY: Facts on File.
Zinsser, J. P. (2007). *Emilie du Châtelet: daring genius of the enlightenment*. New York, NY: Penguin.

ABOUT THE AUTHOR

Helene Götschel is professor for gender in engineering and computer science in the Faculty of Mechanical Engineering and Bioprocess Engineering at Hannover University of Applied Sciences and Arts, Germany. Qualified as a physicist, historian of science, social historian, practitioner in higher education, and researcher in gender studies, she is mainly interested in intersectional and transgressive knowledge transfers. This includes feminist science studies, curriculum development in higher education, and queer theory and new materialism in (science) education. Most of all, her research explores the entanglement of gender and physics in theory and practice. You can reach Helene at helene.goetschel@hs-hannover.de

KRISTIN L. GUNCKEL

6. WHAT DOES QUEER THEORY HAVE TO DO WITH TEACHING SCIENCE IN ELEMENTARY SCHOOLS?

What does queer theory have to do with teaching science in elementary schools and preparing elementary teachers to teach science? This is a question that colleagues and graduate students often ask when I introduce them to queer theory. The purpose of this chapter is to respond to my colleagues' question.

First, however, I want to deconstruct the question to expose its underlying concerns. My colleagues' question is genuine, and I want to respond with care. Effective teachers begin where their students are, connecting to their experiences and perspectives to support them in finding deeper and sometimes different understandings. My colleagues' question is not simple; addressing it requires uncovering deeper concerns and exposing problematic assumptions.

Underlying my colleagues' initial question is a common notion that *queer* is an umbrella term for lesbian, gay, bisexual, transgender, queer (LGBTQ)—the collection of marginalized identities that often get lumped together into the community of the not-straight (Giffney, 2004, 2009; Letts, 2002; Luhmann, 1998). Many people recognize the *Q* in the LGBTQ string of letters as signifying queer, so they assume that queer theory must be about including the experiences of the LGBTQ community in education. As many of my colleagues and graduate students embrace the social justice agenda that calls for inclusiveness and redress of wrongs in education, this is how they assimilate the terms queer and queer theory into their professional work. They can enthusiastically take up the project of fighting homophobia in schools and advocating for the rights of those who have been silenced or legislated as unequal. One interpretation of my colleagues' question is: How do I address the injustices experienced by LGBTQ people in my course on teaching science in elementary schools?

More deeply, the question also belies uncomfortableness. While is it now politically acceptable to acknowledge LGBTQ issues as relevant in teacher education, in what Will Letts (2002) calls the "just-add-LGBT-and-stir" approach, LGBTQ issues are still an uncomfortable subject for many teachers and teacher educators. Because they feel unqualified to talk about LGBTQ-related issues in the classroom, it is not uncommon for course instructors to call in the perceived experts from the campus office of LGBTQ affairs or known LGBTQ-identified and ally faculty to address the topic in multicultural education courses. Many course

DOI:10.1163/9789004331068_008

instructors are uncomfortable addressing this content themselves because they see LGBTQ identities as compartmentalized into boxes defined by "sexual object choice," to use Deborah Britzman's words (2000, p. 41). In the heterogendered discourse that divides the world into the straight/not-straight binary, talking about people classified as the other requires recognition of how the other is constituted in relationship to the normal (Luhmann, 1998). Underlying this construction is acknowledgement that society places people in categories based on the biological sex and gender of the people with whom one engages in sex. Thus, sex is implicated whenever LGBTQ identities are made visible, and many people are uncomfortable talking about sex outside of the bar or the bedroom. The fear of talking about sex in relationship to students and schools makes it even more discomforting to think about how LGBTQ-related content and perspectives might fit into a course on teaching science in elementary schools. What my colleagues may also be feeling when asking their question about queer theory is: Sex is a taboo topic in elementary schools. I don't want to have to talk about sex in my class.

The question gets stretched even further when thinking about preparing elementary teachers to teach science. Science is perceived as logical, objective, and about only certain kinds of content. The topic of LGBTQ issues seems to be outside the rigid boundaries of the standard curriculum categories of physical, Earth, and biological science. Since sex is usually addressed in a separate unit in fifth-grade health class, the whole topic of sexuality, as well, seems outside the standard elementary science curriculum. Given that there is always more to teach than can be fit in a semester, my colleagues' initial question is also asking: How does this content fit into a course about teaching science in elementary schools?

In responding to my colleagues' questions, in this chapter I will first review how queer theory, as a critical theory, is not just about inclusive politics. Rather, it is about disrupting the binaries that get teacher educators into their uncomfortable binds in the first place. Then I will use queer theory to queer three aspects of teaching science in elementary schools and preparing elementary teachers to teach science. To queer something is to ask what it is not, to disrupt normal constructions, and to release the possibilities that normal constructions hide. I will first queer the curriculum standards for teaching science, then learning and teaching science in elementary schools, and finally, preparing preservice teachers to see themselves as teachers of science in elementary schools. I hope to show that queer theory is not a fringe theory that concerns only LBGTQ-identified people, but is a useful tool for reimagining elementary science education and preparing teachers in elementary schools to teach science.

EMBARKING ON A QUEERING

The normal response when answering my colleagues' questions would be to define queer theory in concrete terms. Similarly, in normal academic writing, this section of the chapter would outline the framework for the analysis that follows. Yet, queer theory is not a normal framework because it resists explicit definition (Britzman, 1995; Sullivan, 2003). Instead, one can think of queer theory as a stance that questions what is normal. It takes as a starting point issues of sex, gender, sexuality, and desire to interrogate how "normal" is socially constructed, and aims to disrupt the normal ways of conceptualizing the world to expose new possibilities for ways of thinking and ways of being.

Susanne Luhmann (1998) characterized queer theorizing as the asking of interminable questions. As such, queer theory opens new lines of inquiry that can illuminate alternative ways of conceptualizing how the world is organized and can offer new and fresh possibilities for who we are, how we know, and how we teach. By this definition, inquiry, including scientific inquiry, can be seen as queer (Snyder & Broadway, 2004). This argument, however, is somewhat unsatisfying because it does not allow one to question science or scientific inquiry, and thus puts beyond question the ways that science and inquiry participate in the construction of what is normal. On the other hand, queer inquiry, which I like to call *queering*, asks the questions that science does not dare to ask (Britzman, 1995; Luhmann, 1998). Queering illuminates what we might not otherwise be able to see: much like a new lens on a telescope or microscope makes possible new insights into how the universe is constructed and calls into question prior assumptions, explanations, and theories. Queer theory opens the borders of what counts as scientific knowledge and who can participate in its constructions (Gunckel, 2009). What I find seductive and alluring about queering is the ways that it helps me question the ideas and world that I often take for granted. Queering allows me to explore the unknown territory that this questioning opens. In the process, I find queering to be playful and at times somewhat mischievous as it darts around the edges of sanctioned discourse and pokes holes in essentialized constructions of what is good, what is allowable, and what is known.

An important step in answering my colleagues' questions about what queer theory has to do with anything is to set them straight about what a queer inquiry into education is not. It is not a project of assimilating LGBTQ people into schools. That is a worthy project and one in which I participate, but that is not the focus of queer theory as a lens of critique. The assimilationist project is to raise homosexual, bisexual, transsexual, and transgender identities to be equal to heterosexual and cisgender identities (Fifield & Letts, 2014; Luhmann, 1998; Tierney & Dilley, 1998). In our everyday lives, this is the struggle that LGBTQ people and their allies engage as they work for equal rights in the workplace, in schools, and in society. In terms of what makes a difference in improving LGBTQ people's immediate everyday lives and happiness, the assimilationist project produces tangible results. But it also accepts the socio-cultural-historical structures that define straight and not-straight as a legitimate binary and that privilege heterosexuality and cisgender identity over everything that is not. Through this assimilationist lens, it is difficult to define and articulate the relationship between a struggle for equal rights for LGBTQ people and preparing elementary teachers to teach science. The links are not clear in the same way that starting a gay-straight alliance in a school or assigning students to watch the film *It's Elementary* (Chasnoff et al., 1996) in their educational foundations course feels like one has taken an action that advances the social justice agenda. Instead, embarking on a queering is seeking to expose and release new possibilities for constructing the world and implications for how we think about the relationships among teachers, students, knowledge, and learning in science. Although queering may feel somehow less productive in terms of addressing the immediate inequities that exist for the not-straight, queering calls into question the constructions that legitimize the discrimination in the first place, and thus insists that we (re)imagine and (re)construct our world to eliminate the normal/queer binary altogether.

QUEERING *NEXT GENERATION SCIENCE STANDARDS*: THE MISSING FEELING FOR THE ORGANISM

Since the release of the *Framework for K-12 Science Education* (*Framework*) (National Research Council, 2012) and the *Next Generation Science Standards* (NGSS; NGSS Lead States, 2013), a cutting edge of science education has students engaging in eight practices of scientists. The main idea is that scientists produce knowledge by engaging in specific practices. These practices include asking questions that can be answered empirically; developing and using models; planning and conducting investigations; analyzing and interpreting data; using mathematics and computational thinking; constructing explanations; engaging in arguments based on evidence; and obtaining, evaluating, and communicating information. Rather than memorizing scientific facts or engaging in hands-on experiences that confirm already-known results, students should learn science concepts by engaging in the same knowledge-constructing practices as scientists.

This new framework intertwines scientific practices with core disciplinary ideas (i.e., concepts in physical science, life science, Earth science, and engineering) and seven concepts that crosscut all disciplines (e.g., patterns, cause and effect). One way to think about the intertwining of these three areas is to consider the strands of a rope that are twisted together (Krajcik, Codere, Dahsah, Bayer, & Mun, 2014). Each strand—core disciplinary knowledge, crosscutting concepts, and scientific practices—by itself is relatively weak. But when twisted together the strands form a rope that is stronger than each strand by itself.

Indeed, there are many aspects of the *Framework* and NGSS to celebrate. They bring clarity and rigor to science education. They raise expectations for science literacy to new heights. They challenge traditional teaching approaches that portray science as facts to be mastered rather than as a way knowing about the world. Yet, as much as I like the *Framework* and NGSS, there is a strand that is missing from the rope. This is the strand I call the desire strand or, as I explain below, developing a feeling for the organism (see Keller, 1983).

As portrayed in the *Framework* and NGSS, the eight practices of science are intended to represent how scientists create knowledge. They are envisioned to be less rigid and more representative of the scientific knowledge-producing process than the five steps of the "scientific method" commonly taught in schools and followed faithfully in science fair projects: ask question, form hypothesis, conduct investigation, analyze data, write conclusion about whether your hypothesis was right. Rather than being a linear set of steps, the practices are interconnected as a web of processes that together produce the body of knowledge called science.

Queering these standards begins by queering the phrase *body of knowledge*. Britzman (1995) turned the phrase around to consider knowledge of bodies. A queering of science education asks how bodies are involved in constructing the knowledge of science and thus the learning of science in school. Cognitive linguists broach this idea with their construction of embodied cognition, which recognizes the role that bodies play in producing abstract thought. Lakoff and Johnson (1980) theorized that all people intuitively understand certain concepts because of their sensorimotor experiences with the world. These experiences are then extended through cognitive processes to shape abstract thought and reasoning. For example, humans understand the concepts of up and down through everyday interactions with the physical world. Physical experiences serve as a metaphor for understanding abstract concepts like gravity.

Embodied cognition suggests that our bodies and minds are not distinct. Knowledge of bodies is not just the result of the bodies maneuvering through a physical world that constrains and defines movements. Rather, the movements of bodies are motivated by a bodily desire to satisfy curiosity (Britzman, 2000). Britzman defines this desire as sexuality: the force behind human bodily explorations of the world. The physical, emotional, and psychological pleasure derived from engaging in this quest pushes the body to move, interact with, and connect to the physical world, and it is from these interactions that knowledge is constructed. In remobilizing sexuality in this way, Britzman unchains sexuality from normalizing links to acts of sex and sets free the notion that sexuality is the underlying driver of humans' insatiable quest for knowledge and understanding. By queering sex and sexuality in this way, Britzman both knocks our social preoccupation with sex from its dominant cultural role and exposes the way that sexuality is the foundation upon which all knowledge is built. Thus, knowledge of the world may come from bodily experiences in the world, but the experiences bodies have are inspired and guided by bodily desires.

Observe scientists talking about their work and this underlying drive of curiosity, passion for exploration, and pleasure derived from learning is evident. Whether talking about landing a space vehicle on an asteroid, discovering how to use stem cells to cure diabetes, or identifying the evolutionary mechanism for turning dinosaurs into hummingbirds, scientists' excitement and wonder (see Gilbert & Gray, this volume, chapter 4) reveal the ways that their sexual beings (as reimagined by Britzman) play a role in their inquiry and understanding of the world. Peel back the myths of science as objective, unbiased, and dispassionate, and the underlying sexual foundation of all knowledge and learning is exposed like a dirty secret few dare to talk about.

One of the few who did talk in ways that reveals this sexual foundation for scientific quests and learning was Barbara McClintock. As a scientist, her theories of genetic recombination revolutionized the study of genetics. For her work, she won the Nobel Prize for Physiology or Medicine in 1983. Yet, in her biography of McClintock, Evelyn Fox Keller (1983) described how, at the time of her groundbreaking work, McClintock's fellow scientists, all of whom were *fellows*, asked her about her methodology. McClintock, who used corn in her studies, exclaimed she just had "a feeling for the organism." She said, "I know every plant in the field. I know them intimately, and I find it a great pleasure to know them" (p. 198). She also explained that she found science to be deeply satisfying. "What is ecstasy? I don't understand ecstasy, but I enjoy it. When I have it. Rare ecstasy." (p. 204).

A normal reading of McClintock might interpret these words as evidence that as a scientist who pioneered both the field of genetics and the place of women in science, McClintock was as a loner who had few intimate relationships in her life and so she substituted a relationship with corn and jumping genes for the sexual associations with people that she apparently lacked. A queering of McClintock, however, reveals that it was this feeling for the organism that made McClintock's discoveries possible in the first place. Her feelings for the organisms were not merely a substitution for the pleasure derived from human sexual relations, and science, corn plants, and recombinant DNA were not replacement sexual partners. Rather, McClintock had insights not only into genetics, but also into the work of science. She recognized that her visceral feelings for the natural, physical world allowed her to see what others had until that point missed. While she did engage in the usual practices of science—from questioning to conducting investigations to constructing arguments based on evidence—McClintock also revealed, possibly betrayed, the role that intimacy and ecstasy play in scientific inquiry and discovery.

Knowledge of bodies involves not just knowledge of the world that bodies experience; it also involves knowing the body that experiences the world. Britzman (1995) asked, "What is the knowledge we cannot bare to know?" A queering of science asks how queering the body might produce different knowledge. Queered bodies are not just bodies that do not fit into the heterogendered boxes created to define which bodies are legitimate creators of knowledge (see Slaton, Cech, & Riley, this volume, Chapter 17). Rather, from the queer position, all bodies can be queered by releasing the chains of normalized expectations for bodies and bodily ways of being (Butler, 1999; Halperin, 1995). In the same way that antimatter is the hidden stuff that holds the emptiness of the universe together, a queering of science learning asks if queered bodies produce queer knowledge that is actually the stuff that fills in the voids of our understanding of the world that we are unaware of, or worse, unwilling to examine. In scientific practice, data do not reveal answers; it is the creative process of scientific thought that must interpret the data and construct arguments based on the evidence. Scientific practices are supposed to establish some criteria for what counts as a valid argument about what the data suggest. A queering of this process asks how limiting knowledge to that created by bodies confined to the straightjackets of the normalized has excluded the knowledge, or criteria for what counts as knowledge, that queered bodies know. Queering all bodies and listening to the knowledge that queered bodies might create could lead to frontiers not previously known to exist.

In returning to NGSS, a queering of science education asks where knowledge of queered bodies might fit. How is the knowledge of bodies acknowledged, and where is there room for the knowledge that queered bodies produce? More importantly, where in the eight practices is there room for students to experience the wonders of the world and to be seduced by the pleasure that explorations of that world produce? Barbara McClintock might have argued that engaging in the eight practices produces the ecstasy of science. But in science classrooms, unless knowledge of bodies is nurtured, the eight practices can be easily reduced to steps in a procedure, which is contrary to what the authors of the *Framework* intended. The strand of the NGSS rope that is missing, the strand that could bind the other three strands together, is the strand that inspires passion, makes learning pleasurable, and follows the insights that students' queered bodies, experiences, and perspectives contribute to the construction of knowledge in science classrooms. This strand would leverage students' visceral connections to phenomena and develop students' feelings for the organism.

QUEERING ELEMENTARY SCIENCE INSTRUCTION: INVITING SEXUALITY INTO THE ELEMENTARY SCIENCE CLASSROOM

But, my colleagues and graduate students might ask, how does this relate to teaching science in elementary schools? A sense of fear arises as they consider sexuality and elementary-age children in close proximity. Suggesting that sexuality motivates the construction of science knowledge and that science teaching should nurture visceral feelings for the organism queers the common sense of elementary school as a safe place for innocent children. Much of this fear is grounded in the reduction of *sexuality* to only the act of having sex (Britzman, 2000).

Britzman argued that children are from infancy sexual beings and it is sexuality that drives and permits a baby to learn about itself and its world:

> In the beginning of life, the infant cannot distinguish between the urge for satisfaction and the other whose actions meet the infant's needs. It is an achievement of the human baby to begin to distinguish its body from other bodies. It is an achievement when the baby recognizes the separate existence of the other and of the self. When this occurs, knowledge, although immature in its formulation, begins to be made and used by the baby. This immature knowledge is caught in the baby's desire to be loved and the baby's growing capacity to love as she or he learns to live. (pp. 42–43)

As such, Britzman argues, knowledge is precocious. It is the baby's instinctual curiosity that allows its desires to be met. It is this desire which makes knowledge of bodies possible. Britzman claims knowledge of the world, as experienced by bodies, precedes understanding, and by extension sexuality is foundational to human existence and learning.

Yet, the romantic ideal of children as innately good, fundamentally pure, and needing of protection lest they become damaged by exposure to sexuality has repressed our acknowledgement of the fundamental relationships among sexuality, curiosity, and knowledge construction. Instead, notions of childhood innocence and ignorance conspire to cocoon children in suffocating webs that isolate sexuality from learning. As James Sears (1999) writes, "Childhood innocence is a veneer that we as adults impress onto children, enabling us to deny desire comfortably and to silence sexuality" (p. 9). Rather than protect children, we stifle their curiosity, their desires, and their bodies of knowledge.

In her essay on the relationship between sexuality and knowledge, Britzman (2000) recalled how Anna Freud accused schools of repressing the sexuality of children and warned against the dangers that such repression begets, such as anxiety, unhappiness, loneliness, aggression, and intellectual rigidity. Freud argued that adults should act in the best interest of the child, a phrase that is now corrupted in service of keeping children isolated from their own instinctual desires and the possibilities for knowledge that such desires create. Science is an ideal area for children and teachers to rediscover the learning potential of humans' instinctual desire to satisfy curiosity. Whether exploring similarities and differences between solids and liquids, wondering what happens when seeds are sown in various types of soil, or pondering the mysteries of outer space, science as a field of study can provide the opportunity for children's sexuality to feed their desire to learn.

Yet, elementary school science instruction functions not to feed these desires as much as to continue to repress students' basic bodily connections to knowledge and content. In an age of testing and accountability mania, science is the subject left behind in elementary schools (Banilower et al., 2013). When it is taught, normalized elementary school science instruction relies heavily on prepackaged kits that come complete with classroom sets of materials necessary to perform the activities outlined in the teachers' guide. Science kits are intended to support teachers who feel ill at ease with science content and uncomfortable with children's explorations by removing uncertainties that arise when instruction follows students' questions. In fact, few kits allow students to ask questions at all. Instead, inquiry is boxed and bounded into lessons with expected outcomes, (re)presenting the same already-known and ready-made answers over and over. Furthermore, normal school science often leads to (re)producing heteronormativity and leaves little room for the full range of students' nonnormative bodies, desires, and experiences to be acknowledged, validated, and taken into account in elementary science classrooms (Fifield & Letts, 2014; Letts, 1999, 2001; Snyder & Broadway, 2004). Normalized elementary school science represses opportunities for children to develop their feelings for the organism.

Queering elementary science instruction, then, requires that schools and teachers invite sexuality as reenvisioned by Britzman into the classroom (Shlasko, 2005). This means making learning science in school a pleasurable experience. This does not necessarily mean that science has to be "fun," which is how many preservice and classroom teachers frame their goals for science learning. Rather, science learning experiences that are pleasurable are experiences that satisfy a deep curiosity. What would science instruction look like if teachers acknowledged and leveraged the ways that children's physical, bodily desires to explore shape how and what they know about the world? What would science curricula look like if children were allowed to ask and pursue the questions they most want to ask? What would science in elementary school look like if the full range of children's human bodies were celebrated and the full range of their desires were met with kindness and curiosity?

Pursuing these questions does not necessarily mean that science kits and other provided curriculum materials must be thrown out. Instead, pursuing these questions requires queering their use. That is, rather than fidelity to the curriculum being the goal for science instruction, queering the curriculum and classroom inquiry calls for teachers to use the science kit as a starting point for sparking curiosity. Pursuing children's curiosity, however, will inevitably lead to uncharted territory where ambiguities, contradictions, and uncertainty exist. Paradoxically, uncertainty and ambiguity may lead away from, rather than towards, feelings of pleasure (Fifield, Grusenmeyer, & Ford, 2014). Nevertheless, being true to science and the practices of science, as advocated by the NGSS, means that elementary science instruction must eventually engage students in dealing with uncertainty and becoming comfortable with ambiguity. In other words, acting in the best interest of the child requires opening opportunities for students to experience the full range of feelings for the organism.

QUEERING ELEMENTARY PRESERVICE TEACHER IDENTITIES: DISRUPTING MS. FRIZZLE AND BILL NYE

At this point, colleagues and graduate students should ask: What does this have to do with learning to teach science in elementary school? In response, I ask: What would it mean to prepare elementary teachers to nurture in students a feeling for the organism?

A common normalizing discourse about elementary teachers is that they love teaching students but do not love teaching science. There is some research to support these general attitudes. Elementary preservice teachers tend to express oversimplified notions of caring and emotional connection to students (Goldstein & Lake, 2000). At the same time, many elementary teachers rank teaching science low on their list of favorite subjects to teach (Wilkins, 2009) and feel underprepared to teach science (Banilower et al., 2013). These attitudes get essentialized into an overgeneralized dichotomy: love of science vs. love of students. Elementary teachers, as the saying goes, love their students, while secondary science teachers love science.

One academic response has been to argue that teacher education should support preservice elementary teachers in developing an identity as a teacher of science (Luehmann, 2007). Identity researchers operationalize identity as the stories that one tells about oneself and that others tell about a person (Sfard & Prusak, 2005). From this perspective, changing elementary teachers into teachers who love teaching science requires changing the stories elementary and preservice teachers tell about who elementary science teachers are and what it looks like to teach science in elementary school. Part of the challenge in doing this is that many of the stories teachers are exposed to about teaching elementary school science normalize the love-of-science/love-of-students dichotomy and portray teachers of elementary science as feminine-gendered and sexually sanitized beings.

Two popular elementary science teacher media personalities are Ms. Frizzle, who drives the Magic School Bus books, and Bill Nye the Science Guy of television and Internet fame. Both are portrayed as nerdy, whacky people. Ms. Frizzle takes her students on fantastic journeys to see firsthand the phenomena they are studying (e.g., Scholastic, Inc., 2017). Bill Nye does wild science stunts and demonstrations to reveal the way the world works (see Nye, n.d.). Both Ms. Frizzle and Bill Nye want kids to love science, but as teachers, they stick to the normal discourse about scientists and science teachers as being super-smart encyclopedias of facts. Neither is portrayed as particularly aware of or sensitive to students' needs and inquiries. Ms. Frizzle's students are often reluctant to follow her and question her teaching methods. She ignores these responses and drags her student along on forced trips, never asking for their ideas about what they are experiencing. Similarly, Bill Nye is never shown interacting with students and he doesn't invite them to explore on their own. Their pedagogies, while somewhat unorthodox, emphasize the learning and (re)presenting of science facts. In other words, despite their craziness, they (re) produce traditional school science approaches to teaching science.

Both characters exude a passion and excitement for science that is intended to be contagious to their students. In this respect, they model the ways that passion and desire drive the quest for knowledge. However, as science teachers, they (re) present the repression of sexuality in elementary schools. Ms. Frizzle's dress has a strange ability to change design to reflect the topic of the particular science trips the Magic School Bus takes. Nevertheless, the dress itself and everything else about Ms. Frizzle, from her hair piled on top of her head as a frizzled mess to her prim and painful (although still science-themed) shoes, reflects the problematic notion of teacher as "old maid." Similarly, Bill Nye, dressed in a suit covered by a lab coat with a tight bow tie choking his throat, represents the de-masculinized male elementary school teacher. Furthermore, even in their zaniness, both Ms. Frizzle and Bill Nye reproduce scientists and science teachers as nerdy and sexually sanitized. Erica McWilliam (1999) critiqued the ways that cultural norms for school teachers require teachers to make over their bodies following gendered rules about what counts as the face of knowledge. Ms. Frizzle and Bill Nye both follow these gendered norms that repress sexuality and contribute to the separation of sexuality and the teaching of science.

In this heterogendered discourse, preservice teachers' options for science teacher identities are to become Ms. Frizzle or Bill Nye. The stories about who elementary teachers are and should be—as de-sexualized, feminized/de-masculinized individuals—and about science teachers as crazy people with exhaustive knowledge who love science more than students, provide few opportunities for teacher candidates to see themselves as teachers of science in elementary schools. Thus, if stories about what it means to teach science in elementary school contribute to the construction of elementary teachers' identities, then preparing elementary teachers to teach science requires not changing teachers, but disrupting these heterogendered stories of science teaching that are the basis for developing science teacher identities in the first place.

Preparing elementary teachers to nurture a feeling for the organism might be about releasing the sexual potential of teachers rather than changing people's identities. Lest it is not clear by this point in this chapter, releasing sexual potential is not about bringing sex into classrooms. Instead, I challenge teacher educators, and especially elementary science teacher educators, to consider the ways that their teaching (re)produces heteronormative images of science teachers and (re)enforces the love-of-science/love-of-students dichotomy. What would elementary science teacher preparation look like if teacher educators assumed that everyone has and can nurture a feeling for the organism? What would the science teacher education curriculum look like if rather than focusing on teaching particular methods, routines, and approaches, it refocused on (re)awakening teachers' own desires for knowledge and passion for understanding? What would science teacher education look like if it supported new elementary teachers in connecting their own childhood desires for learning about the world with their passion for teaching children? Such science teacher education might then inspire future elementary teachers to create classrooms where their students can deepen their feeling for the organism.

CONCLUDING WORDS

What does queer theory have to do with elementary science teaching and preparing elementary teachers to teach science? I hope my colleagues now see that queer theory is relevant to everything from the content of the curriculum to the enactment of science teaching to the identities of elementary teachers as teachers of science. Queering the heteronormalized discourse of elementary school science taps the repressed potential of students and teachers to together discover their own curious ways to love and learn science.

REFERENCES

Banilower, E. R., Smith, S. P., Weiss, I. R., Malzahn, K. A., Campbell, K. M., & Weis, A. M. (2013). *Report of the 2012 national survey of science and mathematics education.* Chapel Hill, NC: Horizon Research.

Britzman, D. (1995). Is there a queer pedagogy? Or, stop reading straight. *Educational Theory, 45*(2), 151–165.

Britzman, D. (2000). Precocious education. In S. Talburt & S. R. Steinberg (Eds.), *Thinking queer: Sexuality, culture and education* (pp. 33–59). New York, NY: Peter Lang.

Butler, J. (1999). *Gender trouble*. New York, NY: Routledge.

Chasnoff, D., Cohen, H. S., Bowman, B., Lyman, K., Coates, K., Yacker, F., McCarthy, S., & Thompson, S. (Writers). (1996). *It's elementary: Talking about gay issues in school* [video]. Harriman, NY: New Day Films.

Fifield, S., Grusenmeyer, L., & Ford, D. (2014). Pedagogical change and mourning in elementary teacher education. *Journal of Curriculum Theorizing, 30*(1), 75–86.

Fifield, S., & Letts, W. (2014). [Re]considering queer theories and science education. *Cultural Studies of Science Education, 9*(2), 393–407.

Giffney, N. (2004). Denormatizing queer theory: More than (simply) lesbian and gay studies. *Feminist Theory, 5*(1), 73–78.

Giffney, N. (2009). Introduction: The 'q' word. In N. Giffney & M. O'Rourke (Eds.), *The Ashgate research companion to queer theory* (pp. 1–13). Farnham: Ashgate Publishing Company.

Goldstein, L. S., & Lake, V. E. (2000). "Love, love, and more love for children": Exploring preservice teachers' understandings of caring. *Teaching and Teacher Education, 16*(8), 861–872.

Gunckel, K. L. (2009). Queering science for all: Probing queer theory in science education. *Journal of Curriculum Theorizing, 25*(2), 62–75.

Halperin, D. M. (1995). *The queer politics of Michel Foucault*. New York, NY: Oxford University Press.

Keller, E. F. (1983). *A feeling for the organism*. New York, NY: W.H. Freeman.

Krajcik, J., Codere, S., Dahsah, C., Bayer, R., & Mun, K. (2014). Planning instruction to meet the intent of the next generation science standards. *Journal of Science Teacher Education, 25*(2), 157–175.

Lakoff, G., & Johnson, M. (1980). *Metaphors we live by*. Chicago, IL: University of Chicago Press.

Letts, W. (1999). How to make "boys" and "girls" in the classroom: The heteronormative nature of elementary-school science. In W. Letts & J. Sears (Eds.), *Queering elementary education: Advancing dialogue about sexualities and schooling* (pp. 97–110). New York, NY: Rowman & Littlefield.

Letts, W. (2001). When science is strangely alluring: Interrogating the masculinist and heteronormative nature of primary school science. *Gender and Education, 13*(3), 261–274.

Letts, W. (2002). Revisioning multiculturalism in teacher education: Isn't it queer? In R. M. Kissen (Ed.), *Getting ready for Benjamin: Preparing teachers for sexual diversity in the classroom* (pp. 119–131). New York, NY: Rowman & Littlefield.

Luehmann, A. L. (2007). Identity development as a lens to science teacher preparations. *Science Education, 91*(5), 822–839.

Luhmann, S. (1998). Queering/querying pedagogy? Or, pedagogy is a pretty queer thing. In W. F. Pinar (Ed.), *Queer theory in education* (pp. 141–156). Mahwah, NJ: Lawrence Erlbaum Associates.

McWilliam, E. (1999). *Pedagogical pleasures*. New York, NY: Peter Lang.

National Research Council. (2012). *A framework for K–12 science education: Practices, crosscutting concepts, and core ideas*. Washington, DC: National Academies Press.

NGSS Lead States. (2013). *Next generation science standards: For states, by states*. Retrieved from https://www.nextgenscience.org

Nye, B. (n.d.). *Bill Nye* [YouTube site]. Retrieved from https://www.youtube.com/user/TheRealBillNye

Scholastic, Inc. (2017). *The magic school bus* [website]. Retrieved from https://www.scholastic.com/magicschoolbus/books/

Sears, J. (1999). Teaching queerly: Some elementary propositions. In W. Letts & J. Sears (Eds.), *Queering elementary education: Advancing dialogue about sexualities and schooling* (pp. 3–14). New York, NY: Rowman & Littlefield.

Sfard, A., & Prusak, A. (2005). Telling identities: In search of an analytic tool for investigating learning as a culturally shaped activity. *Educational Researcher, 34*(4), 14–22.

Shlasko, G. D. (2005). Queer (v.) pedagogy. *Equity & Excellence in Education, 38*(2), 123–134.

Snyder, V., & Broadway, F. S. (2004). Queering high school biology textbooks. *Journal of Research in Science Teaching, 41*(6), 617–636.

Sullivan, N. (2003). *A critical introduction to queer theory*. New York, NY: New York University Press.

Tierney, W., & Dilley, P. (1998). Constructing knowledge: Educational research and gay and lesbian studies. In W. F. Pinar (Ed.), *Queer theory in education* (pp. 49–72). Mahwah, NJ: Lawrence Erlbaum Associates.

Wilkins, J. L. (2009). Elementary school teachers' attitudes toward different subjects. *The Teacher Educator, 45*(1), 23–36.

ABOUT THE AUTHOR

Kristin L. Gunckel is an associate professor of science education in the Department of Teaching, Learning, and Sociocultural Studies at the University of Arizona, in Tucson, Arizona (USA). Her research interests include the preparation of elementary and middle school teachers to teach science, the promotion of environmental science literacy, and the use of queer theory as a critical lens on education. After careers as a geologist, environmental educator, and middle school science teacher, Kristin is now a critic of typical school science teaching and an advocate for innovative, exciting, and inclusive science learning opportunities that reimagine science in the classroom.

JOE E. HEIMLICH

7. QUEERING STEM LEARNINGSCAPES

INTRODUCTION

In STEM there is a perceived knowledge base—a literacy, so to speak. Yet the STEM literacy of the population at large is continually challenged as being far below what many in the science education community consider desirable (see, for example, Feinstein, 2011; Liu, 2009; Miller, 2001; Sturgis & Allum, 2004). The counter to this is that people cannot not learn. We are constantly taking in data through our senses and organizing those data in ways that are meaningful to and in our lives. We use the information that we accumulate through the cultural lenses by which we learn it. And STEM content surrounds us in our lives. Whether it's television, radio, movies, shopping malls, museums, zoos, parks, etc.—we are immersed in a world where the products and processes of STEM are always present.

Strauss (1984) convincingly argues that our understanding of learning is framed through a postindustrial, Western orientation. From such an orientation, real learning is often assumed to be limited to the formal arena of schooling. One possible reason for this easy assumption is that since learning is what happens in schools, information, ideas, and feelings that happen or are obtained outside of school are discredited or dismissed as inferior (Coffield, 2000; Hager & Halliday, 2007; Livingstone, 2001). In contrast, this chapter argues that we create *learningscapes* in all aspects of our lives. Learningscapes are the spaces of living through which the many pathways of an individual's experiences, memberships, identities, and activities run and are interpreted through the person's unique frames into meaning or lack of meaning. Further, this chapter posits that these learningscapes must be queered, for it is only in dis-settling normative desires—bringing into consideration the human essence of desire as physical, emotional, and cognitive—that it is possible to achieve the dissonance necessary to make people aware of how they are engaging in STEM as they learn and use STEM content and processes throughout their lives in potentially cumulative and personally beneficial ways. Human learning is not about facts alone; learning is about the blend of facts with emotions and intentions in the totality of conative learning.

LEARNING AS HUMANS

Human learning is not as neat and orderly a process as a postindustrial view might suggest. From the dominant perspective, "learning" is compared against standards

 | DOI:10.1163/9789004331068_009

established for cognitive measurement, usually within a formal educational system. These standards challenge the validity and efficacy of *informal learning*[1] by creating a binary situated on the assumption that learning equals short-term *cognitive* recall. In doing so, we devalue affect-, interest-, and skill-based qualities of all learning by treating them as domains distinct from cognition. Interestingly, what we call "informal" education is actually closer to how people naturally learn (Scribner & Cole, 1973). Most challenging to the view of learning as equal to schooling is to compare how very young children learn with everyday learning processes we encounter across life (Ardoin & Heimlich, 2015).[2]

Most people do not visit a park, go to a movie, read the newspaper, or talk with friends with the primary desire of *learning* science. Indeed, many people likely enter these settings with a very passive view of "I'm not here to learn," which is something those of us who study people in informal STEM settings hear frequently. In one study of zoo visitors, when they were asked, "What did you learn?" I watched them struggle to remember specific cognitive bits related to what they had seen. But when asked, "What did you get out of being here today?" I heard stories about things people learned about others in their group; ideas they had challenged; and how something they saw, learned, or heard built on a preexisting interest. *Learning*, in most people's minds, is normed as code for school. But as someone who studies learning in informal settings, I see how deep, rich, and broad learning can be in these places.

It is well understood in the learning sciences that learning is lifelong, life-wide, and life-deep (Banks, Au, Ball, Bell, & Gordon, 2007; Feder, Shouse, Lewenstein, & Bell, 2009). Lifelong refers, of course, to the humanness of learning as a process of sensory intake, processing, and application that happens at all life ages and stages. Learning is a process which, when the individual is open to it, does not just filter data to that which is "normal" or desired, but constantly challenges prior learning and experiences. This leads to an ongoing articulation of experiences and reinterpretations of what one feels, believes, thinks, values, knows, and does. In other words, lifelong learning is not only about accumulating facts but can be about constructing understandings of oneself and the world in which one lives.

Life-wide learning encompasses two views. One is that learning crosses experiences, and individuals construct meaning through their accumulated experiences (Martin, 2004). The other view of life-wide learning refers to intentional engagement in different experiences, all of which can be considered learning (Malcolm, Hodkinson, & Colley, 2003), which challenges the culturally normed assumption of learning being cognitive only. In this chapter, *learning* refers to the human qualities of what we experience through all our senses and how those experiences inform who we are, how we are, and what we are in all our complexities.[3]

Life-deep learning acknowledges that individuals seek information on specific things: The home gardener often learns much about soil, plant identification, and even plant biology and symbiosis; and some people, when planning a vacation, spend hours reading about where they are going, arranging excursions or activities, and scheduling events. Tough's (1978) groundbreaking work found that adults typically have between five and seven learning projects each year on which they choose to spend hundreds of hours. These projects may be hobbies, vacations, sports or recreation, investigating family history, or any sort of intentional activity even though, in most cases, the individual does not consider it intentional "learning." Livingstone (1999) conducted a national study in Canada that confirmed the ongoing nature of intentional and detectable learning outside schooling. The study found that about 80% of adult learning is informal and averages around 15 hours a week engaged in activities related to community volunteering; work; household learning around such things as home renovations, gardening, cooking, and home maintenance; and other general-interest learnings. These findings did not include the incidental and everyday learning that simply happens or learning associated with cultural institutions like museums, which would have increased the numbers to an even more dramatic level.

I want to focus on one area in which I believe queering could most affect or dissettle informal STEM learning in ways that deepen individuals' understandings of how learning STEM happens. That is to examine the life-wide/life-deep perspective that learning crosses experiences and that individuals selectively construct meaning through their uniquely accumulated experiences. Specifically, I want to examine how STEM learning is influenced through personal activities lived through queer perspectives that intentionally uncover and creatively resist norms around desire that may include but aren't limited to sex, sexualities, and gender. *Queer perspectives* refers to meanings made through learners' own, nonheteronormative lenses and actions that challenge norms around desire to queer learningscapes.

These learningscapes, or pathways moving through the experiences and contexts of a person's life, are created through both intention and accident: by going through life with awareness and oblivion, sometimes being aware of our decisions and other times allowing decisions to be made. For the remainder of this chapter, there is no assumption of a linear, predestined pathway through the learningscapes of a person's life. Because of the sometimes intended, sometimes related, and sometimes unrelated paths a person chooses to take, or merely finds themselves traveling, there are multiple pathways and multiple ways to examine learningscapes. And as noted by Giffney (2009), there are also many different queer theories with sometimes overlapping and sometimes diverging perspectives. Rather than narrow the following narrative, different lenses will be used to understand queer perspectives within the differing STEM learningscapes of individuals.

LEARNINGSCAPES

If learning is constant, then what we experience is a result of conscious and unconscious, intentional and unintentional engagement in learning opportunities, frequently created in the moment by the individual in conversation with the environment and/or with other people. In all these settings, we can choose to attend to information, ideas, concepts, feelings, facts, values, impressions, skills, and all the other sensory-driven information available to us. Some of the settings include places intentionally designed for learning, such as nature centers, science centers, parks, museums, zoos, schools, and the like. A phrase often used to describe networks of such organizations and other potential sites for learning is *ecosystems of learning*. Ecosystems of learning are usually bounded by a focus of the learning, such as an ecosystem of STEM learning, and are rich descriptions of connected enterprises that allow us to examine how the learner is supported (or not) across the community (Uden, Wangsa, & Damiani, 2007). The various ecosystems of learning then combine for a complex web of structured enterprises in which an individual can be supported in their learning.

Yet it is up to the individual to make meaning from and among these many disparate events. A learningscape is comprised in great part of the intentional engagement choices a person makes across the many facets of their life, around which they do (or could) make connections and meaning. By its very nature, a learningscape is ripe for queering. To get to that point, however, it is necessary to first construct the framework on which we build our individual learningscapes.

Individual Learning

In looking at scans across many studies of informal learning (e.g., Conlon, 2004; Feder et al., 2009; Heimlich & Storksdieck, 2007; Marsick, 2009), five basic tenets seem to emerge as central to understanding meaning-making in informal (and coupled formal) STEM learning and its application in an individual's life:

1. What the individual seeks or desires will drive their "agenda."
2. A person has their own motivations.
3. Learning is continual and natural.
4. Learning is cumulative (i.e., contemporary experiences and learning are built on and tempered by prior experiences and learnings, but are not strictly additive).
5. Learning is constructed meaning applied by the learner.

In essence, these tenets provide a base for human learning. To explicate each tenet slightly, we begin with the need individuals have for learning. We know, for example, that people tend to seek out information or situations for personal reasons—to repair something gone awry, plan a vacation, have a wonderful time with friends. Research on informal learning suggests that what one needs to know or what one is interested in in the moment are the predictors of what one will learn and do in informal contexts (Stocklmayer, Rennie, & Gilbert, 2010). I posit that such need, interest, and engagement are often unconscious and sometimes assigned as a need or interest only after the experience. One may not be (and need not be) cognizant of how they respond to, say, a piece of music, but that does not mean the next time the music is heard they will not respond to it differently, as it is no longer new or novel.

Similarly, it is well understood that motivations to engage are closely aligned with social roles. As used here, social roles are not assumed identities or artificial persona, but are, rather, the aspects of one's identity that are dominant in different contexts. Although a person's Identity (upper-case *I*) develops as they age and change, at any given point in time, a person has the same Identity whether correcting children, out socializing with a group of close friends, at a professional meeting, consulting a spiritual advisor, or with a romantic partner. But social role theory suggests that individuals in context choose many of the behaviors they perform (Biddle, 1986), and these behaviors reflect identities (lower-case *i*) expressed in differing discourses, performances, and senses of self. Falk et al. (2007) studied five social-role identity motivations and found that those were closely aligned with the ways people enacted and experienced visits to a zoo. Many other studies have explored how social role, various aspects of identity, and temporal motivations affect those things to which one pays attention. These findings are highly compatible with theories of desire and the complicated dynamics of body identification (Heyes, 2003) and with queer temporalities and motivation (Dinshaw et al., 2007). These temporal selves, or ways we see and interpret our being in different contexts and over time, are all part of the complex Identity of the individual, suggesting these contextual roles are more fluid than what we tend to think of as "Identity." Social role theory suggests that in informal learning, the information one is seeking or may hold or take away from an experience is highly dependent on the role being enacted. If one is in the role of a parent, then there is a different expectation and focus than if one is out with friends or with colleagues from work. Martin (2004) frames expectations about learning in terms of who people are—the complex humanness of the individual—not based on what is/was accomplished. Learning is not about outcomes but is about the individual building experiences around which meanings can be challenged and reinterpreted.

Theories of attention-perception and cognitive schema suggest that people enter informal learning contexts holding a set of prior experiences, beliefs, values, and ideas that shape their expectations and the things to which they give attention (Berlyne, 1951, 1954; Feather, 1982; Read, Vanman, & Miller, 1997). Combining these understandings with motivation and the fluidity of social roles, the complexity of individual learning begins to be revealed. An individual in different contexts with different others will be aware of and give attention to different activities, information, people, ideas, feelings, and the host of what we can learn. It is also well understood that people seek out information that supports their existing beliefs and values, and they often do not notice or attend to experiences and information that do not fit with their entrance narrative—particularly in informal learning environments (Doering & Pekakirk, 1996).

Human learning is not a linear, neat process. As Morris (2000) notes, learning is a function of being human, and personhood is a contradictory and complex phenomenon. We are constantly and continually exposed to vast amounts of stimuli and data, and learning accumulates across experiences and time. Often, because of our selective attention, the learning itself is unnoticed and the specificity of where or how we learned is diminished. This can explain why, for example, Mony (2007) found that visitors to a conservation education setting "learned" facts that were not presented in the exhibits or shared by docents during the experience. The facts came from somewhere else but made sense in this context and seemed to be part of the learning episode to those individuals. Dierking, Falk, Rennie, Anderson, and Ellenbogen (2003) remind us that learning rarely, if ever, occurs and develops from a single experience, even though many STEM education providers and policymakers assume that learning outcomes can be attributed to single or one-off experiences. Learningscapes serve as a tool for integrating the cumulative, intentional, unintentional, and continual aspects of life-wide learning mentioned above.

Social Learning

Learning is a social act. This act includes interactions with people and with biotic and abiotic elements surrounding the individual in the moment. The sociocultural perspective of learning grew from the work of Vygotsky (1962, 1981) that focused on the role of social interaction in creating the environment in which a child learns, not only from the parents but from siblings, aunts and uncles, family friends, and through the larger culture. Vygotsky saw human learning and human development as interwoven with activity—purposeful action mediated by specific tools. Learning in this frame is a social and cultural activity, not an individual phenomenon (Kozulin, 2003). The role of queer communities in sociocultural learning is in challenging heteronormed society not (only) by sexual practices, but through alternative methods of alliance, forms of gender embodiment, subcultural practices, and alternative forms of representation (Halberstam, 2005). Learning that dis-settles the dominant culture is necessarily sociocultural learning.

A more vernacular idea of social learning is central to informal learning. We know that most people visit informal learning settings for the purpose of social engagement with family or friends (Eisenberger, 1999). But sociality is not limited to interpersonal interactions. Dewey (1916/1966) noted that all communication is educative; communication theories inform us that we are in constant states of communication with our environment as well as with specific elements, such as people, within the environment. Conversation itself is not the exchange of knowledge, but is the process of individuals becoming informed about each other's ideas, perceptions, opinions, beliefs, values, and "informings" (or individual insights into understanding how one comes to believe or know something; Scott, 2001). Sharples (2005) suggests that learning is a continual conversation of an individual with the external world and its artifacts: with oneself and with others, including families of choice (Weeks, Heaphy, & Donovan, 2001), communities of identity (Taylor & Whittier, 1998), and the various aspects of identity we create socially (Copes & Williams, 2007).

Most STEM learning occurs outside the formal education system. Integrating the varied STEM-related experiences one has through the 95% of life spent outside schooling (Falk & Dierking, 2012; Feder et al., 2009) with the limited time devoted to STEM in the 5% of life spent in the formal education system demonstrates the complex structure of STEM learning in its fullness. The complex nature of how people learn STEM across lived experiences reveals some the challenges to STEM literacy in society.

It is important to restate that individual learning of STEM across the many personal and social experiences in a person's life cannot be separated within the person's mind. As in the practice of queering, one's learning of STEM cannot be isolated from other information or relegated to a particular place or time. STEM learning can happen over drinks with friends when the discussion turns to the difference between weather and climate, as well as when watching a documentary on space exploration on the Discovery Channel or attending a lecture at a science center. Such STEM learning, and even more importantly the application of this learning, is integrated into the person's life in mostly unconscious ways. Earlier in this chapter, I referred to learningscapes as comprised in great part of intentional engagement and meaning. To queer learningscapes we need to consider how to dis-settle normative desires for the intentional and incidental contexts in which we live our lives. To make STEM learning and application across one's life more visible and understood, it is necessary to discomfort how we move through life in order to be more conscious, critical, and considering of our pathways and the choices we make through them—we need to practice the queering of our learningscapes.

LGBT Learners

Luhmann (1998) notes that queer learning *is* human learning. But being an LGBT STEM learner, as opposed to queerly learning STEM as referenced above, is different from heteronormed experience. Let's consider two ways to view LGBT learners in informal STEM contexts: (a) being STEM learners and (b) challenging others' assumptions.

As individuals, we are constantly looking for an anchor for our sense of self and for personal relevance in learning. Identifying LGBT scientists, mathematicians, engineers, designers, and users of STEM is an important tool to support this search for relevance. Consider the rare exposure to the portrayal of a gay mathematician such as Alan Turing in the 2014 film *The Imitation Game.* Though it is necessary for the informal learner to see others like themselves, there may be a challenge in such visibility reinforcing the heteronormative nature of being the other (Riggs, 2004). I would posit that in informal settings, there is likely an even greater need for inclusion in science, recognizing LGBT learners as part of the learning ecosystem, and going beyond queer discourse (Toynton, 2007) to disrupt the norms of science and learning through rethinking the context in which science learning occurs.

The simple presence of LGBT people in places of informal STEM learning can challenge others' assumptions and alter the dynamics of the learning setting. Halberstam (2005) acknowledges that not all LGBT individuals live lives on the edges of society; nevertheless, larger society tends to interpret each LGBT individual through a dominant cultural view of the subcultural set. The LGBT participant in STEM learning has the potential to discomfort others in (and the content of) the learning exchange (Walker, 2009). Further, the queer participant in STEM learning may make choices because of their identity. For instance, in one study of gay and lesbian parents, I heard male couples share the trepidation of exposing their children in many public settings to comments or reactions that could be hurtful to their family. Because of these concerns, they would often decide to not attend certain events or institutions (Heimlich, 2011). Many of the female couples I spoke to reported thinking the same thing: that male couples with children would likely experience more discomfort in public settings than female couples with children. In this case, wanting to avoid discomfort leads to an outcome that is less than satisfactory for all.

Perhaps the more important queering examines the failure of science institutions, including informal STEM learning sites, to live up to the epistemic virtues of science (Lemke, 2011; Reiss, this volume). There is a need for informal STEM learning providers to push forward discourses of difference, choice, and visibility (Britzman, 1995). The act of giving voice to the other can be surprisingly powerful. Years ago, when I was evaluating the efficacy of a biodiversity curriculum, an educator shared a story about the impact on learners of one of the activities. In that activity, children were ranking the urgency of certain issues. A child shared with their group that their uncle had just been diagnosed with AIDS. The ranking of urgency for AIDS research went from the least important to the most important issue for that group of young people. Forcing the challenge from the "other" facilitates the dis-settling of the individual and can bring appropriate cognitive dissonance to change a passing event into a meaning-making, learning experience.

QUEERING STEM LEARNINGSCAPES

Let's return to the concept of learningscapes and these complex relationships around meaning constructed across, between, and among desires, social roles, social identities, experiences, choice, and attention. The concept of learningscapes suggests that individuals, in their movements through daily life, cumulatively take in sensory data, organize or process the information, and then use the learning in meaningful ways. I argue that across the many aspects of one's identity tied to context and enacted social roles there is a cumulative experience and growing body of held knowledge, and thus the opportunity to synthesize what is learned. (See, for example, Knowles's, 1980, discussion of the characteristics of adult learners.) Whether at a botanical garden on a date, a science center with kids, a natural history museum in a professional role, or in any of the myriad contexts and roles for STEM learning, what an individual gleans from each experience—shaped as it is by the role and perspective of self they took into the context—can begin to be constructed into a body of insight and affect.

As suggested above, queering a learningscape is an aspirational, intentional practice in which an individual critically reflects on how they move through life (in both context and social role), and how they construct meaning among and within the experiences that combine to be their life. Learningscapes provide a framework for the application of queer theoretical lenses to consider the STEM learning of an individual, as there is in queer theory neither fixed content, universal guidelines, nor a stable and singular body of knowledge (Britzman, 1995). Yet queer theories are about uncovering and creatively resisting what are considered normative desires as they are entangled with all the issues associated with sex(ualities) and gendering. Consider the connections across a person's many STEM-related (or potentially related) experiences and among the different aspects of self. These connections can be explored to begin to make sense of how it is one came to acquire and construct what one desires, knows, feels, thinks, believes, and does in the world.

Queering learningscapes involves a kind of mapping. The purpose of a learningscape map is to conceptually (or physically) facilitate examination of the potential meaning of the structures and patterns of engagement across a person's learning through life. Further, the use of a map itself, or the process of trying to create a map of this abstract learningscape, is aligned with queer geography's contributions to "a thoroughgoing interrogation of Cartesian rationality, Euclidean spatial ontologies, and the often norming fixity inherent in cartographic representation" (Brown & Knopp, 2008, p. 40). Mapping the learningscape involves having the individual first consider the many different selves they put forward, the animating desires of those selves, and the experiences in which they engage in each of these roles. They then look across all activities in which they engage through formal and informal systems—the enterprises in the learning ecosystem—which include the spiritual, cultural, entertainment, leisure, physical, avocational, work, and other regular or planned activities of their uniquely constructed life. Conceptually, considering how these activities are clustered and connected, it is possible to scan across many activities and social roles to begin to make connections and meaning. Such a queering of learningscapes would shift the concept from a map of what is, to a map of how normative desires are enmeshed with all the elements of the learningscape: the social roles, the context, the identities, the institutions, the exchanges.

Indeed, this also includes questioning relationships and roles. For example, how do I integrate, or at least hold in tension, the symphony, a work trip to a national park, a solitary walk through a nature center, and the roles I play in each as friend and partner, as professional and colleague, and as a sometimes-moody extrovert who now and then needs to center himself in nature? And as a professional avidly dedicated to the environment and to STEM learning in society, I am deeply concerned with how individuals move through these many episodes and perhaps connect some of the pieces into a meaningful gestalt. Learningscapes can be used to cultivate awareness within an individual of the model by which they are learning, how the learning relates to their prior knowledge, and how that knowledge may be egocentric, protective, compliant, or dynamic (Johnson, 1999, as cited in Conlon, 2004). The nature of the learningscape is such that it should be queered. Not only need we be aware of the many connections between and among our myriad daily experiences, but we should examine how normative and normed desires (perhaps even those learned through routines adopted as an adult) help form how we integrate the pervasive elements of STEM in our daily lives.

It can be dis-settling to one's learningscape to refuse to hold informal learning as the *other* in opposition to formal learning. One struggle within the field of informal STEM learning has been how to claim an identity *other than* not-school (e.g., Falk & Dierking, 1998; Falk, Heimlich, & Foutz, 2009; Marsick, 2009). Learning is attached across all aspects of the identity of an individual and does not reside solely within one fixed mode (such as schooling)—ideally there should be no "other" in relation to which school learning is normal. But in the present reality of how people think about learning and school, informal STEM learning is and should be the queer of STEM learning.

From the Inside

The goal of mapping a learningscape is to shift learning in all its forms from unconscious to conscious. In the context of STEM learning, this would encourage the individual to confront incongruent beliefs and values, integrate cognitive with affective engagement in STEM, and be more intentional in approaching and participating in STEM as they move through life. Queering learningscapes is the intentional dis-settling of the unconscious and bringing the self as a conscious, desiring learner through life to the experiences comprising the learningscape.

Felman (1987, as cited in Luhmann, 1998) suggests that education is the creation of a new condition of knowledge or an original learning disposition. This ties closely to individuals' meaning construction across incidents in their learningscapes, assuming the meaning is critical and challenged. It is the *critical and challenged* aspect of the learning that is problematic. How do we facilitate STEM experiences that lead learners to be truly critical thinkers, starting with critical reflections on the biases and predilections learners bring to the learning exchange? Critical thinking has, for many individuals, become reduced to asking questions of others rather than the important work of asking about the self (Brookfield, 1987). Intentional queering of individual learningscapes by institutions and organizations requires changing how STEM content is presented so that it compels an individual to internalize and question their own desires, assumptions, entry points, prior experiences, values, positions, and beliefs. It then demands that the providers of informal STEM learning rethink not only *how* STEM is communicated, but also *why* certain content is selected and the purpose for the institution and the learner of offering that content. Such reflection by learning providers could facilitate exchanges that better draw learners into critical consideration of content in relationship to their lives, helping learners draw meaning from the many other experiences they have lived.

What might such reflection look like? One winter morning while working on this chapter, the temperature was unseasonably warm for where I live. Lying in bed in the predawn morning, I was suddenly aware of the lack of bird songs. I realized that through much of the year, I rely on those sounds to gently help me awaken. This insight led me to consider what I have learned about migration patterns of birds, understandings I have about climate change, thoughts of friends who are avid birders, and photographs and then art of birds that migrate through my city. Across these connections, I began formulating questions for which I don't have answers. In this moment, I was making explicit my own learningscape and what I desired or needed in that moment. Further, such exploration led to a realization of the gaps in my knowledge and life experiences leading to that moment. In turn, this reflection led me to question why I had not explored those things—what norms had driven my safe, predictable learnings.

How would your thinking about any activity change if you were asked to bring other moments, insights, questions, and more into that moment? Consider your response during a visit to a science center if you are asked to describe other places or experiences different from those you are currently enjoying but that evoke similar insights or emotions. You might be asked when else in your life you encountered something like this and how that moment relates to the present. Or while taking in an amazing natural viewscape, you and those around you are asked to share what music the view brings to mind. Queering the learningscape means pushing beyond any one experience as if it is an isolated event to instead seek to integrate it in the rest of the life (or lives) of an individual.

Queering the STEM learningscape requires discourses that "touch us, move us and leave us unsettled, troubled, confused" (Giffney, 2009, p. 10). In formal schooling and in most informal learning, educators are often afraid of unsettled, troubled, and confused outcomes. It is seen as more important for learners to know predetermined, fixed content than it is to challenge them to question themselves and their providers. Leaving STEM learningscapes unqueered allows people to choose to remain unaware of, or not to challenge, meanings across identities and experiences. An unqueered learningscape does not encourage individuals to confront prior and dissimilar learnings; it fails to push them outside the comfort of absolutes where science is a body of knowledge and *knowing* means looking for answers online; it does not challenge people to deeply question how information, technology, and knowledge are applied within and across their lives. And this is, as Luhmann (1998) calls it, ignorance—not the opposite of knowledge, but a cultivated and in some sense knowing opposition to knowing.

We can consider human learning as the combination of processes whereby the whole person experiences a social situation, the perceived content of which is then transformed cognitively, emotively, and/or actively and integrated into the person's individual biography resulting in a changed person (Jarvis, 2006). Transformative learning is not a result of seeking transformation but an openness to being transformed through what we desire, do, learn, experience, explore, seek, and find. Such a view of human learning transcends context. It suggests that informal learning is actually the realization that learning happens and that we can shape how learning occurs across the life span. But this important view of human learning must embrace the complexity of identity and gender to account for the place of learning, and especially STEM learning, within the social and world views of the larger community.

From the Outside

Power relationships are not hypothetical—they are evident in concrete ways in the real world (Walker, 2009). As with any element of identity, we know that being lesbian, gay, bisexual, trans, or queer influences how an individual engages in society. How safe, welcome, and included does one feel in science centers, nature centers, zoos, or whatever? Am I as a gay man, lesbian, transgender, and/or queer individual discomforted by being in the setting, and does my simple presence discomfort others? How interesting that, in this moment, ambiguous power relations allow the other, by their mere presence, to queer and discomfort heteronormative spaces. This discomfort rests upon a binary power relationship of normal/not normal, that privileges the hetero perspective—a clearly unequal arrangement that reflects what Britzman (1995) called normalcy's immanent exclusions. Pedagogy itself, in any context, is a part of this relationship. In informal settings, as well as in schools, the heteronormed language and images in signs, membership applications, photographs, and marketing materials often reflect binary power relations that position heterosexuality as the norm (Pinar, 1998). Luhmann (1998) also points to alienating and discriminatory experiences within formal education systems. I argue that such discriminatory experiences exist throughout, across, and within the many connections in every individual's learningscape.

BUILDING A STEM LEARNINGSCAPE INFRASTRUCTURE

An individual's learningscape resides within a power discourse—therefore we must disrupt, as we can, the status quo so that individuals can move themselves wholly through all the places and ways in which they encounter, use, and learn STEM across their lives.

How do we discomfort our STEM institutions to help people question what is thinkable? Many informal science institutions play at the edges of the thinkable: problem solving, out-of-the-box thinking, invention, making or tinkering, and innovation, for example. But they stop at that edge and rarely venture into really questioning that which is considered (un)thinkable because they are, I posit, teaching science which is knowable. This ignores the very essence of the scientific process: looking for what is (currently) unknown and sometimes what is, at that point in time, seemingly unknowable. Informal STEM learning experiences that obscure questioning in and of STEM risk homogenizing the diversity of scientific meanings and practices.

For me, the question is not *can* or *should* we queer STEM learningscapes, but rather *how* we queer the many points and contexts at which STEM is shared and taught. Others have and continue to challenge formal education and how this system might be queered. In moving to consideration of learningscapes, it is necessary to look at the schooling as but one piece of an individual's learning and meaning-making pathways. Considering an individual's engagement in STEM at this or that particular point in space and time may serve a provider's perspective, but it does not honor the wider fullness of a learner's life. Rethinking how we convey STEM requires contextualizing STEM learning in the social role and present identity of the individual to help connect the STEM we are teaching to other experiences in that individual's life, including the unlikely connections they may need to bridge for themselves.

Perhaps the most difficult step in dealing with individuals' learningscapes is to radically understand informal STEM learning in relation to its schooling cousin. Considering how and what people need to know in their lives and uncovering and critically working with personal and cultural desires in individuals' engagement with STEM content *is* queering the learningscape. Rather than focusing on STEM content, focusing on the highest levels of critical thinking around the STEM content will be an important change that builds on the movement for incorporating critical thinking in formal and informal learning. Queering the learningscape is not incorporating critical thinking, it is mapping one's journey while embodying critical thinking, casuistic behavior (i.e., deep reflection on one's knowledge and its relationship to actions), and conative understanding. What makes this especially uncomfortable is that STEM providers must move themselves through critical examination of what they do, why they do it, and what they must change in their own practices.

Ultimately, we must queer the STEM informal learning infrastructure to challenge the status quo of STEM in lifelong, life-wide, and life-deep learning in people's lives and learningscapes. This is necessary if for no other reason than to address the dearth of STEM literacy. Indeed, such work might push us beyond obsessions with how much STEM people know and how much more content we can throw at them and into a process for increasing critical science literacy in people's lives. The best reason to queer learningscapes is to enrich and enliven people through STEM learning, awareness, critical consideration, and application *in* their lives.

NOTES

1. There are various definitions and taxonomies for nonschool learning, including the terms nonformal, incidental, everyday, free-choice, and informal. This chapter focuses on *informal learning*, where choice for engagement and choice for what is taken away from a learning episode resides in the individual.
2. A colleague and I once taught a graduate seminar where we spent an afternoon observing an early-childhood education experience and then used that to create finger-paint models of the teaching/learning exchange. The experience was rich. But the following week, the cognitive, adult/professional, role-driven perspective of the graduate students was that the experience was infantile. As we processed

this, the seminarians realized that preconceived and normed assumptions drove their reflections (we watched *kids*? and we did *finger painting*?), and these assumptions distanced their actual learning from their perceptions of the value of the activity as a learning experience.

3 I subscribe to the belief that experiences in one's life are not stable in memory or in their contribution to whom one is, but change as the individual changes—taking on new meaning, being reinvented in recall, and being charged with changing emotions based on the stimulus engaging the recollection.

REFERENCES

Ardoin, N., & Heimlich, J. (2015, October 14). *The importance of the incidental: Environmental learning in everyday life*. Paper presented at the North American Association for Environmental Education Annual Conference, San Diego, CA.

Banks, J. A., Au, K. H., Ball, A. F., Bell, P., & Gordon, E. W. (2007). *Learning in and out of school in diverse environments: Life-long, life-wide, life-deep*. Seattle, WA: University of Washington, Center for Multicultural Education.

Berlyne, D. E. (1951). Attention, perception and behavior theory. *Psychological Review, 58*(2), 137.

Berlyne, D. E. (1954). A theory of human curiosity. *British Journal of Psychology. General Section, 45*(3), 180–191.

Biddle, B. J. (1986). Recent development in role theory. *Annual Review of Sociology, 12*, 67–92.

Britzman, D. P. (1995). Is there a queer pedagogy? Or, stop reading straight. *Educational Theory, 45*(2), 151–165.

Brookfield, S. (1987). *Developing critical thinkers*. Milton Keynes: Open University Press.

Brown, M., & Knopp, L. (2008). Queering the map: The productive tensions of colliding epistemologies. *Annals of the Association of American Geographers, 98*(1), 40–58.

Coffield, F. (2000). *The necessity of informal learning*. Bristol: The Policy Press.

Conlon, T. J. (2004). A review of informal learning literature, theory and implications for practice in developing global professional competence. *Journal of European Industrial Training, 28*, 283–295.

Copes, H., & Williams, J. P. (2007). Techniques of affirmation: Deviant behavior, moral commitment, and subcultural identity. *Deviant Behavior, 28*(3), 247–272.

Dewey, J. (1916/1966). *Democracy and education: An introduction to the philosophy of education*. New York, NY: Free Press.

Dierking, L. D., Falk, J. H., Rennie, L., Anderson, D., & Ellenbogen, K. (2003). Policy statement of the "informal science education" ad hoc committee. *Journal of Research in Science Teaching, 40*, 108–111.

Dinshaw, C., Edelman, L., Ferguson, R. A., Freccero, C., Freeman, E., Halberstam, J., … Nguyen, T. H. (2007). Theorizing queer temporalities: A roundtable discussion. *GLQ: A Journal of Lesbian and Gay Studies, 13*(2), 177–195.

Doering, Z. D., & Pekakirk, A. J. (1996). Questioning the entrance narrative. *Journal of Museum Education, 21*(3), 20–22.

Eisenberger, R. (1999). The museum goer's motives: The social and the sublime. *Visitor Studies Today, 2*(3), 1–5.

Falk, J. H., & Dierking, L. D. (1998). Free-choice learning: An alternative term to informal learning. *Informal Learning Environments Research Newsletter, 2*(1), 2.

Falk, J. H., & Dierking, L. D. (2012). The 95 percent solution: School is not where most Americans learn most of their science. *American Scientist, 98*(6), 486–493.

Falk, J. H., Heimlich, J. E., & Foutz, S. (Eds.). (2009). *Free-choice learning and the environment*. Lanham, MD: Altamira.

Falk, J. H., Reinhard, E. M., Vernon, C., Bronnenkant, K., Heimlich, J. E., & Deans, N. L. (2007). *Why zoos & aquariums matter: Assessing the impact of a visit to a zoo or aquarium*. Silver Spring, MD: Association of Zoos & Aquariums.

Feather, N. T. (1982). *Expectations and actions: Expectancy-value models in psychology*. Mahwah, NJ: Lawrence Erlbaum Associates.

Feder, M. A., Shouse, A. W., Lewenstein, B., & Bell, P. (Eds.). (2009). *Learning science in informal environments: People, places, and pursuits.* Washington, DC: National Academies Press.
Feinstein, N. (2011). Salvaging science literacy. *Science Education, 95*(1), 168–185.
Giffney, N. (2009). Introduction: The "q" word. In N. Giffney & M. O'Rourke (Eds.), *The Ashgate research companion to queer theory* (pp. 1–13). Farnham: Ashgate.
Hager, P. J., & Halliday, J. (2007). *Recovering informal learning: Wisdom, judgement and community* (Vol. 7). Dordrecht: Springer.
Halberstam, J. (2005). *In a queer time and place: Transgender bodies, subcultural lives.* New York, NY: NYU Press.
Heimlich, J. E. (2011). *Parent perception study* (Unpublished internal pilot study). Columbus, OH: COSI.
Heimlich, J. E., & Storksdieck, M. (2007). Changing thinking about learning for a changing world. *Southern African Journal of Environmental Education, 24*, 63–75.
Heyes, C. J. (2003). Feminist solidarity after queer theory: The case of transgender. *Signs, 28*(4), 1093–1120.
Jarvis, P. (2006). *Toward a comprehensive theory of human learning: Lifelong learning and the learning society* (Vol. 1). New York, NY: Routledge.
Knowles, M. S. (1980). *The modern practice of adult education: From pedagogy to andragogy.* Englewood Cliffs, NJ: Prentice Hall/Cambridge.
Kozulin, A. (2003). *Vygotsky's educational theory in cultural context.* Cambridge: Cambridge University Press.
Lemke, J. (2011). The secret identity of science education: Masculine and politically conservative? *Cultural Studies of Science Education, 6*(2), 287–292.
Liu, X. (2009). Beyond science literacy: Science and the public. *International Journal of Environmental and Science Education, 4*(3), 301–311.
Livingstone, D. W. (1999, April). *Exploring the icebergs of adult learning: Findings of the first Canadian survey of informal learning practices.* Paper presented at the meeting of the American Educational Research Association, Montreal, Canada.
Livingstone, D. W. (2001). *Adults' informal learning: Definitions, findings, gaps, and future research* (WALL working paper No. 21). Toronto: Ontario Institute for Studies in Education, University of Toronto. Retrieved from https://tspace.library.utoronto.ca/retrieve/4484/21adultsinformallearning.pdf
Luhmann, S. (1998). Queering/querying pedagogy? Or, pedagogy is a pretty queer thing. In W. Pinar (Ed.), *Queer theory in education* (pp. 141–155). Mahwah, NJ: Lawrence Erlbaum Associates.
Malcolm, J., Hodkinson, P., & Colley, H. (2003). The interrelationships between formal and informal learning. *Journal of Workplace Learning, 15*, 313–318.
Marsick, V. J. (2009). Toward a unifying framework to support informal learning theory, research and practice. *Journal of Workplace Learning, 21*(4), 265–275.
Martin, L. M. W. (2004). An emerging research framework for studying informal learning and schools. *Science Education, 20*, 283–297.
Miller, S. (2001). Public understanding of science at the crossroads. *Public Understanding of Science, 10*(1), 115–120.
Mony, R. S. P. (2007). *An exploratory study of docents as a channel for institutional messages at free-choice conservation education settings* (Doctoral dissertation). Retrieved from https://etd.ohiolink.edu/
Morris, M. (2000). Dante's left foot kicks queer theory into gear. In S. Talburt & S. Steinberg (Eds.), *Thinking queer: Sexuality, culture and education* (pp. 16–32). New York, NY: Peter Lang.
Pinar, W. F. (1998). Understanding curriculum as gender text: Notes on reproduction, resistance, and male-male relations. In W. F. Pinar (Ed.), *Queer theory in education* (pp. 221–243). Mahwah, NJ: Lawrence Erlbaum Associates.
Read, S. J., Vanman, E. J., & Miller, L. C. (1997). Connectionism, parallel constraint satisfaction processes, and gestalt principles: (Re)introducing cognitive dynamics to social psychology. *Personality and Social Psychology Review, 1*(1), 26–53.
Riggs, D. (2004). Resisting heterosexism in foster carer training: Valuing queer approaches to adult learning and relationality. *Canadian Online Journal of Queer Studies in Education, 1*(1). Retrieved from https://journals.scholarsportal.info/pdf/17107598/v01i0001/5_rhifctatalar.xml

Scott, B. (2001). Gordon Pask's conversation theory: A domain independent constructivist model of human knowing. *Foundations of Science, 6*(4), 343–360.
Scribner, S., & Cole, M. (1973). Cognitive consequences of formal and informal education. *Science, 182*(4112), 553–559. doi:10.1126/science.182.4112.553
Sharples, M. (2005, April). Learning as conversation: Transforming education in the mobile age. In *Proceedings of Conference on Seeing, Understanding, Learning in the Mobile Age* (pp. 147–152). Institute for Philosophical Research of the Hungarian Academy of Sciences, Budapest. Retrieved from https://www.researchgate.net/profile/Mike_Sharples/publication/228389244_Learning_as_conversation_Transforming_education_in_the_mobile_age/links/02e7e5277cfa803074000000.pdf
Stocklmayer, S. M., Rennie, L. J., & Gilbert, J. K. (2010). The roles of the formal and informal sectors in the provision of effective science education. *Studies in Science Education, 46*(1), 1–44.
Strauss, C. (1984). Beyond "formal" versus "informal" education: Uses of psychological theory in anthropological research. *Ethos, 12*(3), 195–222.
Sturgis, P., & Allum, N. (2004). Science in society: Re-evaluating the deficit model of public attitudes. *Public Understanding of Science, 13*(1), 55–74.
Taylor, V., & Whittier, N. (1998). Collective identity in social movement communities: Lesbian feminist mobilization. In P. N. Nardi & B. E. Schneider (Eds.), *Social perspectives in lesbian and gay studies: A reader* (pp. 349–365). New York, NY: Routledge.
Tough, A. (1978). Major learning efforts: Recent research and future directions. *Adult Education Quarterly, 28*(4), 250–263.
Toynton, R. (2007). The de-representation of science and queer science students in higher education within the queer/gay discourse. *Teaching in Higher Education, 12*(5–6), 593–605.
Uden, L., Wangsa, I. T., & Damiani, E. (2007, February). *The future of E-learning: E-learning ecosystem.* Paper presented at the IEEE-IES Digital EcoSystems and Technologies Conference, Cairns, Australia. Retrieved from https://www.researchgate.net/publication/4253636_The_future_of_E-learning_E-Learning_ecosystem
Vygotsky, L. S. (1962). *Thought and language.* Cambridge, MA: Harvard University Press.
Vygotsky, L. S. (1981). *Mind and society.* Cambridge, MA: Harvard University Press.
Walker, W. (2009). Adult learning in the queer nation: A Foucauldian analysis of educational strategies for social change. *New Horizons in Adult Education and Human Resource Development, 23*(3), 10–17.
Weeks, J., Heaphy, B., & Donovan, C. (2001). *Same sex intimacies: Families of choice and other life experiments.* New York, NY: Psychology Press.

ABOUT THE AUTHOR

Joe E. Heimlich, PhD, is director of research for the Center of Science and Industry (COSI) in Columbus, Ohio (USA), co-director for COSI's Center for Research and Evaluation, and a researcher with the Center's Lifelong Learning Group. Joe's work focuses on informal learning in and capacity-building for zoos, nature centers, science centers, and other museums. He is professor emeritus with The Ohio State University, where he was Extension leader in environmental science, and specialist in museums and organizational capacity building. He also held appointments in Ohio State's School of Environment and Natural Resources, the Environmental Science Graduate Program, and the College of Education and Human Ecology.

TOMMYE HUTSON

8. WHAT'S IN A NAME?

Reflections on Learning and Teaching in Central Texas

For as long as I can remember, I've always been keenly aware of and sensitive to the power of naming things and places—and most especially, people.

A GIRL NAMED TOMMYE

I've been a science teacher for over 30 years and, I now realize, tangled up in queering my whole life. To start with, my parents were a very young Air Force couple in the early 1950s. As such, they never dreamed they would produce anything other than a firstborn son. Understanding that there were to be no daughters, the name they had agreed upon before my arrival was a combination of their middle names, Thomas and Lynn. Hearing, "Sargent—you have a girl," I am quite sure caused my father to, as a friend and colleague recently suggested, "Lose his shit." I was in my 40s before I discovered that my parents had also selected a girl's name, but only as a joke. And obviously, it was discarded.

Country-western music was the only kind allowed in our home, and, at one point, I thoroughly identified with Johnny Cash's hit song, "A Boy Named Sue." Even as a child, I had had my share of disputes over my real name in schools and other places around the world. When I asked my dad the reason for my name, he assured me that he thought it quite clever to give me a name that was unique—it would cause "them" to consider my identity. I wasn't quite sure who "them" was. The "them" that I ran into weren't always willing or able to engage in the mental gymnastics it might take to look past the name. Although I identified heavily with the Sue in the song, I just hoped it wouldn't result in too many bar fights. Here's some of what I did run into:

- There was a time (long before medical privacy laws) when you could call a doctor's office for results from routine tests and procedures. Women, for example, were told to call their gynecologist's office for PAP test results. In the late 1970s, a couple of weeks after a PAP test, I called for my results and was told that they would have to talk to the patient—to "him." I was not a new patient, but the person I was speaking with was new to the office. I told her I was not aware that my gynecologist performed PAP tests on men but, regardless, to please check the patient information more closely, as I was indeed "him."

 | DOI:10.1163/9789004331068_010

- When applying for a high school teaching job, I was told that my credentials, references, and classroom experience were all exemplary and quite badly needed. Too bad, though, that I wasn't a man. They had assumed, in fact, that's what I was until I arrived for the interview. The campus principal—a man, and accompanied by three other male assistant principals—asked why my resume didn't indicate I was a woman. "What are you trying to pull?" he asked. I asked him why he had assumed I was a man. I got the job but soon regretted it. He never let me forget the favor he'd done by allowing me to teach on his campus.
- Shortly after September 11, 2001, a young administrator verifying background information on school district employment records called me in to provide my legal name with valid documents, as nicknames were not acceptable for official records. I told her that my name was indeed listed correctly, but she insisted that wasn't possible. With a Spanish-language birth certificate (from Puerto Rico), teaching certificate, college transcripts, state driver's license, and Social Security card in hand, I assured her that my name was indeed legal, correct and, further, that I was eternally grateful my parents had been kind enough not to name me after a piece of dried-up bread. Her name was Melba.

The truth is, I don't know why my parents called me Tommye. At one point, my mother said it was because, when I was several hours old, my father declared that I didn't look like anything else. I didn't know what to do with that answer. But here are some things that I feel are true. My father was genuinely disappointed and remained so for the rest of his life. I was never sure how my mother felt about much of anything. She was from that generation of very supportive women who made the lives of professional soldiers as comfortable as possible. She was alone, suffering, and silent much of the time. After my arrival, there were three more children—all in possession of the "correct" external plumbing. My dad ended up with one daughter and three sons. It's also worth noting, not one of those boys got named Sue or anything even remotely gender neutral.

CEDAR HACKERS

Several years ago, I heard then U.S. Senator Barack Obama say something about eighth-grade graduation. He thought it was premature to mark the passage from middle to high school with pomp-and-circumstance-style celebrations when the real work of completing college and competing for a good job was still ahead. After hearing that, I wished I could introduce him to Clay, a young man from my second year of teaching when I was the middle and high school science teacher in a small school in rural central Texas. I was Clay's middle school science teacher, and eighth-grade graduation was all he ever got.

It was 1987, and Clay was a wonderful young man. He worked hard and was a straight-A student. He was the kind of student who makes you a better teacher because he would constantly raise questions that occurred to him. He was truly curious about, well, *everything* to do with science. He used to come to my classroom after school before the bus left to ask me questions he had had that day. And he would bound off the bus and come to my room before classes started to ask me about something he'd read the night before. He read with the same intensity that cedar trees burn—voraciously and tenaciously. Every weekend during the summer, his mother took him to town where he checked out the maximum number of library books, only to finish them before the week was over. He pestered her to go back to town before the weekend, but they couldn't afford two trips in one week. He was, in a word, precious.

Clay's twin sister, Carrie, was quiet and not nearly so driven to study or conversation. I taught them both in seventh- and eighth-grade science. She copied his homework and sat quietly in the back of the room while he and I chatted before and after school. She was his opposite and his shadow. They were poor. Their daddy and all the men in their extended family were day laborers who worked for farmers clearing cedar trees from marginal, hay-producing land—they were cedar hackers.

After spring break of their eighth-grade year, Clay was a different person. He had lost his spark. He stopped reading. He didn't stop by to see me before or after school. He wasn't interacting with his classmates. He just stopped talking, period. After a few days, I thought he might be sick and asked his sister what was wrong. She said, "He graduates in May and is going to work this summer." Misunderstanding her answer, I said, "But, he'll be back in the fall." "No, Miss," she said, "he's graduating. He's finished with school." Their father had explained this to Clay just before he returned to school from spring break. As a "man" now, Clay was expected to start earning for the family. He had two older brothers who had not done so well in school. Clay had hoped his father would make an exception for him. But he was needed to work. Carrie evidently felt like she had to justify the family tradition. This was normal in her family's world—after the eighth grade, boys went to work and girls went on to high school.

At the afternoon graduation, Clay wore an overly starched, short-sleeved white dress shirt that was too big for him and a clip-on tie. His sister was wearing … I don't remember what. My eyes were on Clay. There he sat on stage with his class, a 13-year-old child crying quietly while his peers and everyone else were reminded that next fall they would begin their best years yet. His life was, in his estimation, over. He was out of time. It was his turn to accept the role his family and society had assigned him as part of a necessary but marginalized population—people who were used up for their less-skilled functions making agriculture possible on hostile land. And if he and his folks couldn't come and cut cedar, there were always other strong backs to hire, like the "Messikuns." What else could he do? He was in no position to buck the system into which he was born. Clay was a kind young man; hopefully, that kindness was so deeply rooted that it was not destroyed by the life that lay ahead. He would be close to 45 years old now. If he stayed with his family past age 18 and followed what were considered acceptable norms, he would be married and have kids. If I could wish anything for him (as I could change nothing), I would hope he had children with whom to share his love of reading for sheer pleasure and discovery—daughters and, particularly, sons who were all privileged to continue school beyond eighth-grade graduation.

THE BIRTH OF ZELDA

This is 42-year-old John, my hairdresser, describing his high school principal—who was my boss at the time:

> He used to hit me in the back of the head as he walked by, calling me a "disgusting fag." Oh, he was a great role model! Turned out, five years later, his son, the basketball player, came out. I felt sorry for his son. You know the principal had that boy committed, right? Yep—wanted the boy "fixed." But I bet that boy's happy now, though. He lives with a professor, teaches, and coaches girls' basketball. He's been real successful—got district championships a couple of times.

I love John. He's a gentle giant who is a magician when it comes to hair, calling the magic he works "stress highlights." I don't truly care how he does it, but somehow the gray isn't as noticeable as it grows out. He's always delighted to see me and to talk about anything and everything. I know all about John's miserable experience at high school, his failed attempts at working other "more appropriate" jobs (his words, complete with fingers making air quotes), and his being lucky enough to have a devoted partner and the blessing of parents who embraced him literally and figuratively as he navigated his adult path to "become whole" (again, his words). He and his partner live in his grandmother's house, across the backyard from his childhood home. Like me, they have several shih-tzus—we kid that every other breed is "just posing." John and his partner have a relationship that has been thoroughly tested by moving from their social family in Dallas back to John's (less than accepting) smaller hometown to deal with the needs of aging parents, intermittent loss of jobs and income, injuries, and their own aging and illnesses. Their care and compassion for each other and their extended circle of friends are at the heart of lasting commitments. And, not wanting to admit envy, I wish I was as grounded and as lucky.

I met John through his sister, an elementary teacher with whom I worked. John arrived in my life at the end of another exhaustive search for a hairdresser. I was just about ready to shave my head and had already invested in wigs—seriously. As I have moved with my career from state to state over the years, I have found that it isn't easy to just find somebody who can do my hair. Let me start by saying I'm not that picky. I was never allowed to have long hair; early on I got used to short blow-and-go styles. I used to be easy to please and didn't take much time in the chair. No one needed to fuss over me and my hair—until I became conscious of my own gray hair. Now I understand that covering up the gray takes a hairdresser's time, more than it takes if I do it myself at home.

Hairdressers, as I'm sure everyone knows, serve lots of roles: like teachers, they are often counselors, mediators, and pseudo-parents. Good hairdressers engage customers in conversation to create a friendly, embracing space where life is a bit more relaxed. At least that's why my mother said she went to the beauty shop. But that has not been my general experience. More than once, I have called a new establishment to make an appointment and given my name to the person answering the phone. "Ask him if 2:00 would be OK." Then I have to go through the "my dad had a weird sense of humor" explanation, or just say, "It's OK—I'll let him know." Either way, it's the same old song and dance—"My name is Sue. How do you do?" As a new customer, saying I taught high school science was met with stony silence, or "Wow, you must be really smart," or "Gee, I wish I'd paid more attention to chemistry when I was in high school," or the ever-famous, "Gee, how is being so smart as a woman gonna help you when Jesus comes back?" I hate to say it, but these comments always came from female hairdressers, never from men.

Such utterances told me a lot about what those women saw as normal, perhaps immutable, about learning for girls. In their homes, the pervasive culture of school intertwined with the Bible belt to discourage exploring, or even dreaming about, things other than "what normal girls do": homemaking, cheerleading, and hairdressing. Fear of being an outlier is often stronger than innate curiosity and the desire for something more. By assuming roles as working women that were safe, or at least acceptable, they were secure and securely trapped in a system that, like Clay's, guaranteed them little chance of change. From that perspective, their comments stung less and caused me to surely wonder what their daughters were being taught. I wondered if their daughters might happen to be in my classroom. Could I possibly offer a different perspective that might change their trajectories? I couldn't even be sure they would be interested in *having* options.

These days I'm not willing to risk another bad hairdo because someone is hurrying to get me out of her chair. One poor soul got so flustered at the prospect of working on a chemistry teacher's hair that my color, normally a dark blonde, was black as pitch. I looked like I'd decided to go goth—except I'd have been about a decade ahead of the trend. That hairdressing disaster ushered in the year of the wig. After about a year, Zelda was born. I made an appointment for Zelda, a kindergarten teacher who needed a very short haircut. Being Zelda made life more comfortable for everyone involved. Zelda's birth gave me permission to allow someone a bit lighter and carefree out into the world. It was several years and a couple of moves later that I found John. With John, I get to be myself—which on any given day could be Tommye or Zelda. John really likes Zelda better. They hug each other when they meet. And I'm always happy with my hair.

TEACHING K–12 SCIENCE

> *Texas Essential Knowledge and Skills for Science: Subchapter A. Elementary.* §112.11. (b)(2) Scientific investigation and reasoning. The [kindergarten] student develops abilities to ask questions and seek answers in classroom and outdoor investigations. The student is expected to: (A) ask questions about

> organisms, objects, and events observed in the natural world. (Texas Education Agency, 2007–2015)

I have to admit, for much of my career I never thought much about the role that state-mandated curriculum plays in what teachers do when they are behind closed classroom doors. I owe that to a certain degree of competence and self-assurance in my teaching ability, perhaps because I had an advanced degree in science before I sought teaching certification. One of my education methods professors told me, "You know the science, and you know what kids need to know in their everyday lives—just teach it." I took that advice as permission to ignore what were, at the time, curriculum *suggestions*. But with time and great attention to detail, folks representing various political agendas were appointed to the Texas State Board of Education and placed in charge of developing the state curriculum. Over a period of years, the work of various Boards yielded the now massive Texas Essential Knowledge and Skills documents that serve as the state's public school curriculum.

It's well documented that the Board members who produced these documents fall into several categories. Some of them hold to notions about students that are truly out of touch with current classrooms and the students in them. Others have never attended public schools nor did their children attend public schools—yet they draft curriculum for public schools and their decidedly "less-privileged" students. The final group are those without children at all. They have no direct knowledge of children and rely on what they remember, what they have heard, and perhaps what they choose to watch on TV (and it's probably not *Modern Family*). As an example, consider the kindergarten science standard excerpted above that says students are expected to learn how to ask questions. Let's think about a room full of garden-variety 5- to 6-year-olds. Do you know, have you ever known, or were you ever, a kindergarten student? How many of them are less than curious about their world and how it works? A 5- or 6-year-old person doesn't do anything but ask questions! They can't help it. For most of them, the world is still a pretty interesting place to be. Now, imagine a room with about 22 individuals who do nothing but ask questions. Kindergarten, I would submit, is all about teaching children to take turns and raise their hands *before* asking questions. Kindergarten doesn't teach children to ask questions; its goal is to contain and control students and their questions.

And how exactly would you measure and hold teachers accountable for this standard? What if you, as a teacher of kindergartners, realized that a singular child never asked a question or had any interest in going outside to explore? Would you have failed in your teaching? Does attempting to assess—to quantify—an obviously qualitative standard erase the very essence of that standard? To my way of thinking, not everything worthwhile is quantifiable, nor should it be. If it can't be measured, does it then need a rubric? Can an assessment instrument be developed that could determine such a competency?

If you truly want high-quality science questions, hand a group of kindergartners magnifying glasses and a box of rocks—it's that simple. And then stand back. They will generate questions—probably more than most teachers are able to answer. Being able to answer questions is important, but not nearly as important as the ability to accept that you may have to live with being unable to answer them all, and, further, to realize that the answers you offer may generate even deeper questions that may exhaust your limits as a teacher on any given day. If a kindergartener asks about the reason for the differences between rocks, an acceptable response would be qualitatively different than it would be for an eighth-grade earth science student. Providing the best answer for each student in a way that addresses their cognitive capacity is a daunting task. More than once, I've seen kindergarten teachers answer their students with, "because God made it that way." Some elementary teachers present such garbage for answers because science has (in my opinion) been given short shrift when reading and math are so heavily emphasized. In middle school, teacher preparation is so variable that ensuring highly qualified teachers is, for all practical purposes, impossible. I've seen eighth-grade teachers point their students to a textbook and tell them, "Find the answer yourself, that way you'll remember it." In each case, an opportunity to seriously investigate interesting questions and engage in science talk is squandered. Teachers don't have to know everything! That is something I strive to make sure preservice teacher candidates understand: You can enjoy the process of discovery with your students.

At the end of the day, much of the touted standards-based science platform deals poorly with children's thinking and learning. It seeks to teach children *how* to think and, by fiat, *what* to think. Should that be the goal of a curriculum that claims to value investigation and reasoning? Perhaps a better goal is to help children develop questions that science can adequately address (with better answers than "because God made it that way"). We can help students discover that natural processes could be responsible for what we see. Children need to understand the value of good questions and how to develop questions that are worth answering. They should understand that answers take time; throughout history, those who have asked questions haven't always been around for the answers.

While current curriculum may appear sound, training students to score well on high-stakes tests is quite possibly the worst reason to include science (or any other content) in the curriculum. So who should be in charge of this very impressionable and often fragile population, understanding that they will most likely be taught how to think and what to think? Knowing that curriculum (particularly science) is continuously revised, which standards are the correct ones? Is the term "standard" even appropriate in science education? Isn't that what the teacher who says "because God made it that way" would declare—that that response is a standard that never fails to provide the answer? Could this be one reason we see so many parents pressing their children into home schooling and other alternatives to public schools: a profound lack of understanding about what schools are teaching and a basic mistrust where their children are involved? These troublesome issues make for restless sleep. I often wonder if the "standard" will fail to have any significant meaning. This makes for great debate over late-night wine tastings. But such exchanges render no truly workable and just solutions for what I've come to consider the victims of the system. I worry about the long-term consequences when children no longer enjoy learning because they don't understand it can be fun. They are worn down by a world of multiple-choice items where no one cares to involve them in meaningful discussions. I should probably just get more sleep.

MY MOTHER AND LUCAS

My mother is now 82 years old (going on 14), a retired special education diagnostician and first-grade teacher. She's still called almost daily to substitute teach, even though she has requested repeatedly that she no longer be contacted. When she's called, she goes—she can't help it. I both can and cannot believe her drive and enthusiasm—they are what keep her going. A child of the Depression—she's still working because "I should," she says—she teaches primarily in special education classrooms where other people refuse substitute assignments. "Those sweet little ones are God's gifts too," she says. She retired three times, finally staying home to take care of my dad while he was stricken and eventually died from Lewy body disease, a particularly cruel form of dementia. She never let him be taken to a nursing home; she was responsible for his day-to-day care for nearly six years. The only time they were separated was when he was hospitalized with heart problems. All of this is to say, she's a tough cookie.

After 56 years together, they were the type of couple that when one inhaled, the other exhaled. After he died, she took about a year to recover her breath. Once she was breathing on her own again, she said she needed to be busy. She initially proposed returning to elementary schools as a volunteer. When she called the school district office to ask about volunteering, a secretary she had known for years talked to her about substitute teaching. Mom jumped at it. Once word got out, the phone started ringing. One of the first requests was for a long-term replacement for a young woman who was having a baby. Twelve weeks in a fourth-grade math/science classroom. I was worried that we'd carry her out on a stretcher, but I needn't have.

She called me nightly with the latest antics of her class. About two weeks in, she said she could do everything with them in math, but admitted to not being as sure about science. I could not believe my ears. My mother didn't feel confident teaching science? This was weirdly similar to preservice teacher research I was working on about low science self-efficacy in elementary candidates. She further told me that she never liked science when she was in school, not as a student and not as a teacher. Now I was completely flummoxed! How on earth could I have not known how my mother felt about this field of study I found so engaging? She mentioned a chemistry teacher in high school who could never make it clear "where the energy went" when balancing equations, and a physics professor who graded on a curve and gave her an A, even though she never scored above 60%. She went on about her bad experiences in a science course she tried to take by correspondence while living on Guam and other equally frustrating stories from high school and college.

I was her daughter, the science teacher, and I'd never heard about any of this before! Why had she withheld this from me? Had I betrayed her somehow by becoming a science teacher? No answers were forthcoming, so I dropped that line of inquiry to focus on her immediate problem. She sounded kind of desperate. I promised that after completing my own student teacher observations, I would visit her classroom twice a week to present science activities to help her students keep up with the other fourth-grade classes. Then she told me about Lucas. "That kid will ask the most bizarre questions. I don't know what to say to him or, half the time, even what he's talking about." I asked her what she meant and she said, "He's just weird." I asked her how weird. Then she paused, "No, actually, I think he may be unidentified—as gifted." I couldn't wait to meet Lucas.

When the fourth graders entered the classroom on my first visit, everything was already set up on their desks for the lesson on the water cycle. As I introduced myself and asked them to look carefully but not to touch their displays (clear plastic cups with lids containing a hot saltwater solution dyed blue to represent the ocean), Lucas immediately started asking questions. What I didn't know was that Lucas had difficulty pronouncing *r*'s and *l*'s. Water cycle was watuh cykoh. The rest of the students were practically directed through the lesson, quite nicely, by Lucas's questions. He asked everything I anticipated a student of that age would ask: very few abstract questions, lots of concrete concepts. It was almost as if Lucas was ensuring that his peers all understood the concepts and vocabulary. I was impressed: Lucas was animated, personable, and smart. But I wasn't convinced he was "unidentified" as my mother put it, a gifted learner not yet recognized as such by the system. Then he asked, "Doctoh Hutson, when ocean watuh evapowates and comes togethuh to make cwouds, is that watuh vapoh fweshwatuh or sawtwatuh?" My mother smiled and winked as if to say, "Told you so!"

I couldn't believe it! This kid asked me a question that had never been proposed before in all my years teaching the water cycle to sixth graders on up to high schoolers. Did water leave the salt behind when it evaporated? This 9-year-old hadn't lost his curiosity. Schooling hadn't yet managed to dull his enthusiasm for discovery! Of course, no one had ever been as interested or thorough in their questioning, either. I wasn't sure if Lucas was thinking this was a "stump-the-teacher" question or what.

I answered him back, "Lucas, I am going to let you discover the answer. Very carefully, take the lid off your cup and flip it over really fast." I pantomimed the action he should perform. He did it, just as I modeled. The other students were watching; the room was very quiet. "Now," I told him, "that lid has water on it that has evaporated, and I want you to taste it—however you want to—and then tell me, is it freshwater or saltwater?" I thought he might put his finger on it, but Lucas licked it—he was nine, after all. "That's incwediboh! It's fweshwatuh! So, you mean, when it evapowates, it weaves the sawt behind? That's just incwediboh!" He was so excited. He couldn't wait to tell his mother, dad, brother, and whomever else was around. I wanted to stay the rest of the day and just listen to Lucas make more connections to things he found curious. I was in love with discovering science, all over again, through Lucas's eyes.

The magic of the moment soon disappeared as my mother started barking orders. "Don't take those lids off! If you don't get quiet, you'll lose time for recess! Sit down in your seats! Why do I always have to get onto you boys? Why can't you stay in your seats and listen like the girls do?" I quickly gathered up the displays and supplies to save them from at least some of the hostility she was so prone to displaying. And I realized that the real reason she didn't like science was that it's messy. Kids get excited. If you are used to imposing external control, science appears to be chaotic. But in chaos are discovery, magic, and wonder that are marvellously contagious. Knowing that (and remembering my mother's love of "quiet classrooms" and "well-behaved children"), I was secretly glad that she didn't like to teach science. Why would I share that with someone who saw "keeping the boys under control" as the primary indicator of a "good day" at school?

As I had promised, I returned a few more times. We played with magnets to predict and discover how many paperclips they could hold in a chained fashion, energy balls to discover that we could conduct electricity if we held hands to create a complete circuit, and magnifying glasses to dissect and examine pinto beans that had been soaked overnight. I was very careful not to let students get "too loud" (my mother's words) in the science talk time. And to model a different type of classroom management for my mother, I used a timer to be sure the children weren't out of their chairs for more than two minutes as they moved around the room.

I wouldn't have done well teaching in the same setting as my mother. We are too different in our philosophies. And, while I love her dearly, I understand that to expect her to change is not realistic. She is mostly kind, considerate, and fair, but only within a range of possibilities that appear increasingly narrow as society is rapidly evolving to embrace more than she can accept. She worries that children are becoming too worldly and not having true childhoods any more, as if there is a single acceptable childhood standard. When I'm in her home, I don't share with her some of the books I've read, movies I've seen, television shows I enjoy, or friends I have. I choose not to out of respect for her. Hoping to live up to the honor-thy-mother standard, I do my part to ensure that she enjoys peace in her own home. At the age of 82, what else could be more important?

SINGLE-SEX SCIENCE

From the start, I've wondered why science is not well received by many middle and high school students. At one point, I was "all in" where single-sex classrooms were concerned. I had moved to yet another small rural school where middle school students were assigned to single-sex classes when life science and human reproduction were taught. At my interview, the principal claimed that this was a great arrangement that produced real gains in understanding in that age group. I wondered if he was seeing what he wanted to see or telling me what he expected to see if I became the (one and only) middle school science teacher. But after two years of teaching single-sex seventh-grade science, I was convinced that the principal was correct. Based on results of the statewide eighth-grade science test, our single-sex arrangement seemed to be a success.

Then something more important than state tests changed things—football. This was, after all, a Texas middle and high school: a farm club for college athletics. It turned out that the reason for single-sex classrooms had nothing to do with effective learning, teaching human reproduction, or boosting test results. It was, instead, a concerted effort to make it easier for the athletic director to schedule a weightlifting program for the middle school boys to get them ready for the high school football program. When the boys didn't develop the hoped-for muscles, the scheme was scrapped.

The students, however, weren't so ready to go back to things as usual. Those coming into the seventh grade were sorely disappointed because they'd been looking forward to girls- and boys-only classes. And many of the kids who had been in the single-sex science classes in middle school wanted single-sex classes in high school—and not just in science! This was a small campus with fewer than 130 students in grades 6 to 12, so it would have been entirely possible to honor their request. But the administration was not at all prepared for this pushback by students.

On the whole, the students seemed to experience the single-sex science classroom as a place where they could explore content with fewer concerns about being proper girls and boys. For many of the girls, the normal gender role meant that it was not cool to be smarter than or to challenge the boys. For the boys, it typically meant it was not cool to be interested in academics beyond whatever was needed to pass the class. When I asked why they felt the need to perform as expected, they spoke first of parental and family pressure, next of expectations from the coaching staff, and finally of peer pressure. This confounds the idea that kids don't listen to adults. These students were not unusual in their desire to please adults, especially those who held tremendous power in their lives—the power to grant privileges, such as driving or playing in the next game. These were students who graduated in 2010, not the 1950s.

I think the lesson here is not that sex-segregated classrooms are necessarily a good thing. We should continue to seek ways to disrupt the tight fabric of what is considered typical or conventional in sex and gender roles and relations. In this case, perhaps a single-sex learning space was not the real issue. Instead, maybe it was about (unexpectedly) discovering how to lower the background noise that is typical of adolescents just long enough for true discourse and learning to take place. Certainly, the idea that diversity is more than merely either/or (racially, politically, or sexually) remains a new idea for many people. But when we are open to greater discussion and understanding, we cannot help but discover unexpected benefits. One environment where this can occur is the classroom. My grandfather (a deacon in the Southern Baptist church) used to say something like, "This world doesn't take all types, but we've sure got 'em." I'm sure he was quite wrong. I think it takes all types, and we're just beginning to realize what each has to offer.

IS IT SOMETHING IN THE WATER?

A funny thing happened in the years following the single-sex seventh-grade science classes. Several students from two consecutive graduating classes of fewer than 50 students combined came out as lesbian or gay during and shortly after senior year. When we talked about it, they said their classmates encouraged them to come out. I'd like to think this is indicative of how the idea of what is "normal" is being tested, experienced, and redefined by younger generations. Maybe they are less interested in the normal/abnormal divide that folks often create—the us versus them dichotomy. Or maybe this was just an exceptional group of students. But I'd like to believe that, on the whole, students are discovering that there are more ways of being human than their parents were allowed to experience. I find students are less threatened by the diversity that was hidden in previous generations and still lies hidden in many parts of the world. They are often able to experience each other without the labels their parents found, and still find, necessary to explain their world: good/bad, gay/straight, rich/poor, black/white, and so on. As children seek to understand their parents' points of view, they are quick to point out that labels are questionable and don't reflect the value of an individual's life.

By students' own reckoning, the single-sex classroom experience had given them opportunities to think about their identities. As one young woman put it, this gave her the strength to acknowledge that she might be better accepted if she moved away from the small rural community she had known all her life so she could "figure out some things." So she joined the Navy. Five years later, she returned for the first time to visit her family and friends for Christmas. She also brought her fiancé, a young woman who was also in the Navy. She said on Facebook that she missed her family and was excited about the visit, but didn't say anything about moving back home. Another student moved to a large university town in north-central Texas and, as his mother laughingly told me, became a perennial student of the arts. Two other students moved to the nearest large city seeking educational and career opportunities while settling into relatively comfortable lives with partners.

Did the seventh-grade science class, where students had time and space to consider and process their identities, make such a difference? Or am I looking too closely, too much involved to be objective? I taught middle and high school science in that town for another five years. There were no more single-sex science classes (but plenty of single-sex athletics), and the number of students "going gay" seemed to disappear as well. Perhaps they went unnoticed in an era of don't ask, don't tell. Or, as one parent hoped, maybe it was just a fad—something kids had seen too much of on the Internet and would eventually grow out of.

Or—was it something in the water? The town's drinking water came from four or five local wells. During that time the wells were tested by state inspectors, and one was found to have unacceptably high arsenic content. The amount of water drawn from that well was significantly reduced (but not eliminated!) to lower the arsenic level in the overall water supply. A grandfather of one of the gay students thought he was onto something. Cornering me at the school's fall festival, he quietly shared his thoughts about "this unholy mess." "Could it be that arsenic in them wells?" he asked. "What about it, Doc? Could that make 'em go gay?" He wanted me to know how miserable he was thinking about that "queer" grandson of his off at college "doing God knows what." After all, he reasoned, if too much arsenic could kill you, wouldn't you know that something was wrong before you died? Wouldn't it affect your brain and your thinking? Couldn't it affect how you thought about yourself and what was "right and wrong"?

I patiently listened, marveling at the tangled motivations for his ranting. He loved his grandson, and he was genuinely worried about the likely damnation of the boy's eternal soul to hell and the general demise of Western civilization. But he failed to appreciate his grandson's contributions to his classmates as a friend and track team state qualifier. He failed to understand that his grandson didn't have to suffer humiliation working in the same school office for the same principal who had 15 years earlier called John (now my hairdresser) a "disgusting fag." Bless his heart. It was all I could do to keep from offering to drown him in that arsenic-laced water, being as how he was so miserable and all. Sometimes it pays to be patient, or not.

REFERENCE

Texas Education Agency. (2007–2015). *Texas essential knowledge and skills for science: Subchapter A: Elementary*. Retrieved from http://ritter.tea.state.tx.us/rules/tac/chapter112/ch112a.html#112.11

ABOUT THE AUTHOR

Tommye Hutson has taught about science in rural middle and high schools for 35 years. She has held full-time and adjunct positions in a variety of teacher education programs in Texas. When not working with students, her time is spent caring for her mother, playing with her grandchildren (there are currently five), and walking her two dogs. When not spoiling them (kids and dogs), she enjoys country life on the family farm in central Texas. From that vantage point, she can testify that the stars really are clear and bright, deep in the heart of Texas—a well-used telescope sits on her back porch.

KATHRYN L. KIRCHGASLER

9. STRANGE PRECIPITATE

How Interest in Science Produces Different Kinds of Students

> Race and sex differences in students' interest in, and affinity for mathematics and science have been reported in a number of studies. ... Black and female students express lower levels of interest and affinity for mathematics and science than white and male students.
>
> —Gail E. Thomas (1986, p. 31)

> Educational experiences designed to leverage the personal interests of learners have been used to increase the participation of girls in middle school, of urban high school youth of color, and of elementary school children from immigrant families.
>
> —National Research Council (NRC) (2012, p. 286)

The idea that demographic groups exhibit different levels of interest in science is treated as a matter of fact in U.S. science education. Decades of research have taken *interest in science* as an object of analysis. Despite variation in how it is studied, it is generally assumed to be a measurable quantity of the mind.[1] That is, a student has a more- or less-developed interest in science than someone else or than a population average. This comparative aspect makes it possible to contrast not only individuals but also subgroups, such as the race and sex differences reported in research from the 1980s (see the first epigraph).[2] When the problem of inequity is formulated in this way, the solution seems clear: Teachers must connect science to students' personal interests. This is regarded as good for all, but particularly important for those assumed to lack an affinity for science: identified in a 2012 U.S. policy framework (see the second epigraph) as girls, urban youth of color, and children from immigrant families. Knowing and leveraging the interests of these subgroups is promoted as a method of achieving equity and inclusion.

Because underrepresented groups are diagnosed with different interests—and in certain cases, less overall interest in science—it has become reasonable to prescribe different teaching strategies for them. Building on the 2012 framework above, the U.S. *Next Generation Science Standards* (NGSS) present case studies for how to make science "accessible" to diverse subgroups (NGSS Lead States, 2013a, p. 359). Interest is a key factor marking those subgroups as pedagogically distinct. What defines a subset of students as gifted and talented is, in part, their "intense interests

 DOI:10.1163/9789004331068_011

and motivation in science" (p. 365). In contrast, girls' interest in science is said to "dramatically decline" compared with boys' (NRC, 2012, p. 281), and "non-dominant" racial and ethnic groups are identified as having "traditionally been alienated from science" (NGSS Lead States, 2013a, p. 366). These claims have acquired such a high degree of facticity that specific studies are no longer cited to back them up. They are now matters of fact (Latour & Woolgar, 1986).

Borrowing from chemistry, it is possible to think of interest in science as something that only becomes real through a reaction—interest in science is not a permanent object, but a solid precipitate. In chemistry, a precipitate is a solid substance that falls out of solution during a reaction under a certain set of conditions. I use the notion of precipitate to suggest that while interest in science has acquired solidity as an object of research and reform, it did not always exist in its present state as a psychological attribute differentiating students by gender and race. Treating interest as a precipitate is to shift the analytical focus from a matter of fact to the processes of its materializing. It calls attention to the fluid reactions taking place just before this precipitate solidified. This chapter investigates the swirling set of historical conditions and practices out of which *interest in science* precipitated as a solid attribute of persons—one that has repeatedly crystallized around notions of difference in sex and race.

I explore how interest in science got configured and reconfigured as a pedagogical concern at two moments: the 2013 release of the NGSS, and the early 1900s rise of the general science course in U.S. high schools. General science was a course launched to adapt science to perceived differences in girls and the "masses" entering high school at the time—including their disparate interests. Almost a century later, interest in science shows up in the NGSS as a strategy to promote equity by adapting instruction for girls and non-dominant groups. Rather than treat these deployments of interest, gender, and race as self-evident, I draw on scholarship from queer theories and science studies to examine how such distinctions have been historically produced. I ask: How did interest in science become a measurable attribute that some gendered and racialized groups appear to possess less of than others? And how has it operated to produce different kinds of students who appear to require distinct forms of science education?

HISTORICIZING THE PRESENT—A QUEER THEORIES APPROACH

I juxtapose past and present to question a particular common sense: that attempting to connect science to the interests of underrepresented groups is an unproblematic good. Efforts operating within this common sense tend to analyze power in terms of identities and representation. These studies ask: Are the interests of students from diverse groups being included in the science classroom? Neglected in these analyses, however, are the power relations that operate through the production and regulation of different kinds of people (Foucault, 1990; Hacking, 2007). Categories of diversity are neither natural nor neutral. As scholarship in queer theories explores, what becomes seen as diverse depends on the cultural production of a norm against which difference, or abnormality, can appear (Somerville, 2007). This can occur through what Foucault (1994a) calls "dividing practices" (p. 126), by which people become objects of scientific study and come to see themselves and others as particular types of subjects. Delineating a domain of students who are interested in science simultaneously abjects a domain of others cast as uninterested and in need of regulation. By *abjection* I mean that locating these students' desires as the source of their presumed distance from science reinscribes their difference as deviation and authorizes the need for its correction.

Just like categories of diversity, the scientific concepts we use to describe their distinct attributes are not natural (Danziger, 1997). Interest in science is not the same thing in different times and places. A century ago, U.S. science educators spoke of interest in science as a democratic ideal that defied measurement (e.g., Kilpatrick, 1921). Today, it is quantified in the individual, mapped in the brain, disaggregated by subgroup, and used to rank nations around the world (see Bang & Valero, 2015). Rather than assume interest in science constitutes a timeless dimension of human nature, I approach it as a contingent product of human history that changes based on how it is conceptualized, measured, and deployed.

My methodological approach, called a history of the present (Foucault, 1977), asks about the confluence of conditions and practices that allowed today's common sense to materialize (Butler, 2011). By destabilizing notions of linearity from past to present, the aim is to upset the prevalent chronology of continual progress, which assumes that research yields an expanding knowledge base about student diversity, thereby ensuring greater equity. Instead, historicizing illustrates how past distinctions and fears have solidified in contemporary categories of those who must be motivated to take an interest in science and apply it to their daily lives. Historically, interest in science has emerged in relation to other dividing practices of sex, gender, and race, which were mutating and mutually constitutive over the past century (Somerville, 2000).

My analysis focuses primarily on (a) the NGSS and their associated *Framework for K–12 Science Education* (*Framework*), and (b) the journal *General Science Quarterly* (*GSQ*) published from 1916 to 1929. *GSQ*, one of the earliest science education journals, initially emphasized research on how to adapt instruction to the pupils entering so-called comprehensive high schools. (These high schools were by no means comprehensive, since separate institutions existed for those classified as "feeble-minded," "Indian," and "Negro," among others.) I bring together *GSQ* and the NGSS because they each warn science teachers to attend to differences in students' interests, and in both cases this concern was linked to national anxieties about a "changing population." I read both *GSQ* and the NGSS as collections of statements that display historically particular modes of reasoning (Stoler, 2009). While some of the language of *GSQ* reads as bizarre and of the past (e.g., eugenics terms like "race betterment"), many of its criteria for seeing and sorting difference bear resemblances to those of today. As such, I treat the NGSS as equally strange and use similarities and differences with *GSQ* as a strategy to denaturalize the present.

At issue throughout this chapter are *not* the individual intentions of the authors, but the scientific techniques, schooling practices, and epistemological grounds on which differences are classified. Scientific ways of seeing and sorting do not belong to the realm of private individuals, but reflect collective and historically contingent norms (Daston & Galison, 2010). These norms generate interest in science as a working object that social scientists and science teachers are trained to identify and act upon. Invented as a conceptual resource to guide the work of researchers and educators, interest in science does sociopolitical work by dividing children and curricula. It can be understood as a fabrication in both senses of the word (Popkewitz, 2012b): It is *made up*, a fiction imposed on the world, but it is simultaneously *made into* reality. That is, it is not just a social construction in terms of language and ideas, because it acts on the world in material ways.

I organize the analysis as follows. First, I consider how interest in science emerged in the early 1900s, as researchers drew on evolutionary theories to infer natural differences in interest by "sex" and "racial stock." Second, I highlight an important shift over the next few decades from inferring differences in interest by nature to calculating interest by the numbers through empirical techniques. Third, I explore how interest in science became linked to the problematic of underrepresentation, so that it now seems common sense for the NGSS to identify girls and non-dominant racial and ethnic groups as requiring distinct pedagogies to boost their affinity for science. I conclude by considering how interest in science depoliticizes questions of justice—erasing historical processes and installing the barrier to equity as an attribute of gendered and racialized subjects: their perceived lack of desire for science.

HOW INTEREST PRODUCES DIFFERENT KINDS OF STUDENTS

By Nature: Inferring Interest from Theories of Racial and Sex Differences

At the turn of the 20th century, interest in science had already solidified as a working object discussed by psychologists and teachers. To see how interest in science is not just naturally there, then, requires going back to a time before it became a commonsense way to describe students.

Education in the United States has long been concerned with managing pupils' desires (Baker, 2005). During the 19th century, this inner desire was not conceptualized as interest but as the child's will. In the mid-1800s, science education sought to promote moral character and to bring each pupil's will into alignment with the Divine Will through learning to obey God's natural laws as revealed by science (e.g., Mann, 1867). Within this logic, what held back the so-called "savages" was a difference in their souls, as indicated by their use of heathen incantations rather than Christian medicine (p. 147). At this time, interest in science was not yet thinkable as a psychological trait existing in the mind and varying by degree. Nevertheless, the notion of a pupil's will located in the soul bears some resemblance to the later notion of a student's interest located in the mind. In both cases, the goal was to harness some motive force within the individual child to promote an idealized mode of conduct.

By the early 1900s, the popular understanding of humans as religious beings with varying virtues was gradually substituted with a biological model of humans as natural organisms with distinct degrees of intelligence (Carson, 2007; Wynter, 1995). Normal human nature was defined scientifically by studying those imagined to be less evolved races (Gould, 1977). Recapitulation theory, pervasive in U.S. social sciences at this time, posited that the individual human develops through all of the stages of the human race—from primitive Negro to civilized European, and from feminine emotion to masculine reason (Fallace, 2015). In science education scholarship, these principles made it possible to reorganize science education from a "recapitulatory point of view" (Downing, 1925, p. 74). If the natural stages of mental development follow the evolution of man, then

> it seems to be the function of education to expedite the child in his passage through these primitive stages and to bring him as rapidly as possible to a comprehension and appreciation of the cultural heritage of mankind. (p. 54)

Recapitulation theory made it possible to configure the educational space as a trajectory of hierarchical stages. As shown in Figure 9.1, the first stage was *Appreciation* (bottom step), a term used interchangeably with interest. Only once this stage was acquired would pupils be ready to move up to *Skill in Thinking* (middle step), and finally up to *Knowledge* of the sciences (top step). As pupils moved from the early grades to senior high school (horizontal axis), it was assumed that not all would develop at the same rate, nor have the capacity to reach the higher aims (vertical axis). By following this sequence, science education scholars argued, the natural hungers that drove the child and primitive man could be developed into the appreciation for abstract knowledge that distinguished the mature scientist and civilized man (Clark, 1921; Quickstad, 1917).

Expediting the child's passage up this trajectory would require knowledge of which pupils already possessed an interest in science, and which were still driven by "emotional attitudes rather than reasoning" (Downing, 1925, p. 74). To acquire this knowledge of the pupil, teachers were urged to study psychology, sociology, and pedagogy (Elhuff, 1916). These fields were understood as having accumulated "a rapidly growing mass of fact" (Downing, 1925, p. *ix*) regarding the "important factors which must be recognized by the modern teacher," including sex and racial heredity (Hunter, 1920, p. 385).

First, interest in science was to be distinguished by sex. Anthropological studies drew on evolutionary theories to assert that in savage races, men and women were more similar to one another in anatomical features and societal roles (e.g., Delauney, 1881; see Russett, 2009). Greater sexual differentiation became a sign of civilizational development, a principle informing educational efforts to reinforce the proper roles of men versus women (see Lesko, 2012). In a modern civilization, science education scholars explained, boys and girls ought to display different

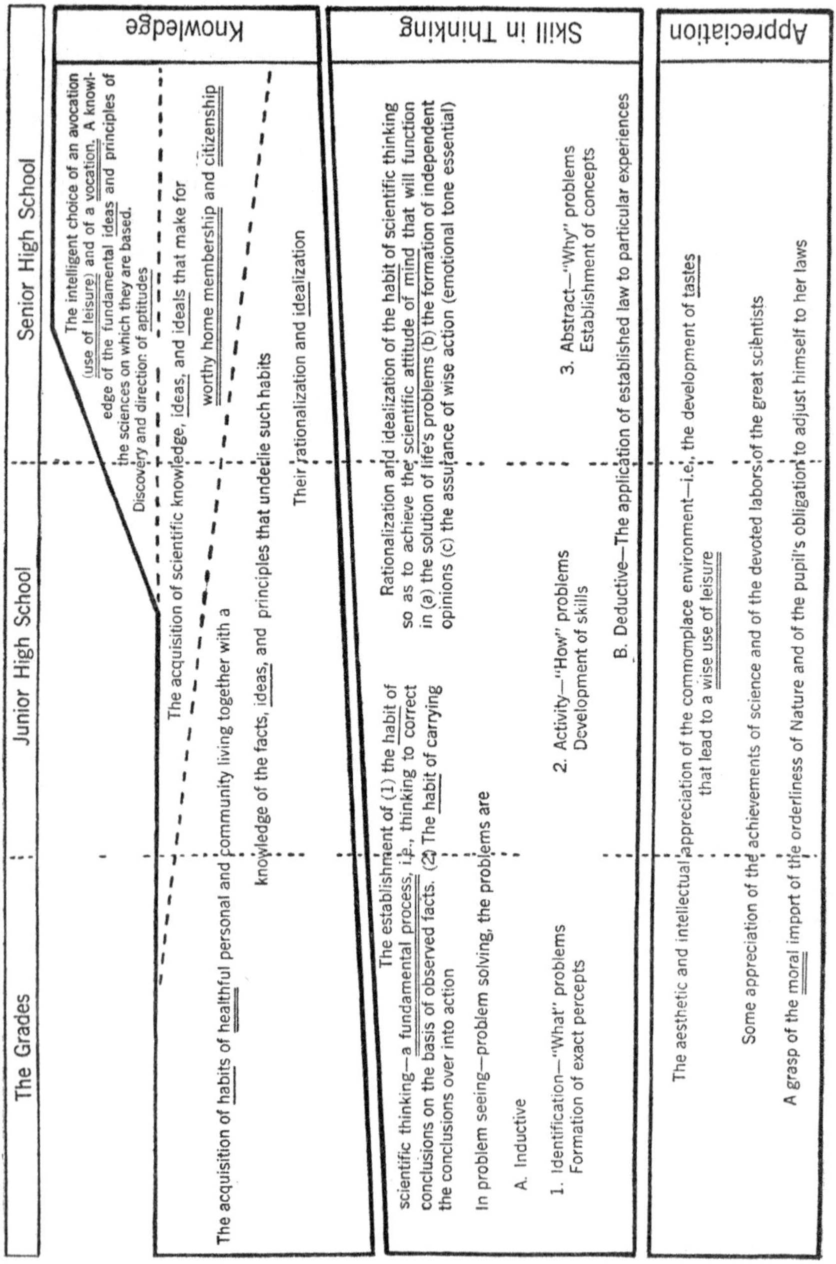

Figure 9.1. A developmental sequence of science education aims (Downing, 1925, p. 83)

interests by nature. Where boys should be interested in the great industries, outdoors, sports, and the affairs of the world of man, the "predominant interests of girls are in the home and in the activities within the woman's sphere" (Massachusetts State Committee on General Science, 1916, pp. 89–90). *GSQ* provided teachers with separate course outlines by sex, recommending industrial chemistry for boys and household chemistry for girls (Waterhouse, 1917; Gerry, 1920). As the title of one article put it, "Peggy Learns that Housekeeping is Science" (Hanson, 1919, p. 219). By recognizing and responding to those divergent interests, the general science course aimed to further the sexual differentiation understood as essential to ensure that America remained an advanced civilization.

Second, interest in science was to be distinguished by nationality and race. Science education scholars expressed concern that the U.S. population was changing and losing its interest in science (e.g., Bowden, 1924). In an era of immigration quotas and Jim Crow segregation, national fears of urban disorder and racial degeneration were coupled with hopes that science could provide knowledge of and solutions to societal problems (Chamberlin & Gilman, 1985). Sociological studies classified immigrant groups' mental and moral traits to identify what separated each from the "Anglo-Saxon ideal" (Jones, 1904, p. 133). Resulting typologies linked racial heredity with dispositions toward scientific reason, such as the "impulsiveness of the Italian" (p. 133) that made them less receptive to science than to "superstition" (p. 76). Similar taxonomic principles emerged in racial science. In an influential eugenics text, the pursuit of scientific discovery was identified as a mental trait of the Nordic race, associated with the moral traits of Protestantism and self-reliance, and with the physical traits of fair skin and blue eyes (Grant, 1916). Across the social sciences, the notion of interest in science as a hereditary and evolutionary trait was produced through (and productive of) racial distinctions. Recapitulatory principles, coupled with empirical practices of inferring inner differences from external features, allowed for the possibility of reading interest in science off of racialized bodies.

By identifying the racial stocks of the school, the child's interests were assumed to reveal themselves to the teacher as scientific observer. One *GSQ* article drew on sociological studies to classify the intellectual desires of the English, the industrial ambitions of the Germans, and the common interests of Italians and Serbs (Caldwell, 1917). In another article, a science teacher inferred pupils' interests based on the section of the city contributing to the school's population:

> [The school] consists largely of various types of foreigners: Italian, Polish, German, Hungarians. ... A few are bright and alert, some give evidences only of a passing interest in things educative, while a majority perhaps are rich with family tradition and superstition. (Carpenter, 1916, p. 47)

Here racial hierarchies are not only reinforced but reformulated. Within a given type of foreigner, pupils could be imagined as arrayed along a scale, from a majority interested in superstition up to a bright few with an interest in science.

Out of the evolutionary theories and social science typologies circulating at the time, interest in science solidified as a site for ordering both children and curricula. It simultaneously helped to fabricate different kinds of pupils and the distinct tiers of instruction to which those categories were matched. For the "mongrel lot of pupils of all races" (Woodhull, 1917, p. 224), developing an appreciation for science, as seen in Figure 9.1, would be the primary aim and the prerequisite for any other goals, including scientific knowledge. Instruction for pupils from southern and eastern Europe would focus on "relevant material from sources in their every day [*sic*] life" (Richardson, 1916, p. 33), "rather than making the futile attempt of imposing upon people the special aspects of science which are properly of interest to special students" (Caldwell, 1917, p. 135). In contrast to general science, physics and chemistry became designated as specialized sciences. These courses would now be reserved for the few—"all American born"—seen as uniquely interested in quantitative abstractions (Woodhull, 1917, p. 138). Interest in science thus played a crucial role in the emerging stratification of U.S. high school science, with a lower level offering relevance to daily life and a higher level promising the rigor of college preparation.

General science aimed not only to respond to pupils' natural interests, but also to direct those desires toward proper ends. Regulating the supposedly private realm of sentiments, dispositions, and domestic relations has long served as a strategy of governance (Stoler, 2009). Based on the premise that girls and inferior racial stocks were driven by emotion rather than reason, the general science strategy was to reach these pupils "through sentiment" by appealing to their existing interests (Grier, 1920, p. 48). The ultimate aim, though, was not to entertain but to incite an "interest in the science of right living" (Bowden, 1924, p. 481) in terms of physical, mental, and sexual hygiene. Returning to Figure 9.1, access to higher tiers of science education became dependent on demonstrating an appreciation for science, defined as "a grasp of the moral import of the orderliness of Nature and of the pupil's obligation to adjust himself to her laws" (Downing, 1925, p. 83). To take an interest in science meant following the rules of hygienic living now attributed not to Divine Will (as seen in the mid-1800s) but to Nature's laws as revealed by scientific experts.

Bodies gendered and racialized as lacking an interest in science were especially targeted for regulation. General science's emphasis on applying science to everyday life permitted greater surveillance of the nation's "unassimilated material" (Hunter, 1922, p. 526) that must be Americanized and trained in "straight thinking" (p. 530). This included proper (i.e., heteronormative) attitudes towards sex and sexual restraint for "race betterment" (i.e., avoiding miscegenation) (Clark, 1921, p. 129). General science was also concerned with policing desires for certain kinds of knowledge and societal roles. As we have seen, "Peggy" must learn that for her, housekeeping *was* science, just as the "masses" must come to appreciate that physics and chemistry were properly of interest only to the few. In *GSQ*, then, interest in science functions as a "dense transfer point for relations of power" (Foucault, 1990, p. 103)—a site for monitoring the boundaries of proper desires, sexual spheres, racial belonging, and American citizenship.

So far, I have considered a few of the conditions out of which interest in science precipitated in the early 1900s. I have argued that in science education scholarship from the 1920s, interest in science was constituted *by nature*—where differences in interest were not directly assessed but inferred based on theories of race and sex differences. Biological and social scientific theories provided teachers with knowledge of their pupils to classify their interests and determine what pedagogy they would require. In the next section, I trace key shifts in how a student's interest in science could be seen and sorted. By the 2010s, interest in science would no longer rely on evolutionary theories widely rejected as sexist and racist. Rather, as I explore below, it is now constituted *by the numbers* through quantitative data. In a strange inversion, it is now racial and gender differences that are not assumed *a priori* but are nonetheless inferred to exist based on numerical comparisons of these subgroups' interest in science.

By the Numbers: Inferring Racial and Gender Differences from Data on Interest

Interest in science does not refer to the same thing in the 2010s as it did in the 1920s. In the intervening century, there has been a shift from explicit theories of natural difference to numerical data as the arbiter of educational reality (Popkewitz, 2012a). Nevertheless, as evident in the NGSS case studies, interest in science still functions as a distinction-making measure. Before turning to the NGSS, however, I want to highlight three events that made possible today's notion of interest in science. The first is the use of survey techniques to visualize presumed differences between categories of pupils, such that the data embody implicit theories of human difference. The second is the emergence of *attitude* and *motivation* as measurable psychological constructs. The third is the linkage of these constructs to the problem of underrepresented groups. Despite these shifts in how interest in science is verified and deployed, it continues to produce and regulate distinct kinds of students in ways that potentially undermine its use as a strategy for equity.

At the turn of the 20th century, efforts to generate scientific evidence of human difference were moving from exterior features (e.g., cranial measurements) to the interior of the mind (e.g., mental testing). By the 1920s, IQ tests entered U.S. schools where they promised to objectively reveal pupils' mental capacities. Science education scholars lamented that there was not yet an instrument like the IQ test to measure pupils' attitudes toward science (Kilpatrick, 1921). An early attempt adopted sociological survey techniques to determine the science topics of interest to boys versus girls (Lyon, 1918). In this study's design, natural differences in interest between the sexes were presupposed and empirically verified as a seemingly automatic inscription of self-reported differences. For instance, the survey results were said to reveal that "removing stains naturally held the girls," while boys were "naturally ahead of the girls" on magnetism and electricity (p. 389). This study demonstrates a comparative mode of reasoning—still prevalent today—wherein individualizing techniques display distinctions in interest as statistical and stable attributes of pre-given categories. Though the existence and salience of these categories may only be hypothesized at the outset, the study's procedures of classifying, counting, and correlating data solidify them as distinct types and link them to new attributes and evidence of difference—in effect, making them up as different kinds of people (Hacking, 2007). While this 1918 interest survey indicated sex differences in preference for some science topics over others, it was not yet possible to calculate interest in science as a general quantity of mind.

Not until the 1930s were interest, motivation, and attitudes routinely studied as individual attributes. The emergence of the construct of attitude promised a direct link between what was visible on the surface (i.e., written responses) to what was presumed to be going on underneath (i.e., within the subject's mind) (Danziger, 1997). An early study of attitudes analyzed the content of Polish immigrants' letters to map each individual along an evolutionary trajectory of Americanization: from Polish attitudes (e.g., beliefs in religious magic) to American attitudes (e.g., a theoretical interest in the natural sciences) (Thomas & Znaniecki, 1920). Since attitudes were conceptualized as shaped by one's home environment and yet amenable to external intervention, interest in science could operate as a site for monitoring individuals' progress toward cultural assimilation.

Soon after, the Likert scale made it possible to measure attitudes on a quantitative scale and along a bell curve (Thurstone, 1928). This statistical technique reconstituted attitude as a trait, like IQ, that would now appear to vary in degree and to be unevenly distributed in the population. From the 1930s on, Likert scales proliferated across domains of schooling, marketing, and political campaigns, as they offered to objectively reveal which subgroups should be targeted in efforts to induce new desires and behaviors (Danziger, 1997).

The constructs and techniques emerging during these early decades of the 20th century performed key roles in later reforms to help underrepresented groups in science. To be sure, a significant break occurred between the 1920s and the 2010s. In the 1920s, differences by sex and racial stock were taken as natural—as in resulting from nature and also reflecting the normal state of society. By the 2010s, differences in science education by gender and race are now seen as unnatural—disparities that demand research and intervention. Yet this may be less a case of rupture than of reconfiguration. When gender and racial disparities appeared on national research agendas in the 1970s (e.g., Crowley, 1977), the search for explanations would not begin in a vacuum. It would draw on historically available theories and techniques, including decades of research to acquire knowledge of pupils' differences in psychological terms. According to a national report on underrepresented groups (Marrett, 1987), the tendency to seek individual-level explanations for gender and racial disparities in science "may be the consequence of the theoretical models used to study mathematics and science education," particularly "various motivational theories" (p. 26). Already-circulating notions of interest in science thus participated in forging the research agenda about underrepresented groups, asking: Why are girls and minority students *less interested* in studying science and *less motivated* to pursue scientific careers?

During the 1970s and 1980s, psychologists developed numerous Likert-style instruments to measure interest, motivation, and attitude toward science (Newcombe et al., 2009). By constraining the self-report of attitudes to a range from 1 to 5, the Likert scale generated new equivalencies and hierarchical distinctions. Given the ordering capacity of numbers (Cline Cohen, 1982), it would now be possible to (a) conceptualize interest in science as varying strictly in degree, (b) average individual scores using demographic categories, (c) rank subgroups by these average scores, and (d) establish statistical norms against which an individual or subgroup could be judged. Within this statistical style of reasoning, massive data sets permitted generalized truth claims, such as that "women and racial minorities" have less interest in science (e.g., Thomas, 1986, p. 31). Such data can be understood as neither passive readings of reality nor deliberate misrepresentations. Rather, they refer to "a reality in which they are themselves heavily implicated" (Danziger, 1997, p. 192). Through social science techniques, historical assumptions of difference are quantified and materialized, reconstituting the problem and its possible solutions (Rose, 1991).

Despite inconclusive and contrary findings, such as higher "positive attitudes evidenced by minority groups" (Kahle, 1982, p. 545), an effect of this research was to crystallize a common sense about female and minority students' negative attitudes toward science. By the early 2000s, unsuccessful attempts to ensure science-for-all were attributed to the fact that women and minorities "opt out of a subject that they see as incommensurate with their interests" (Alsop, 2005, p. 4). This logic circulates in recent U.S. science education policy as well (see Knaier, this volume, chapter 10).

The NGSS *Framework* formulates the problem of inequity around differences in interest. While acknowledging that one source of inequity has been "biased stereotypical views about the interests or abilities of particular students or demographic groups" (NRC, 2012, p. 281), the *Framework* goes on to assert:

> Students' own motivation and interest in science and engineering can also play a role in their achievement. ... Thus attention to factors that may motivate or fail to motivate students from particular demographic groups is important to keep in mind when designing instruction. (p. 281)

The *Framework* treats interest in science as a key factor explaining achievement gaps, insisting that—separate from biased stereotype—certain demographic groups do require instruction to be made more motivating in targeted ways. Such claims, made possible by decades of numerical data, have become so matter-of-fact that numbers are no longer cited to back them up. Interest in science has materialized as an invisible attitude in an individual's mind, one that can be probabilistically inferred from a student's gender and racial classification. Interest in science is now co-constitutive with gender and race—each inferred from the other.

In the NGSS case studies, interest in science acts as an indicator that teachers can use to recognize distinct types of students and their science learning needs. Individuals marked as "gifted and talented" are depicted as entering the classroom with fully developed scientific interests: "Jerry's naturalist interests were immediately apparent in the first week of school" (NGSS Lead States, 2013d, p. 2). In contrast, for girls and non-dominant groups, interests are identified at the level of the subgroup rather than the individual. Girls are described as having some interests—"literacy," "life science," and "aesthetic aspects of science"—that can be leveraged to spark an interest in more problematic areas of physics, engineering, and technology (NGSS Lead States, 2013c, pp. 11–12). Non-dominant groups are assumed to have an interest in their home cultures, local communities, and social activism (NGSS Lead States, 2013b, p. 12). Within the logic of the case studies, interests in pit vipers and robot characteristics are prototypical of gifted and talented students, while interests in fuel use at the local gas station or food items in one's home are prototypical of non-dominant racial and ethnic groups. A danger is that these vignettes serve as primers, training the "classificatory eye" (Foucault, 1995, p. 147) of the teacher to read particular expressions of interest in science as a sign of a certain "kind" of student: one with greater or less talent.

Though certainly not the intention, the effect of this focus on interest in science is that racial and gender distinctions are simultaneously reinscribed and transformed. Girls and non-dominant racial and ethnic groups become conceived of as distinct types of science learners that appear to demand different forms—and in some cases, tiers—of science education. Not only does the gender and racial segregation of tracked science courses not appear as an explanation for inequity in the NGSS, these tracking practices are potentially reinforced as a more efficient means to address the divergent interests assigned to each demographic group. This risk is underscored by the resemblances between the interests assumed to appeal to the masses in the 1920s and to non-dominant groups today—their own everyday lives, local communities, and concrete projects—as opposed to the abstract and quantitative knowledge seen as of interest only to the few and the gifted. Consequently, the stratification of science education is naturalized as a reasonable response to perceived differences in students' levels of interest in science.

Moreover, interest in science demands dividing students by whether their current interests should be freely encouraged on the one hand, or closely monitored and transformed on the other. For those designated as gifted and talented, the teacher is to "encourage autonomy by allowing the student to follow and cultivate her/his interests" (NGSS Lead States, 2013d, p. 7). The premise is that Jerry's interests are already aligned with science, a disposition qualifying him to conduct independent investigations with students of "similar interests and ability" (p. 3). But not all students' interests are considered equally scientific. Non-dominant groups are said to require strategies that connect science to daily life, such as recording water use in the home, in order to "increase interest among both parents and students" (NGSS Lead States, 2013a, p. 366). The premise is that this subgroup's allegedly negative attitude toward science has roots in the home. For non-dominant groups, gaining knowledge of their prior interests is a means to an end—a "bridge for student engagement" (NGSS Lead States, 2013b, p. 8)—so that they "embrace and further investigate what they are learning, instead of being resistant to learning science" (NGSS Lead States, 2013a, p. 367). This logic bears a concerning resemblance to arguments made by *GSQ* scholars for appealing to the interests of the masses: "If they are to be elevated, they must first be reached" (Rowell, 1916, p. 55). Interest in science, then, continues to function as a dense transfer point of power, simultaneously monitoring the desires of individual students and holding the family and community responsible for cultivating proper attitudes.

Interest in science, as materialized in *GSQ* and the NGSS, only becomes visible through a comparative mode of analysis. Its historically provisional criteria for perceiving and evaluating human difference constitute what Butler (2011) calls an exclusionary matrix. In reiterating the boundaries around the normative science student, interest in science helps to "produce—demarcate, circulate, differentiate—the bodies it controls" (p. xii). The bright interests of the few only appear against the passing interest of the masses, while the intense interests of the gifted and talented students stand out in contrast to the alienation and resistance ascribed to non-dominant subgroups. The subject who desires science learning, then and now, remains "constituted through the force of exclusion and abjection, one which produces a constitutive outside to the subject" (p. xiii). In this sense, interest in science resembles positive psychology notions like curiosity and grit. Each depends on a "founding repudiation" (p. xiii) through the production of the less interested, the incurious, and the ungritty (see Slaton, Cech, & Riley, this volume, Chapter 17). Moreover, the positivity of these constructs results in "normalizing a proximal subset of the Other" (Fifield & Letts, 2014, p. 396)—those who express interest, curiosity, or grit in culturally recognized and sanctioned ways.

HOW TO IDENTIFY A DEPOLITICIZATION REACTION

> Depoliticization involves construing inequality, subordination, marginalization, and social conflict, which all require political analysis and political solutions, as personal and individual, on the one hand, or as natural … on the other. (Brown, 2006, p. 15)

> All students can profit from [instruction that builds on prior interest and identity], but the benefits are particularly salient for those who would *feel disenfranchised* or disconnected from science should instruction neglect their personal inclinations. (NRC, 2012, p. 287, italics added)

Interest in science depoliticizes questions of why some students advance in science education and others do not. The call for teachers to know their diverse students' interests elides the fact that both the construct of interest in science and the pre-given categories of diversity have been "socially and historically constituted and are themselves the effect of power" (Brown, 2006, p. 16). Instead, different groups with distinct interests and degrees of motivation in science become what teachers are exhorted to see, sort, and act upon in targeted ways. In *GSQ* and the NGSS, the approaches of "knowledge of the pupil" (Quickstad, 1917, p. 159) and "making diversity visible" (NRC, 2012, p. 288) conceptualize underlying differences in terms of problems to be solved, gaps to be filled, and deviations to be rectified. The hope is that by seeing these distinctions more precisely, science education can more effectively bring about their improvement. For those groups feared to possess subnormal levels of interest, this approach subjects them to attempts to monitor their affect and to shift their identities and attachments toward more socially desirable ones.

The political nature of the processes discussed in this chapter—the production and regulation of gendered and racialized bodies—is disavowed by the fact that interest in science is treated as natural, individual, and ahistorical. Interest in science operates normatively to produce enthusiastic and apathetic learners, while this normativity is made virtually invisible through scientific techniques. Discussions of ability can be substituted with interest to avoid more politically fraught questions of IQ, for which the history of gender and racial discrimination is more widely known (e.g., Gould, 1996). Analysis of systemic inequities can be replaced with interest, obviating the need to examine these disparities in any detail (see Lewis, 2003). Further, as evident in the final epigraph above, disenfranchisement is rendered into a personal feeling that must be managed, rather than a set of political relations. When interest in science becomes the primary lens for seeing equality and difference, the sociopolitical dilemma of how to redress science education's exclusions is replaced with an agenda of attitudinal change. That agenda—rather than accounting for the ongoing history of marginalization—is targeted at those very groups identified as only "historically" marginalized.

What then to do with this precipitate called interest in science? Can it be recrystallized to remove biases and obtain a more objective measure? This chapter has argued that interest in science is not a timeless trait of nature, but only materializes as an object of research and reform. Attempts to measure "it" rely on historically available theories and techniques that help produce the distinctions they purport to discover.

Then should interest in science be added to a list of hazardous research products, alongside IQ or cultural deprivation, and removed from discourse altogether? The problem is not the words "interest in science." Consider that the NGSS, in shifting from a deficit- to an asset-based approach (NGSS Lead States, 2013a, p. 366), never use the term "unmotivated," and yet the logic remains that certain subgroups require strategies to motivate them. The problem is not with interest in science "itself," but with the dividing practices that have made it appear. These practices embody a comparative mode of reasoning that continues to fabricate distinct kinds of students in need of different tiers of science education. Even if interest in science were removed, other precipitates would persist as normalizing modes of seeing and sorting difference: achievement, talent, worldview, identity, disciplinary dispositions, and so on.

What if science education research were to shift from using these precipitates to classify diverse subgroups to instead identifying the reactions through which they materialize? The aim would be to investigate how the theories and instruments we inherit—e.g., developmental stages, standardized tests, and home surveys—have social and political histories of their own that may subvert the intentions we bring to their use (Kirchgasler, 2017). To be clear, this is not a call to abandon all such theories and methods of science education research. Butler (2011) makes a similar case regarding terms like woman or queer: The argument is not *against* using these identity categories, but to insist on an "inquiry into their formation" (p. 174). This boils down to sensible lab practice—to insist on knowing what is floating around in solution before extracting the precipitate for practical use. It is to treat the tools of research and pedagogy not as natural givens but as products of human history, and to demand care in considering the "exclusionary operations of their own production" (p. 174).

The point of historicizing interest in science and its related precipitates is to "refuse prescriptive closures" (Chen, 2012, p. 237), opening space for new questions and possibilities. As Foucault (1991) argues, destabilizing the common sense does not paralyze the practitioner. On the contrary, it provokes the political and ethical deliberation that is anesthetized by treating equity as a technical problem. On what grounds is it decided what type of education a student or demographic group requires or deserves? In making diversity visible, what are the theories and techniques by which difference is classified? And finally, what other purposes might animate a science education freed from the injunction to respond to presumed differences in students' interests and needs for rigor versus relevance? These new possibilities will not evacuate power relations from the science classroom, since the stabilization of what counts as school science is already an effect of power. Still, it is important to work to transform those relations by "detaching the power of truth" from the forms of stratification, normalization, and exclusion it currently authorizes (Foucault, 1994b, p. 317). This will first require rethinking the premise that making science education more relevant to students' interests will necessarily result in their inclusion and empowerment. It is to recognize the precipitate as strange.

ACKNOWLEDGMENTS

Research for this chapter was supported by the University of Wisconsin-Madison Holtz Center for Science & Technology Studies and the Department of Curriculum & Instruction. I gratefully acknowledge this support, as well as invaluable feedback from Bernadette Baker, Durell Callier, Karishma Desai, Noah Weeth Feinstein, Steve Fifield, Malathi Iyengar, Christopher Kirchgasler, William Letts, Thomas Popkewitz, Brenda Nyandiko Sanya, and Lei Zheng. Any faults are my own.

NOTES

1 There is significant variation in the operationalization of interest and the science toward which that interest is directed (e.g., science in general versus school science). For a literature review on attitudes toward science in which interest is one of several related constructs, see Osborne, Simon, and Collins (2003). For a synthesis of interdisciplinary research on interest in science and mathematics learning, see Renninger, Nieswandt, and Hidi (2015). For an analysis of key shifts in the discursive formation of interest in science, including its recent emergence as a tool of international ranking, see Bang and Valero (2015).

[2] Throughout this chapter, I cite the precise classificatory terms used by historical sources, such as presumed differences in "sex" and "racial stock" in the early 1900s, or in "gender" and "non-dominant groups" in the early 2000s. Tracing these shifting classifications is a strategy to denaturalize these categories and examine their historical production within distinct grids of scientific and schooling practices.

REFERENCES

Alsop, S. (2005). Bridging the Cartesian divide: Science education and affect. In S. Alsop (Ed.), *Beyond Cartesian dualism: Encountering affect in the teaching and learning of science* (pp. 3–16). Dordrecht, The Netherlands: Springer.

Baker, B. (2005). State-formation, teaching techniques, and globalisation as aporia. *Discourse: Studies in the Cultural Politics of Education, 26*(1), 45–77.

Bang, L., & Valero, P. (2015). Chasing the chimera's tails: An analysis of interest in science. In T. S. Popkewitz (Ed.), *The "reason" of schooling: Historicizing curriculum studies, pedagogy, and teacher education* (pp. 131–149). New York, NY: Routledge.

Bowden, G. A. (1924). Vitalizing the problems of good citizenship by means of the general science course. *General Science Quarterly, 8*(3), 477–478.

Brown, W. (2006). *Regulating aversion: Tolerance in the age of identity and empire*. Princeton, NJ: Princeton University Press.

Butler, J. (2011). *Bodies that matter: On the discursive limits of "sex"* (2nd ed.). New York, NY: Routledge.

Caldwell, O. W. (1917). An interpretation of the new point of view in science teaching. *General Science Quarterly, 1*(3), 131–136.

Carpenter, H. A. (1916). General science in the junior high school at Rochester. *General Science Quarterly, 1*(1), 46–53.

Carson, J. (2007). *The measure of merit: Talents, intelligence, and inequality in the French and American republics, 1750–1940*. Princeton, NJ: Princeton University Press.

Chamberlin, J. E., & Gilman, S. L. (Eds.). (1985). *Degeneration: The dark side of progress*. New York, NY: Columbia University Press.

Chen, M. Y. (2012). *Animacies: Biopolitics, racial mattering, and queer affect*. Durham, NC: Duke University Press.

Clark, B. M. (1921). General science a factor in race betterment. *General Science Quarterly, 5*(3), 123–129.

Cline Cohen, P. (1982). *Calculating people: The spread of numeracy in early America*. Chicago, IL: University of Chicago Press.

Crowley, M. F. (1977). *Women and minorities in science and engineering*. Washington, DC: U.S. Government Printing Office.

Danziger, K. (1997). *Naming the mind*. London, England: Sage Publications.

Daston, L., & Galison, P. (2010). *Objectivity*. New York, NY: Zone Books.

Delauney, G. (1881). Equality and inequality in sex. *Popular Science Monthly, 20*, 184–185.

Downing, E. R. (1925). *Teaching science in the schools*. Chicago, IL: University of Chicago Press.

Elhuff, L. (1916). Relation of general science to later courses in physics and chemistry. *General Science Quarterly, 1*(1), 17–22.

Fallace, T. D. (2015). *Race and the origins of progressive education, 1880–1929*. New York, NY: Teachers College Press.

Fifield, S., & Letts, W. (2014). [Re]considering queer theories and science education. *Cultural Studies of Science Education, 9*(2), 393–407.

Foucault, M. (1977). Nietzsche, genealogy, history (D. F. Bouchard & S. Simon, Trans.). In D. F. Bouchard (Ed.), *Language, counter-memory, practice: Selected essays and interviews* (pp. 139–164). Ithaca, NY: Cornell University Press.

Foucault, M. (1990). *The history of sexuality, Vol. 1: An introduction*. New York, NY: Vintage Books.

Foucault, M. (1991). Questions of method. In G. Burchell, C. Gordon, & P. Miller (Eds.), *The Foucault effect: Studies in governmentality* (pp. 73–86). Chicago, IL: The University of Chicago Press.

Foucault, M. (1994a). The subject and power. In P. Rabinow & N. Rose (Eds.), *The essential Foucault* (pp. 126–144). New York, NY: The New Press.

Foucault, M. (1994b). Truth and power. In P. Rabinow & N. Rose (Eds.), *The essential Foucault* (pp. 300–318). New York, NY: The New Press.

Foucault, M. (1995). *Discipline and punish: The birth of the prison* (A. Sheridan, Trans., 2nd ed.). New York, NY: Vintage Books.

Gerry, H. L. (1920). Natural science in the secondary school: A digest of recent literature. *General Science Quarterly, 5*(1), 1–15.

Grant, M. (1916). *The passing of the great race or the racial basis of European history*. New York, NY: Charles Scribner's Sons.

Grier, N. M. (1920). The present day status and the future of public school physiology. *General Science Quarterly, 5*(1), 43–48.

Gould, S. J. (1977). *Ontogeny and phylogeny*. Cambridge, MA: The Belknap Press of Harvard University Press.

Gould, S. J. (1996). *The mismeasure of man*. New York, NY: W.W. Norton & Company.

Hacking, I. (2007). Kinds of people: Moving targets. *Proceedings of the British Academy, 151*, 285–318.

Hanson, T. (1919). Peggy learns that housekeeping is science. *General Science Quarterly, 3*(4), 219–224.

Hunter, G. W. (1920). The relation of general science to biological science in the secondary school. *General Science Quarterly, 4*(3), 381–389.

Hunter, G. W. (1922). Elementary science as a preparation for citizenship. *General Science Quarterly, 6*(4), 522–530.

Jones, T. J. (1904). *The sociology of a New York City block*. New York, NY: Columbia University Press.

Kahle, J. B. (1982). Can positive minority attitudes lead to achievement gains in science? Analysis of the 1977 National Assessment of Educational Progress, attitudes toward science. *Science Education, 66*(4), 539–546.

Kilpatrick, W. H. (1921). The wider study of method. *General Science Quarterly, 6*(1), 277–284.

Kirchgasler, K. L. (2017). Scientific Americans: Historicizing the making of difference in early 20th-century U.S. science education. In T. Popkewitz, J. Diaz, & C. Kirchgasler (Eds.), *A political sociology of educational knowledge: Studies of exclusions and difference* (pp. 87–102). New York, NY: Routledge.

Latour, B., & Woolgar, S. (1986). *Laboratory life: The construction of scientific facts*. Princeton, NJ: Princeton University Press.

Lesko, N. (2012). *Act your age: A cultural construction of adolescence* (2nd ed.). New York, NY: Routledge.

Lewis, B. F. (2003). A critique of literature on the underrepresentation of African Americans in science: Directions for future research. *Journal of Women and Minorities in Science and Engineering, 9*(3–4), 361–373.

Lyon, H. (1918). Student interest in subject matter. *General Science Quarterly, 2*(3), 387–389.

Mann, H. (1867). *Life and works of Horace Mann*. Boston, MA: Lee and Shepherd Publishers.

Marrett, C. B. (1987). Black and Native American students in precollege mathematics and science. In L. Dix (Ed.), *Minorities: Their underrepresentation and career differential in science and engineering* (pp. 7–32). Washington, DC: National Academy Press.

Massachusetts State Committee on General Science. (1916). General science bulletin. *General Science Quarterly, 1*(1), 37–46.

National Research Council (NRC). (2012). *A framework for K-12 science education: Practices, crosscutting concepts, and core ideas*. Washington, DC: The National Academies Press.

Newcombe, N. S., Ambady, N., Eccles, J., Gomez, L., Klahr, D., Linn, M., Miller, K., & Mix, K. (2009). Psychology's role in mathematics and science education. *The American Psychologist, 64*(6), 538–550.

NGSS Lead States. (2013a). *Next generation science standards: For states, by states*. Washington, DC: The National Academies Press. Retrieved from http://www.nextgenscience.org

NGSS Lead States. (2013b). *Next generation science standards: For states, by states*. Case study 2: Students from racial and ethnic groups and the *Next Generation Science Standards*. Retrieved from https://www.nextgenscience.org/sites/default/files/%282%29%20Case%20Study%20Race%20and%20Ethnicity%206-14-13.pdf

NGSS Lead States. (2013c). *Next generation science standards: For states, by states.* Case study 5: Girls and the *Next Generation Science Standards.* Retrieved from https://www.nextgenscience.org/sites/default/files/%285%29%20Case%20Study%20Girls%206-14-13.pdf

NGSS Lead States. (2013d). *Next generation science standards: For states, by states.* Case study 7: Gifted and talented students and the *Next Generation Science Standards.* Retrieved from https://www.nextgenscience.org/sites/default/files/%287%29%20Case%20Study%20TAG%206-14-13.pdf

Osborne, J., Simon, S., & Collins, S. (2003). Attitudes towards science: A review of the literature and its implications. *International Journal of Science Education, 25*(9), 1049–1079.

Popkewitz, T. (2012a). Numbers in grids of intelligibility: Making sense of how educational truth is told. In H. Lauder, M. Young, H. Daniels, M. Balarin, & J. Lowe (Eds.), *Educating for the knowledge economy? Critical perspectives* (pp. 169–191). New York, NY: Routledge.

Popkewitz, T. (2012b). The sociology of education as the history of the present: Fabrication, difference and abjection. *Discourse: Studies in the Cultural Politics of Education, 34*(3), 439–456.

Quickstad, N. J. (1917). Some phases of the general science problem. *General Science Quarterly, 1*(3), 153–161.

Renninger, K. A., Nieswandt, M., & Hidi, S. (Eds.). (2015). *Interest in mathematics and science learning.* Washington, DC: American Educational Research Association.

Richardson, H. A. (1916). Some experiences in teaching general science and physiography. *General Science Quarterly, 1*(1), 30–35.

Rose, N. (1991). Governing by numbers: Figuring out democracy. *Accounting, Organizations and Society, 16*(7), 673–692.

Rowell, P. E. (1916). Why science in the grades. *General Science Quarterly, 1*(1), 53–55.

Russett, C. E. (2009). *Sexual science: The Victorian construction of womanhood.* Cambridge, MA: Harvard University Press.

Somerville, S. B. (2000). *Queering the color line: Race and the invention of homosexuality in American culture.* Durham, NC: Duke University Press.

Somerville, S. B. (2007). Queer. In B. Burgett & G. Hendler (Eds.), *Keywords for American cultural studies* (pp. 187–191). New York, NY: New York University Press.

Stoler, A. L. (2009). *Along the archival grain: Epistemic anxieties and colonial common sense.* Princeton, NJ: Princeton University Press.

Thomas, G. E. (1986). Cultivating the interest of women and minorities in high school mathematics and science. *Science and Education, 70*(1), 31–43.

Thomas, W. I., & Znaniecki, F. (1920). *The Polish peasant in Europe and America: Monograph of an immigrant group.* Boston, MA: The Gorham Press.

Thurstone, L. L. (1928). Attitudes can be measured. *American Journal of Sociology, 33*, 529–554.

Waterhouse, R. H. (1917). General science in Amherst junior high school. *General Science Quarterly, 2*(2), 318–336.

Woodhull, J. F. (1917). The high school situation. *General Science Quarterly, 1*(3), 137–140.

Wynter, S. (1995). 1492, a new world view. In V. Hyatt & R. Nettleford (Eds.), *Race, discourse, and the origin of the Americas: A new world view* (pp. 5–57). Washington, DC: Smithsonian Institution Press.

ABOUT THE AUTHOR

Kathryn L. Kirchgasler is a lecturer in the Department of Curriculum and Teaching at the University of Kansas in Lawrence, Kansas (USA). Her research intersects curriculum studies, science education, and teacher education, especially related to issues of marginalization and exclusion. She studies how pedagogical practices divide children into hierarchical categories that appear to require distinct forms of instruction. At stake is how these dividing practices embody cultural normativities and historical assumptions of difference at odds with current commitments to equality and justice.

MICHELLE L. KNAIER

10. WHAT MAKES *GIRLS* AND *BOYS* SO DESIRABLE?

STEM Education beyond Gender Binaries

INTRODUCTION

Throughout my childhood and into my adult life, I willingly took on many nonconforming labels: tomboy, science major, science educator, childless by choice, outspoken, independent, skeptic, risk-taker, and queer. I express a variety of masculine and feminine traits, and therefore do not fit into the normative gender binary—hence my *genderqueer* identity. I have not always identified as a genderqueer woman, but ever since I can remember I have pushed the boundaries of what it meant to me to be a "girl."

As a child, I did not play with dolls. Instead, I played tackle football with the boys from the neighborhood. I created zoos with blocks and little plastic animals. I designed racetracks in the sandbox for my matchbox cars. During my adolescent years, I recognized that, unlike my sister, I did not have an affinity toward babies. I never had the desire to become a mother, and so many times my mom and aunts would say to me, "Wait until you're older. You'll change your mind." Then they would laugh. This always upset me. I felt like they were not listening to me.

Being a queer *girl* in a heteronormative household—a place where the assumption was that the children were heterosexual and conformed to the societal norms of masculine/feminine gender binary categories—was not easy. My response to expectations of gender conformity was to push back. *It was always a struggle to get me to wear a dress.* I chose to do, say, and believe what made sense to me—fortunately with few repercussions.

My mother stressed the importance of education. She never once tried to sway me away from studying science (like many girls experience), or from teacher education for that matter. However, once I became a science teacher, members of my family repeatedly asked, "How can you be a teacher when you hate children?" Again, I found myself angry. *How does my lack of desire to have children equate with me hating children?* I felt misunderstood—like that unheard child.

My personal experience illustrates that normative gender-binary assumptions (like masculine/feminine) hurt those who do not conform—especially when they are applied to children. These assumptions may also narrow the minds of those who do conform—especially when they try to understand nonconforming children.

 DOI:10.1163/9789004331068_012

Gender-conforming students are taught that they are "normal," and these students may, in turn, think nothing of rejecting nonconforming students who are "weird," "strange," or "queer." Students whose gender identities or expressions are questioned or policed due to nonconformity or perceived nonconformity are thus silenced. *How many of our students feel that they are not being heard?*

This chapter demonstrates and challenges how normative gender-binary assumptions are still being perpetuated in STEM (science, technology, engineering, and math) classrooms. To support this claim, I highlight gender-restrictive examples from the *Next Generation Science Standards* (NGSS) (NGSS Lead States, 2013a), focusing on its "all standards, all students" approach. And though I recognize that there has been advancement for girls (particularly cisgender girls; see Enke, 2012, for a history and critique of the term *cisgender*), we also need to offer support to gender-nonconforming students (e.g., girls who present masculine traits, boys who present feminine traits, and trans students). I consider a curriculum that purposefully recognizes and includes various gender identities. However, I then contemplate that even this approach is too narrow. I conclude by arguing that STEM education should move beyond the gender binary—and other conveniently constructed labels—through implementing queer theory to push the boundaries of and deconstruct what it means to be a girl (or a boy or other gendered ways of being) in STEM classrooms.

THE COLLAPSE OF SEX AND GENDER

Before discussing how the normative gender binary is being preserved in STEM classrooms, I first address the collapse of sex and gender (Lee & Buxton, 2010) that occurs when discussing gender identity within classrooms. I am concerned about the way *sex* and *gender* are often used interchangeably within education research and policy. Lee and Buxton (2010) explain:

> While gender and sex are often thought of as synonymous and representative of a clear biological distinction, social scientists usually make a distinction between the two. Sex generally refers to the biological aspects of being male or female, while gender generally refers to the behavioral, social, and psychological characteristics of men and women or boys and girls. Thus, while sex is biological, gender is a social construct. While sex is generally fixed (at least in human beings), gender is more fluid and is largely the result of socialized behavior patterns. For the purposes of demographic distinctions and how terms are used in relation to educational data, however, gender is almost always used to represent the biological dichotomy between males and females; in other words, it is collapsed with sex. (p. 13)

By collapsing sex and gender, many of our students' identities and attributes are not recognized or addressed. Boys and girls are "'essentializ[ed]'" (NGSS Lead States, 2013c, p. 2)—homogenized into the normative gender binary—in much of education research. This is problematic. Letts (1999) describes the serious consequences of the collapse of sex and gender as follows:

> The issue I want to raise here for consideration is the automatic slippage that superimposes dichotomous notions of gender—femininity and masculinity—onto "biological" sex in such a way that makes gender also seem dichotomous, discrete, polarized, and "natural"—in effect, making it heterogender. There are more ways to think critically about gender than the broad categories "male" and "female" provide us, such as the importance that gender plays in an individual's life. To ignore this complexity and multilayeredness of gender makes it seem as if gender is naturally, or biologically, occurring (essentializes it), and as if a person is some comprehensible embodiment of either "masculine" *or* "feminine" (oversimplifies and reduces it). (p. 103)

This chapter does not explicitly grapple with biological *sex* identities. It also does not collapse gender and sex—but recognizes the relationship between them, for the terms are not synonymous or prescriptive. Moreover, it is worth noting at the outset that the commonly seen normative biological sex binary (male/female) is as much in doubt as the normative gender binary (masculine/feminine). Biological sex identities, like gender, occur on a spectrum (e.g., intersex) (Butler, 1990, 2004; Fausto-Sterling, 2000; Killermann, 2013). Nevertheless, as discussed below, the common assumption of a *sex* binary appears to do much of the (troubling) work underlying the assumption of a *gender* binary. Thus, these identities, and their complexities, should be recognized when considering who is represented in our classrooms.

NORMATIVE GENDER BINARY IN THE NGSS

Considerations about teaching "all students," and indeed of my own childhood memories, (re)surfaced as I read the NGSS, specifically "Appendix D—'All Standards, All Students': Making Next Generation Science Standards Accessible to All Students" (NGSS Lead States, 2013c); and "Case Study 5: Girls and the Next Generation Science Standards" (NGSS Lead States, 2013d). These texts purport to address "what classroom teachers can do to ensure that NGSS are accessible to all students" (NGSS Lead States, 2013c, p. 1). "Appendix D—'All Standards, All Students'"

> highlights the practicality and utility of implementation strategies that are grounded in theoretical or conceptual fremeworks. ... First, it discusses both *learning opportunities and challenges* that the NGSS present to student groups that have traditionally been underserved in science classrooms. [Author's note: The text later defines these groups as economically disadvantaged students, students from major racial and ethnic groups, students with disabilities, students with limited English proficiency, girls, students in alternative education programs, and gifted and talented students.] Second, it describes

> effective strategies for *implementation* of the NGSS in classrooms, schools, home, and community. Finally, it provides the *context* of student diversity by addressing changing demographics, persistent science achievement gaps, and education policies affecting non-dominant student groups. (pp. 1–2, original italics)

This portion of the NGSS includes seven case studies written by members of the NGSS Diversity and Equity Team. They are meant to "illustrate science teaching and learning of non-dominant student groups as they engage in the NGSS" (p. 2):

> The terms "dominant" and "non-dominant" groups are used with reference to student diversity (Gutiérrez & Rogoff, 2003). The dominant group(s) does not refer to numerical majority, but rather to social prestige and institutionalized privilege. This is particularly the case as student diversity is increasing in the nation's classrooms. Even where the dominant group(s) is the numerical minority, the privileging of their academic background persists. In contrast, non-dominant groups have traditionally been underserved by the education system. Thus, the term "non-dominant" highlights a call to action that the education system meets the learning needs of the nation's increasingly diverse student population. (p. 1)

"Case Study 5: Girls and the Next Generation Science Standards" (NGSS Lead States, 2013d) tells the tale of a third-grade engineering project focused on forest restoration. Throughout the case study, the authors highlight teaching strategies (based on research including Baker, 2013; Milgram, 2011; and Scantlebury & Baker, 2007) that, they contend, work particularly well for girls. They claim that the proposed strategies are specially geared toward the way girls behave, think, and learn. For example, "*Ms. G. chose a curriculum topic that had relevancy and real-world application, an angle that would purposefully interest and engage the girls in the class*" (NGSS Lead States, 2013d, p. 3, original italics). However, these claims provoke such questions as: What makes this a "girl" interest? What is the difference between a "girl" interest and a "boy" interest? Are these interests based on gender or sex? Are these interests relevant only to gender-conforming girls? Who determines these gender or sex differences?

Another example reads, "*Ms. G. previously had brought in a female expert in science, a strategy that enhances girls' confidence in science*" (p. 4, original italics). While I do not disagree that this strategy boosts girls' confidence, it also shows boys that women are scientists too, which may start to breakdown misogynistic views held about girls within STEM classrooms. There is a reason why girls are considered the nondominant gender and sex in these scenarios, and this underlying problem of feminine and gender-nonconforming oppression is not being addressed within the classroom—nor is there a prescription to do so. Furthermore, there is little or no evidence of constructive interaction between girls and boys demonstrated within the scenario described in "Case Study 5." This is disconcerting because to begin the breakdown of the male- and masculine-orientated practice of science—which would clearly benefit girls—girls, boys, and students of *all* genders ought to be encouraged to engage in and collaborate on scientific tasks.

Here are the three other examples of teaching strategies recommended for girls:

> *Ms. G. carefully planned the partners and groupings in her class, an organizational structure that encourages participation for the girls in science.* (p. 4, original italics)

> *Having students generate the problems and possible solutions that will drive class activities, an important scientific practice, is also very motivating for all students, including girls.* (p. 6, original italics)

> *The school's after-school club represents a structural decision that positively impacts girls and their attitude toward science.* (p. 6, original italics)

The writers of "Case Study 5" state that the strategies proposed will help all students, but insist that they work "particularly [well] for girls" (p. 2). It is still not clear to me, however, how or why these strategies might do so. Indeed, much of the research on which the strategies rest, such as Baker (2013) and Milgram (2011), is immersed in heteronormativity, grossly collapses sex and gender, and unreflectively promotes gender conformity. Perhaps just as the NGSS need to move beyond gender and other binaries, so does the underlying research on "girls" (and all other genders and sexes) in STEM.

Letts (1999) realized that "the types of language being used and the ways in which it is used seems essential to exposing and disentangling the heteronormative nature of school science" (p. 105). Notably, the language used in the NGSS assumes that one is working within heteronormative classrooms with heteronormative, cisgender students, which are normatively masculine spaces (Letts, 1999, 2001a, 2001b). Although one might consider it progress that the NGSS attempt to remedy the disadvantages for or exclusion of girls in STEM classrooms by offering teaching strategies that will benefit them, these texts may undermine that purpose by reflecting detrimental heteronormative assumptions about students. They do this, in part, by crudely collapsing considerations of gender into simplified sex categories. Indeed, the words *gender* and *girls* were used interchangeably throughout the April 2013 online version of "Appendix D" (NGSS Lead States, 2013b)—though in the June 2013 version of the NGSS the word *gender* had been replaced with *girls* (NGSS Lead States, 2013c). And while "Case Study 5" (NGSS Lead States, 2013d) is supposed to enlighten teachers on how to teach science to include the nondominant gender—girls—the instruction comes across as demeaning to girls *and* boys as it reflects normative, binary male/female and masculine/feminine beliefs and stereotypes.

In reflecting on my own "masculine" and "feminine" actions, thoughts, and traits, I found myself, a girl who had a passion for science, upset by the invisibility of queerness—in this case, gender-nonconforming identities—within the NGSS documents. I again felt like that unheard child from so long ago. I pondered: *What about students like me?*

INVISIBILITY IN STEM CLASSROOMS

The scenarios and strategies presented in the NGSS (2013a) are steeped in heteronormative assumptions (e.g., our female students are cisgender girls who are not interested in learning science). Moreover, the NGSS send the message that scientists, students, and community members with gender-nonconforming practices do not warrant recognition—and this is certainly not true. Consequently, I am not the first scholar to address such assumptions in the NGSS and earlier STEM reforms (e.g., National Research Council, 2007, 2008, 2009, 2011) (see Fifield & Letts, 2014; Letts, 1999, 2001b; Rodriguez, 1997). Fifield and Letts (2014) affirm that "'science for all' entails a discourse of invisibility (Rodriguez, 1997) in which the lives of those we wish to include in science education are homogenized and reduced to those who already dominate science education and the broader culture" (p. 399), i.e., straight, white, middle-class, cisgender males marked by heteronormative masculinity.

To illustrate this masculine, heteronormative domination of STEM, I offer a personal anecdote. Throughout my teaching career, at the start of a science class, I would ask my middle school students (though I also found this useful at the elementary school, high school, and undergraduate levels) to draw a "scientist." Most of my students (at all levels), year after year, would draw a white man with crazy hair wearing a lab coat and holding a test tube—the normative image of a scientist. In my experiences, I have found that there are girls who draw women scientists (usually in a lab coat with test tubes), but very few—and, in fact, I have witnessed only one student draw a scientist of color (and she, too, was dressed in a lab coat). But the reality is that scientists come in many sexes, genders, and ethnicities—and they don't all work in a laboratory. How we approach STEM education will either perpetuate biased societal norms or allow students to envision a plethora of hand-drawn scientists—scientists who perhaps look like them.

In addition to heteronormative assumptions, the portrayal of girls in the NGSS also omits other attributes. Identities such as race, ethnicity, sexual orientation, socioeconomic status, and ability remain widely invisible—or are addressed separately in the other NGSS case studies. Furthermore, the fact that there already are girls (and boys) with deep desires to learn about and do science is not significantly addressed—at least not with the help of prescribed strategies. Indeed, the absence of nonconformity led me to wonder whether the NGSS should be more sensitive to the nuanced diversity of gender identities.

GENDER INCLUSIVENESS IN STEM CLASSROOMS

More robust gender inclusiveness thus seems, at least at first blush, to be a program worth considering (even if, for reasons explored below, it is ultimately inadequate).

In other words, *if* gender is a relevant consideration in how education is performed, then the masculine/feminine binary is simply too blunt an instrument. *If* educators need to tailor their methods along gender lines, then they should consider a broader range of gender identities (e.g., androgynous, genderqueer, transgender, two spirit).[1] *Is being cisgender the only way to define girls and boys? What about transgender or genderqueer students? How do we teach all students—even our nonconforming students?*

What if we consider other gender identities? One result of this might be that we spend time developing strategies for the many categories that fall outside of normative binaries—and doing so may at least be a useful interim step toward inclusive STEM instruction. Indeed, in addition to the relatively singular identities of gender, educators must address the complexities of the intersectionality of identities (e.g., gender, sex, race, ethnicity, socioeconomic status, sexual orientation, religion, exceptionality) of all students. After all, the goal of "Appendix D—'All Standards, All Students'" is to provide guidance to educators on how to make science "accessible to all students" (NGSS Lead States, 2013c, p. 1). And the NGSS warn us that "students could belong to multiple categories of diversity" (p. 2). Furthermore, they assert that "as there is a wide variability among students within each group, 'essentializing' on the basis of a group label must be avoided" (p. 2). Where the NGSS fail to follow their own advice, however, is in essentializing gender labels by excluding queer and nonconforming identities. Ultimately, by gendering (or sexing, as the case may be) teaching strategies along simplistic, binary lines, the NGSS perpetuate hegemonic and misogynistic ideologies and practices—thus compounding the intersectionality of racism, classism, homophobia, heteronormativity, and ableism—within K–12 STEM classrooms and teacher education.

One solution to this problem might be to take account of the more multifaceted variety of gender identities currently recognized—at least among individuals attuned to such things. Perhaps, in the words of the NGSS, there are "effective strategies" particularly suited to teaching not just "boys" and "girls," but also those of our students whose gender identities are different from the norm. In other words, just like the NGSS attempt to address a nondominant gender (girls), curriculum developers could consider and use many teaching strategies that may be geared toward, and more inclusive of, transgender and other gender-nonconforming identities. Although cooperative learning strategies should prove beneficial to all students, perhaps teachers could be persuaded against grouping students based on gender or sex, to avoid "outing" or causing discomfort to students with nonconforming genders. Or, more specifically, perhaps the varied categories of gender identity require their own tailored teaching strategies. Is there a way to "*purposefully interest and engage the* [*genderqueer students*] *in the class*" (NGSS Lead States, 2013d, p. 3, original italics)? Is there "*a strategy that enhances* [*androgynous students'*] *confidence in science*" (p. 4, original italics)?

We should recognize, of course, that this is a responsibility classroom teachers may not be aware of and may be ill-prepared to fulfill. As Barton (1998) explains:

> Teaching a gender-inclusive science in schools is a daunting task, for it requires a change in the ways that teachers are prepared to teach science. Teachers need to enter the teaching profession with a special knowledge of subject matter and the student body to engage students from diverse backgrounds in meaningful and participatory learning. … Teachers need to have a thorough understanding of science, including its content, history, culture, and discursive practices, as well as an understanding of student populations and educational processes, so they can provide opportunities for personally relevant engagement in science by a wide variety of students. Without this extensive knowledge, teachers will lack the ability needed to help students conceptually understand the knowledge base of science and to engage in scientific discourse and practice. (pp. 11–12)

But perhaps this, too, is the wrong approach. Perhaps neither girls nor students with nonconforming genders need special, or even different, treatment; perhaps they simply deserve equitable and just treatment. Curriculum should not be packaged in little pink, blue, or purple boxes—and, contrary to suggestions made in the research on which the NGSS rely (e.g., Milgram, 2011), neither should our students. Even if we were to add specific strategies and content directly relevant to students with nonconforming genders, we would almost certainly still exclude identities. And more to the point, what authority or curriculum developer would we empower to define and suggest teaching strategies for reaching the nonconforming, genderqueer child? Is that even possible? Is it the first step in a never-ending (and perhaps futile) Balkanization of education along ever-more complex gender lines? It may be that attempting to tailor education to such identities is doomed from the start.

So perhaps the solution to the problem of a restrictive two-category system (i.e., boy/girl, masculine/feminine) is not simply adding more categories. Perhaps *categories* themselves are the problem. What if we move beyond gender identities? Instead of focusing on the assumed "girls" and "boys," why aren't we teaching beyond cisgender heteronormativity? Why aren't we teaching to all students? What makes the labels *girl* or *boy*—or even more finely tuned categories—so *desirable*?

MOVING BEYOND GENDER

I am not proposing that we ignore or erase gender identity. Just as U.S. Supreme Court Chief Justice John Roberts has been rightly criticized for contending that "the way to stop discrimination on the basis of race is to stop discriminating on the basis of race" (*Parents Involved in Community Schools v. Seattle School District No. 1*, 2007), the way to solve the problem of excluding students with nonconforming genders is not to ignore gender. But by curbing the *desire* for gender conformity within STEM education, by using queer theory to challenge the idea of normative gender and sex binaries (Pinar, 2009), and by queering our approach to teaching STEM, we may come closer to meeting the challenge of teaching all our students.

One way to avoid silencing our nonconforming students is to do away with restrictive assumptions about, and desires for, familiar gendered and sexed behaviors. Sears (1999) warns:

> When it comes to sexuality and gender, many educators mold children into curriculum cookie-cutter identities. Students are socialized into this make-believe world of self and the other: male/female (biological sex), heterosexual/homosexual (sexual orientation), man/woman (gender identity), masculine/feminine (gender roles), opposite/same (sexual behavior), straight/gay (sexual identity). (p. 5)

But—pushing beyond gender and other prepackaged labels—"teaching queerly demands we explore taken-for-granted [or essentialized] assumptions about diversity, identities, childhood, and prejudice" (Sears, 1999, p. 5). Student identities are complicated, and by "urging [teachers] to open our minds and hearts so that we can know beyond the boundaries of what is acceptable, so that we can think and rethink, so that we can create new visions," we might "celebrate teaching that enables transgressions—a movement against and beyond boundaries" (hooks, 1994, p. 12).

I concur with Letts (2002) that an "add-and-stir" approach (which I considered above) is not sufficient. By avoiding a "laundry list" of gender identities, and by advocating for "a more critical examination of the assumptions that infuse current conceptualizations of equity in education," queering STEM education

> extends beyond a concern for the identities of the students who sit before us in our classrooms. Instead, this stance explicitly uncovers and works against heteronormativity, deconstructs the powerful allure of dichotomies (such as male/female, [masculine/feminine, boy/girl]), and originates from a unique epistemological standpoint. (pp. 119–120)

By queering our approach, we move beyond gender, maybe to a place that includes all students.

Through queering the approach to STEM education by "examining one's beliefs about sex, gender, and sexuality and coming to terms with one's comfort and knowledge (or lack thereof) about these topics and the epistemological implications that all of this has for each of us" (Letts, 2002, p. 124), teachers will move toward teaching all students. To be clear, "the point of queering is not to fit the 'all' into one or more norms, but to mark, deconstruct, and pass beyond the regulatory impulse to fit being(s) into finite categories" (Fifield & Letts, 2014, p. 395). Teachers who expect girls to conform to feminine stereotypes and assumed behaviors will continue to curb girls' performances in masculine-oriented spaces of STEM learning. Similarly, boys who do not conform to assumed masculine norms (e.g., "sissies," "babies," "fags") will struggle in science curricula that embed those norms (Letts, 1999, 2001a, 2001b). Letts (2001a) asserts that

> we need to start to view schools, and specifically in this case school science, as sites of intervention, and as offering resources for the refashioning of gendered, racialized, classed, and heterosexualized identities by explicitly taking up these issues in our instruction. (p. 193)

The suggested classroom strategies put forth by the NGSS (2013a) are, in fact, effective strategies for *all* students. But categorizing these strategies by gender seems haphazard—and essentialist. Instead, perhaps educators could implement strategies untethered to gender or any essentialist notions. For example, Lyon, Jafri, and St. Louis (2012) contend that "a growing body of research shows that students who do not find personal meaning or relevance in STEM will not pursue STEM beyond what is required in school" (p. 49). Indeed, this is one strategy suggested in "Case Study 5." And although some of the research on which they base this claim found it to be true for girls (Jeffe, 1995), making science relevant to students' lives also benefitted other groups of minority students (Basu & Barton, 2007; Campbell, Denes, & Morrison, 2000; Zacharia & Barton, 2004).

Why must we gender or sex such teaching strategies? How is homogenizing the diversity trapped within these labels useful for us in STEM education? Is it so we can differentiate the conformers (the dominant groups) from the nonconformers (the nondominant groups)? At a minimum, we need to be asking ourselves these questions before recommending strategies that have a direct effect on students. And at our best, we should be recommending strategies that recognize the fluid, and perhaps uncategorizable, nature of student identities.

CONCLUSION

Being a queer student in a heteronormative STEM classroom is not easy. I do not see students like me recognized in the NGSS (NGSS Lead States, 2013a). How do we nurture the students who are pushing the boundaries of what it means to be a girl (or a boy) in STEM classrooms (Letts, 2001a, 2001b)? Ultimately, perhaps by implementing queer theory—frameworks that deconstruct binaries like masculine/feminine, male/female, and boy/girl (Fifield & Letts, 2014; Sullivan, 2003)—we might advance toward an equitable educational experience that considers all students, *even students like me*.

When I was a STEM teacher, I did my best to create a safe learning environment for *all* my students. I made a conscious effort to secure a safe space for nonconforming, or queer, students—as I am aware that these students make up a "silent minority" in classrooms (Casper, Schultz, & Wickens, 1992; Hulsebosch, Koerner, & Ryan, 1999)—by pushing beyond normative gender assumptions. Atwater (2010) reminds us that "it is in their science [technology, engineering, and math] classrooms that teachers get to decide what voices are heard, when they are heard, and under what circumstances they are heard" (p. 104). *I will always be grateful for the science teachers who heard my voice.*

Brickhouse, Lowery, and Schultz (2000) "wonder whether [girls'] willingness to violate gender norms might not make them better able to withstand the sexism of the scientific workplace (Eisenhart & Finkel, 1998) than women who conform more to gendered societal expectations" (p. 456). I wonder whether my gender-nonconformity contributed to my success in science classrooms. Until or unless we address the gendered and sexed issues and assumptions that shape opportunities in science classrooms, "girls" will not succeed—*unless they are students like me.*

NOTE

1 As a queer educator, it would be irresponsible of me not to offer descriptions of the following terms:
- Someone who is *androgynous* may not be obviously male or female at first glance because they merge what are stereotypically masculine and feminine characteristics in many different ways.
- *Genderqueer* describes people whose gender identities exist outside of the traditional masculine/feminine binary.
- When you are *transgender*, you have a gender identity or gender expression that is different from your assigned biological sex.
- Certain Native American cultures describe transgender people as having *two spirits* (Huegel, 2011).

REFERENCES

Atwater, M. M. (2010). Multicultural science education and curriculum materials. *Science Activities, 47*, 103–108.

Baker, D. (2013). What works: Using curriculum and pedagogy to increase girls' interest and participation in science and engineering. *Theory into Practice, 52*(1), 14–20.

Barton, A. C. (1998). *Feminist science education.* New York, NY: Teachers College Press.

Basu, S. J., & Barton, A. C. (2007). Developing a sustained interest in science among urban minority youth. *Journal of Research in Science Teaching, 44*(3), 466.

Brickhouse, N. W., Lowery, P., & Schultz, K. (2000). What kind of a girl does science? The construction of school science identities. *Journal of Research in Science Teaching, 37*, 441–458.

Butler, J. (1990). *Gender trouble.* New York, NY: Routledge.

Butler, J. (2004). *Undoing gender.* New York, NY: Routledge.

Campbell, G., Denes, R., & Morrison, C. (2000). *Access denied: Race, ethnicity, and the scientific enterprise.* New York, NY: Oxford University Press.

Casper, V., Schultz, S., & Wickens, E. (1992). Breaking the silences: Lesbian and gay parents and schools. *Teachers College Record, 9*, 109–138.

Enke, A. F. (2012). The education of little cis: Cisgender and the discipline of opposing bodies. In A. Enke (Ed.), *Transfeminist perspectives: In and beyond transgender and gender studies* (pp. 60–77). Philadelphia, PA: Temple University Press.

Fausto-Sterling, A. (2000). *Sexing the body: Gender politics and the construction of sexuality.* New York, NY: Basic Books.

Fifield, S., & Letts, W. (2014). [Re]considering queer theories and science education. *Cultural Studies of Science Education, 9*, 393–407.

Gutiérrez, K., & Rogoff, B. (2003). Cultural ways of learning: Individual traits or repertoires of practice. *Educational Researcher, 32*, 19–25.

hooks, B. (1994). *Teaching to transgress: Education as the practice of freedom.* New York, NY: Routledge.

Huegel, K. (2011). *GLBTQ: The survival guide for gay, lesbian, bisexual, transgender, and questioning teens.* Minneapolis, MN: Free Spirit Publishing.

Hulsebosch, P., Koerner, M. E., & Ryan, D. P. (1999). Supporting students/responding to gay and lesbian parents. In W. Letts & J. Sears (Eds.), *Queering elementary education: Advancing the dialogue about sexualities and schooling* (pp. 183–193). Lanham, MD: Rowman & Littlefield Publishers.

Jeffe, D. (1995). About girls' "difficulties" in science: A social, not a personal matter. *Teachers College Record, 97*(2), 206–226.

Killermann, S. (2013). *The social justice advocate's handbook: A guide to gender*. Austin, TX: Impetus Books.

Lee, O., & Buxton, C. A. (2010). *Diversity and equity in science education: Research, policy, and practice*. New York, NY: Teachers College Press.

Letts, W. J. (1999). How to make "boys" and "girls" in the classroom: The heteronormative nature of elementary-school science. In W. Letts & J. Sears (Eds.), *Queering elementary education: Advancing the dialogue about sexualities and schooling* (pp. 97–110). Lanham, MD: Rowman & Littlefield Publishers.

Letts, W. J. (2001a). Boys will be boys (if they pay attention in science class). In W. Martino & B. Meyenn (Eds.), *What about the boys? Issues of masculinity in schools* (pp. 186–198). Philadelphia, PA: Open University Press.

Letts, W. J. (2001b). When science is strangely alluring: Interrogating the masculinist and heteronormative nature of primary school science. *Gender and Education, 13*(3), 261–274.

Letts, W. J. (2002). Revisioning multiculturalism in teacher education: Isn't it queer? In R. M. Kissen (Ed.), *Getting ready for Benjamin: Preparing teachers for sexual diversity in the classroom* (pp. 119–131). Lanham, MD: Rowman & Littlefield.

Lyon, G. H., Jafri, J., & St. Louis, K. (2012). Beyond the pipeline: STEM pathways for youth development. *Afterschool Matters, 16*, 48–57.

Milgram, D. (2011). How to recruit women and girls to the Science, Technology, Engineering, and Math (STEM) classroom. *Technology and Engineering Teacher, 71*(3), 4–11.

National Research Council. (2007). *Taking science to school: Learning and teaching science in grades K–8*. Washington, DC: National Academies Press.

National Research Council. (2008). *Ready, set, science! Putting research to work in K–8 science classrooms*. Washington, DC: National Academies Press.

National Research Council. (2009). *Learning science in informal environments: People, places, and pursuits*. Washington, DC: National Academies Press.

National Research Council. (2011). *A framework for K-12 science education: Practices, crosscutting themes, and core ideas*. Washington, DC: National Academies Press.

NGSS Lead States. (2013a). *Next generation science standards: For states, by states*. Washington, DC: The National Academies Press. Retrieved from http://www.nextgenscience.org

NGSS Lead States. (2013b). *Next generation science standards: For states, by states*. Appendix D – "All standards, all students": Making the *Next Generation Science Standards* accessible to all students. Retrieved from http://www.nextgenscience.org

NGSS Lead States. (2013c). *Next generation science standards: For states, by states*. Appendix D – "All standards, all students": Making the *Next Generation Science Standards* accessible to all students [June 2013 version]. Retrieved from http://www.nextgenscience.org/sites/default/files/Appendix%20D%20Diversity%20and%20Equity%206-14-13.pdf

NGSS Lead States. (2013d). *Next generation science standards: For states, by states*. Case study 5: Girls and the *Next Generation Science Standards* [June 2013 version]. Retrieved from http://www.nextgenscience.org/sites/default/files/%285%29%20Case%20Study%20Girls%206-14-13.pdf

Parents Involved in Community Schools v. Seattle School District No. 1, 551 U.S. 701. (2007).

Pinar, W. F. (Ed.). (2009). *Queer theory in education*. New York, NY: Routledge.

Rodriguez, A. J. (1997). The dangerous discourse of invisibility: A critique of the national research council's national science education standards. *Journal of Research in Science Teaching, 34*(1), 19–37.

Scantlebury, K., & Baker, D. (2007). Gender issues in science education research: Remembering where the difference lies. In S. K. Abell & N.G. Lederman (Eds.), *Handbook of research on science education* (pp. 257–285). Mahwah, NJ: Lawrence Erlbaum Associates.

Sears, J. T. (1999). Teaching queerly: Some elementary propositions. In W. Letts & J. Sears (Eds.), *Queering elementary education: Advancing the dialogue about sexualities and schooling* (pp. 3–14). Lanham, MD: Rowman & Littlefield Publishers.
Sullivan, N. (2003). *A critical introduction to queer theory*. New York, NY: New York University Press.
Zacharia, Z., & Barton, A. C. (2004). Urban middle-school students' attitudes toward a defined science. *Science Education, 88*(2), 197–222.

ABOUT THE AUTHOR

Michelle L. Knaier uses queer theory to advocate for LGBTQ-inclusive teacher education and K–12 curriculum. She is a doctoral candidate at Purdue University, in West Lafayette, Indiana (USA), enrolled in the Department of Curriculum and Instruction curriculum studies program. Her research focuses on multicultural education. Michelle also is a graduate of SUNY Cortland, holding a master of education in childhood education, with an emphasis in educational technology; and a bachelor of science in education in elementary and early secondary education, with an emphasis in biology. She often reflects on her experience as a former middle and high school science teacher while incorporating critical theories into her research on queer(ing) curriculum.

SHERI LEAFGREN AND SCOTT SANDER

11. CHILDREN, NOMADS, AND QUEERING

Desire and Surprise in a Wiggly World

INTRODUCTION

While visiting a new city, a friend cautioned Sheri, "You can't walk to your hotel by yourself—it's not safe; you're always looking sideways and up and down, touching buildings and dogs, and wandering after whatever catches your eye! You walk around like a 5-year-old!" He further admonished, "You should look straight ahead and act like you know where you're going!" At first, possibly feeling that she should feel afraid (but didn't) or offended (but wasn't), Sheri carefully examined his concern. She thought about the two kinds of traveling she might have undertaken on her trek back to the bargain hotel near the bus station: walking like an adult with straight-ahead looking and determined purpose, or "like a 5-year-old" with round-about looking and indeterminate destination(s).

There are benefits of straight-ahead looking: efficiency, relative safety, an appearance of competence, the security of knowing your destination, and confidence in knowing that the place you end up is the place you intended to go. But one cannot help but wonder what is missed along the way. In this particular walk, Sheri might have missed the following:

- the enormously tall building completely backlit with a turquoise-y glow;
- three people—out of about 15 passers-by—whose top button on their pants was undone (from which she surmised a very good restaurant was on her route);
- the smoothness of a marble column;
- the slick bumpiness of a polished stone wall;
- a tree growing out of a mesh trash receptacle attached to a street light pole (yes, growing! she pulled on it and it was rooted to the mesh!);
- the unpredictable nature of the uneven sidewalk; and
- hearing the story of the man who needed and received from her exactly 68 cents, "no more, no less."

But she didn't miss those sights, sounds, surfaces, and stories. She chose to take the path that a 5-year-old would travel: one with misdirections and surprise—a path that suggested that perhaps her adult self had not yet settled down into the proper adult actions.

 | DOI:10.1163/9789004331068_013

In this chapter, we liken moving through the geographies of a city to that of moving through the spaces of the early childhood (science) classroom and curriculum. We see the classroom as Deleuze (as cited in Roy, 2003) saw the city: as having the potential to unleash uncontained elements "that are not fully captured or controlled by the machinery of the State[,] … [elements that serve as forces] of 'deterritorialization' that elude capture and regulation by the State[,] … evok[ing] powerful subversive energies and intertexts, and revers[ing] the usual dismal picture of urbanity" (pp. 4, 37–38). These uncontained elements act with a nomadic tendency toward a "violence in thought" that "state apparatuses dislike because it is a violence in thought that makes us less predictable, and for Deleuze, a new thought is born of a violence in thought" (pp. 37–38). Through an explanation of Deleuze and Guattari's urban nomad, Bond (2007) advances the concept of disrupting predictable routes through geographies of ordered space:

> The urban nomad lives in a thoroughly striated environment, yet, since there are dissymmetrical passages between smooth and striated spaces, the urban nomad may, in fact, "live smooth" and engage in "nomadic transit in smooth space." In short, the urban nomad can become even in the most striated spaces. But the urban nomad's becomings … are never easy but rather are "difficult" and "uncertain" … given that the passages and reversals between smooth and striated space always imply concomitant deterritorializations and reterritorializations of physical and mental spaces. (p. 9)

Britzman (1995), too, constructs a geographic concept of the classroom, resistant and elusive, raising the possibility of "articulating pedagogies that call into question the conceptual geography of normalization," perhaps even remapping normalcy as part of an ethical project. Remapping normalcy may require 5-year-old-like ways of moving through spaces—uncontained and unpredictable ways of negotiating landscapes of city or classroom.

We liken ways of moving—"like a 5-year-old" versus "as one should"—to, respectively, nomadic and sedentary relationships with negotiated movements in smooth and striated spaces. Deleuze and Guattari (1987/2004) contrast nomad life with sedentary life in ways that reflect ways of moving and being in the spaces of city and classroom. As Aldea (2014) explains in Deleuzoguattarian terms, sedentary and nomadic are used to "consider the use of space and people's relationship to the land they inhabit" (para. 7). She describes the striated space of the sedentary order as similar to an aerial view of agricultural land: "divided and demarcated[,] … defined by borders and boundaries[;] … as you move from one distinct place to another[,] … roads and walls determine the route you have to take" (para. 8). In striated spaces, destination points A and B are intractable, with little attention afforded to the traits of the spaces travelled in between. Aldea portrays the smooth space of the nomadic order as the desert—"an expanse of land, without clear borders or exclusive ownership … [where] stopping places are subordinated to the journey itself" (para. 9). While traveling nomadically, one is fully attuned to the traits of the space—traits such as uneven sidewalks, unbuttoned pants, and trees growing from trash receptacles along a road—and the journey becomes something other than what it might have been. This attunement and authentic noticing as a way of living and knowing with the world is a hallmark of science. And yet in classrooms—spaces ostensibly of science learning and knowing—children are told which way to look and what to look for, to disregard the traits of the spaces they journey. We adopt Broadway's (2011) vision of a queered curriculum of, as Fifield and Letts (2014) put it, "storied journeys of self, other, and the world that arc from the familiar to the strange" (p. 398). In this way, those familiar scenes—"in which nomadic selves were long ago disciplined, forced to settle down, to behave properly in proper time and space" (S. Fifield, personal communication, December 14, 2014)—of classroom and city become strange and queered.

We imagine a queering/ed pedagogy as adopting a route in which we do not look straight ahead to predetermined points defined by static roads and walls. Instead, we like thinking about traveling along an "impossible geography—[one that] exists and does not exist at exactly the same time" (Golding, 1993, as cited in Britzman, 1995, p. 155). This is a messy and unpredictable route, just the kind that allows a smoothing of the striations of State pedagogy, not leaving a tracing of a route, but making and taking one toward recovering possibilities foreclosed by straight-ahead looking and straight thinking. In response to her titular question, Is there a queer pedagogy? Britzman (1995) wonders, "What is required to refuse the unremarked and obdurately unremarkable straight educational curriculum?" (p. 151). Here we suggest that children (some of them actual 5-year-olds) play a role in doing what is required: resisting the normalizing practices of the classroom, queering the curriculum, unfixing identities, and smoothing the striations that hold us gridlocked and frighteningly still.

WALKING AROUND LIKE A 5-YEAR-OLD: NOMADS AND QUEERS

> A child of five would understand this. Send someone to fetch a child of five. (attributed to Groucho Marx)

Explorimenting

Willie, a kindergartner, asked his teacher if they were going to be doing any more "exploriments" in school. From his joyful fusion of *experiments* and *explorations*, emerged a portmanteau that was neither original nor new to the larger world—but it was to him. His animating struggle with their words—grown-ups' words—for his work allowed for the generation of a fresh expression that perfectly captured the sort of relationship he and the other children enjoyed with the environment and with one another—in spite of their location in the sedentary, partitioned, and calmed-down spaces of School.

A kindergarten classroom is like a city where one might notice that walking around like a 5-year-old is, ironically, not often the expected or accepted means of moving through space. Deleuze and Guattari (1987/2004) write that, "one of the fundamental tasks of the State is to striate the space over which it reigns" (p. 385). The teacher, serving the State, does so by arranging the

> milieu to communicate clearly the striations and order that s/he sees as necessary to the classroom territory[:] … placing desks in … columns and rows, assigning places to sit on the carpet, and publishing documentation of compliances with stars on graph paper and non-compliances with color-coded behavior cards. (Leafgren, 2013, p. 278)

These structures make clear to anyone who enters the space what is expected and where to look (straight ahead) and what to do and not do. And so, when children arrive with their side-ways looks and their desire to touch, to talk, to intersect with other children and with things—to compose storied journeys that make the ordinary strange—the nomad/State opposition comes into play. The basic Deleuzoguattarian principle is that nomad space is smooth and heterogeneous, allowing movement to flow in infinitely various directions, while State space is striated and homogeneous. To better control its territory, the State stratifies and divides, thus ruling over interior movement, imposing external limits, and seeking to capture nomads as they follow their Stateless instinct (Beaulieu, 2009).

Willie's playful negotiation of language served in small ways to nomadically elude capture by the State as he disrupted the defined scientific terms posted on his classroom walls. Via a dissymmetrical passage between the smooth and striated spaces, Willie's *explorimemts* joined the State terms on the classroom walls, a crayon-drawn poster that included not only the coined word itself, but the "definition" as scribed by Willie, and drawings of children (by children) in the process of explorimenting. We wonder if pinning up the word neatly between commercially-made posters of scientific terms such as *experiments*, *hypothesis*, and *conclusion* served to capture it and make it static. However, as noted in Leafgren (2013), "explorations of the classroom as a Deleuzoguattarian place—always changing, always mutating—involve the interaction of smooth and striated spaces … and … serve as nomadography allowing a de-institutionalized experience with what is traditionally an institutionalized space: the classroom" (p. 278). Tally (2010) explains that in nomadography, fresh and unexpected territories are coconstructed in associations between nomad and State, "projecting an alternative line of flight—a line that will allow us to re-imagine the classroom in enlightening and productive new ways" (p. 15). Willie's violence in thought led to the construction of not only a word, but of the concept represented by his word, and so produced an alternative line of flight—one without the limits that had been in place before he had asked his question.

We liken the rhizomatic nature of the nomad and nomadography to queer theory's study of limits and suggest that as children act as nomads, they queer the curriculum.

And vice versa. Children are not alone in acting nomadically and queering the limits, and we are not suggesting a dichotomy that places children as nomads and adults as State. However, as human beings who haven't settled down yet, who perhaps experience the world differently than do adults at times and places, children's

> living, breathing disobedience—their resistance to domination—implies a rhizomatic, deterritorializing interaction within the enclosed and partitioned structures of the classroom space and interactions, and therefore manifests a nomadic penchant for resist-ing the restrictive techniques … of the sedentary, partitioned and designated spaces of the dominant culture's [adult's] classroom environment. (Leafgren, 2009, p. 101)

Thus, we share children's, and sometimes their teachers', storied journeys in navigating classroom spaces—following them as they flow every which way, disrupting and reimagining those "blessed-by-the-State trajectories" (Roy, 2003, p. 21) that so many adults feel compelled to follow. In doing so, we share a hope for possibilities that queering maintains.

The Roof

There was a day when people were making repairs on the roof of a school. Noises galore. Ladders striking the side of the building, materials hoisted onto the roof, men's voices calling out orders and cautions, tools scraping, pounding—all sounds reaching the ears of the teachers and children, but nothing could be seen from inside the building and classrooms. The sounds created a mystery for those inside—a discrepant event.

In one classroom, 27 kindergartners were occupied with making budget decisions for their next shopping trip to a nearby grocery store as their classroom filled with strange scraping sounds from above their heads. These rough noises assembled in relation to the children as traits of the classroom to attend to and be with—a storied journey arcing from the familiar to the strange. Thus, even while continuing their work, the children's budget discussions were punctuated with expressions of interest and wonder about the unexpected noises, and several of the children offered theories fanciful and pragmatic:

- "I think it's a witch up there!"
- "No, it rabbits running around! I hear their fingernails!"
- "I bet it's rain and snow, the hard kind!"
- "My roof does that when our house is settling down."
- "The roof is falling!"
- "It's a robber!"

In collaborating with the world—the unexpected sounds—the children became fascinated with this curiosity and spontaneously generated reasonable theories to explain the sounds. Absent direct(ed) STEM-inquiry instruction, the children aligned the evidence of the mysterious noises to their experiences. Some connected the sounds with what they know about roofs and the noises they make: weather, wood settling, the fragility of human-made objects. Others connected the sounds to recent classroom experiences: the rough sounds resembled the sound of the classroom rabbit scratching across the hard floor. And one child noted that the noises were eerily similar to the sound effects Ganiva made when acting out Strega Nona's (a witch characterized in a book the children read and acted out together) stirring pasta in her magical pot. Gilbert and Gray (this volume, Chapter 4) allude to children's science as inquiry personalized. In the case of the roof sounds, the children's nascent theories were joyfully offered as a means of communicating and sharing their fascination with a phenomenon—evoking collaboration among the sounds, their own theories laid next to and over the theories suggested by their classmates, their teacher, and the day-to-day workings of their classroom and school.

In the more typical School/State striated space of the classroom, the expected response would have been to maintain straight-ahead looking, staying true to the course of the destinations predetermined by gridded routes to the next point. (Consider—even a point of intrigue and authentic interest, such as planning a budget for a grocery store visit, can become locked into the grid of State-sponsored striated space if children are not permitted to attend to the unexpected, interesting traits along the route, thus subordinating those traits to the route itself). And, in fact, for most of the children in the school, this is what happened: The blessed-by-the-state trajectories were safely maintained. However, these 27 children and their teacher followed the path of the 5-year-old, nomadically attending to the traits of the path, not the destination. They left their budget plans and the classroom to see what they could see—to investigate an unknown (and thoughtfully theorized) phenomenon.

As they walked together through the quiet School hallways, passing the open doors of other classrooms, they heard teachers' clipped voices—"Pay attention!" (But not to the mysterious sounds!) "Up here!" "I need your attention." "It's just a noise!" "Stop it."—and took in snapshot views of children sitting still at their desks placed in rows with eyes facing front or with heads bent to paper. This despite the mysterious noises that were impossible to ignore! Once outside, the kindergartners and their teacher walked around the school until they found the ladders leaning sturdily (yet somehow precariously) against a wall of the building that had no windows—the side where the children surmised the cafegymatorium would be.

By happy-stance, just as they came to the ladders, two men who had been on the roof were scrambling down and joined the children on the ground. One man remained on the roof and as the children stood below—absolutely transfixed—he tossed down one object after another: first, the workers' long-handled scrapers and smoothers, then three frisbees, a basketball, three playground balls, a football (inflated), a football (deflated), a single shoe, a hula hoop, and a pair of very stiff and dirty blue jeans. The workmen on the ground laughed and shielded the children from the falling detritus. The new information led the children to new questions and thoughts, and they asked workmen the following: What's up there? What were you doing? What made the noises? What about the noise that sounded like [imitates it with eerie accuracy]? How'd you get up there? Were you scared? What's wrong with our roof? What's that do [points to tool]? Which is harder, climbing up or climbing down? How'd you get your stuff up there? Is that our hula hoop? How'd it get up there? Whose pants are those? The workers were willing to answer questions—as many as they could.

The children's curiosity slaked for the time being, they thanked the roofers for their time and for explaining the noises to them, and they began their journey back to their classroom. And here is where the very best part comes: As the children were returning to their classroom, the teacher overheard two children discussing whether the witch would want to cook the rabbit if she managed to catch it on the roof and whether or not the men had tried to save it. The awe and fantasy were not pushed away by the reality of the men on the ladder, only elaborated through driving questions that would lead to further inquiry.

Honoring children's science—with its qualities of acute attention to stimuli, being fully present in the moment, embracing surprise and contradictions, and willingness to take risks and share vulnerable and fragile hypotheses with others—led the teacher to unanticipated ends. The discussion that followed was driven by the children as they processed what they had observed outdoors in alliance with their inside theories. Some children saw the theories of witches and rabbits debunked (e.g., "Those men would have said something … or screamed!"); others accommodated a theory with new evidence (e.g., "I'll bet those pants were the robber's pants—the work-guys scared them off of him!"). As children queer the Schooled curriculum path, they act as nomads.

Curriculum is a space the State desires to limit and control via the straight-ahead looking of the narrowed viewpoint inherent in School's scientifically-based quest for certainty. "Seen through a Deleuzian lens, a curriculum that follows narrow goals, that attempts to homogenize and limit the signs and processes of learning, … [that denies] difference and complex multiplicity, runs the risk of locking us into oppressive grids" (Roy, 2003, p. 4). In the anecdote described above, the 5- and 6-year-olds and their teacher smoothed the schooled/gridded path, actively engaging in STEM-identified actions such as scientific observation, thinking, communication, and problem-solving, all driven by their own attunement and curiosities. As the child is nomad in the school milieu, it follows that s/he would resist the restrictive implications of such a curriculum, "manifesting a nomadic penchant for creating lines of flight and resisting the restrictive techniques of power that the school-State serves to impose" (Leafgren, 2013, p. 276). It is this resistance that serves to smooth the rigid striations of the kindergarten classroom. Wonder smooths. Awareness smooths. Watchfulness smooths. Verve smooths. Enthusiasm smooths. Kinship smooths. Curiosity smooths. Spirit smooths. Inquisitiveness smooths. Initiative smooths. Excitement smooths. Desire smooths.

DESIRE IN A WIGGLY WORLD

> You're only making a mess by trying to put things straight. You're trying to straighten out a wiggly world and no wonder you're in trouble. (Watts in Watts & Putterman, 1971b, video 8:59–9:07)

Aah, desire. What animating desires lie in the hearts of schoolchildren and what deadening aversions dwell in the State apparatus of School? Within notions of desire one can find the simultaneous actuality of oppressive State instrumentality and liberating nomadic freedom. Would desire be so sweet without the State? As Gao (2013) explains:

> Desire is usually understood as something abnormal, avaricious and excessive, the opposite of rationality, to be controlled and suppressed in [hu]man[s]. Deleuze's desire is much wider, referring not only to [hu]man[s], but also to animals, objects and social institutions. In Deleuze's view, desire is not a psychic existence, not lack, but an active and positive reality, an affirmative vital force. Desire has neither object, nor fixed subject. It is like labour in essence, productive and actualisable only through practice. (p. 406)

If School operates in ways to control and suppress the perceived-as-irrational desires of school children—trying to put the wiggly straight, as suggested by Watts in the epigraph above—Kesson (2005) makes clear those same desires are an affirmative vital force, a "dynamic momentum for human activity [and the] foundation for authentic engagement with the world" (p. 43). Zembylas (2007) theorizes a pedagogy of desire that arouses "creative, transgressive and pleasurable forces within teaching and learning environments" catching students and teachers in "powerful flows of desire" that subvert representations and significations that have been normalized and stratified by the State and its instruments (p. 331). We, along with the editors of this volume, argue that queering attends to these desires and to the odd (im)possibilities that are necessary to unsettle normalized and institutionalized significations. "Queer necessitates insisting that queer (v.) is … how to make schools educative" (Broadway, 2011, p. 301). Teachers and children, in queering, may "'awaken in another human being powers, dreams beyond one's own; to induce in others a love for that which one loves; to make of one's inward present their future (Steiner, 2003, pp. 183–184)'" (Broadway, 2011, p. 301). Below, we share an example of young children's storied journeys as they interact with their teachers, the School, and the world in desirous and nomadic ways toward a Deweyian-like goal of nurturing mutual dispositions and abilities for purposes of generous and generative living: "generous, in being sensitive to others, and generative in being in love with life and a desire to learn and grow" (Leafgren, 2009, p. 243).

A Flower Desires to Be Given as a Gift

> Animals can desire, ants, the heart can, and fingers, and hands, everyone can desire. (child quoted by Paola Barchi in Cardy, 2011)

There were days in which 10 children, ages 4 to 5 years old, went walking with their teacher, Paola Barchi in Reggio Emilia, Italy. The following account of their walks is constructed from notes taken by a colleague of ours, Terri Cardy, during a presentation of children's work by their teacher at an International Study Group at the International Centre Loris Malaguzzi in Reggio Emilia, Italy (Cardy, 2011).

While on their walks, the children travelled a wiggly path through the city, thus enacting a pedagogy of desire described by Zembylas (2007) as "a landscape of becoming, in which forces, surfaces and flows of teachers/students are caught up in a desiring ontology" (p. 331). The storied journeys were transformed as the children and their teacher attended to traits along their route: in this case, a tall wooden wall that obstructed a view of construction and renovation. The children couldn't see what was happening behind the perplexing enclosure, and they became fascinated by the mystery of it. Following their path of fascination, the children's conversations took them to an unexpected place: desire.

> The children compare and create theories around the theme of desire, which to their mind is first and foremost an attitude, the tension and vaulting of knowledge toward encounter with the world. The children attribute values such as universality, immaterialness, and generosity to the idea of desire. (Cardy, 2011, p. 2)

In their encounter with this surprise, the children considered the relationship of the new wall to their city. They wanted to give the city a gift of perhaps making the wall beautiful; they saw that the wall wanted to be beautiful. The children understood that desires were not only involving humans, but also about things having desires, desires that emerge and live in interactions.

> *Teacher*: How can a wall feel our desires?
>
> *Child*: Hold desires in hand and then place my hand on the wall and push for a long time. Then the wall will feel desire.

Paying attention to what happens in the moment on wiggly walks through the world is a practice of close observation and knowing and feeling. The children were able to see things in their journey—in the traits of their path, in moments of surprise, in the everyday-ness of a wall that had been assembled by workers to protect passers-by—that were significant and worth attending to.

On a different walk on a different day, a child noticed a plant—a flower—growing from the pavement. This flower held lovely significance: "That plant wants to tell us it has a perfume." And so, the walk led them to a new story in their journey. "A genuine purpose always starts with an impulse [the wall! the flower!]; … the impulse becomes a desire" (Dewey, 1938, p. 67). The desire flows into purpose as the child's observations of what surrounds them derive significance in the context of the larger whole—assemblages of places, people, objects, self, movements, time, thoughts. Children are thus, in Deleuzoguattarian terms, "'desiring machines' or 'assemblages[,]' … desire flows through and between (human and non-human) machines/assemblages/bodies in complex ways" (Ringrose, 2011, p. 600).

Child: The flower desires to be given as a gift!

Desire happens without positing a desiring agent or desirable object; it is emergent in relations and not the property or act of any one being or thing. This gives desire back to the relational universe of being—being scent, flower, city, human, wall in relation. In this instance, the child saw the flower growing from a crack in the pavement in the city—a city she cared about—and she shared her observation with other children who also loved the city.

Child: With my nose, I desire to smell the perfume of flowers. Everyone can smell perfume and it makes the city happy.

The children's love + city + the flower + the flower's scent + their desires + the city's desires assembled in qualities of generosity and generativity.

In their work together, the children and the teacher as part of desiring assemblages relax contentedly into the moment, becoming a communion of mutual curiosity and wonder—a collaborative inquiry of the world.

> The person is the product of the endless play of desire and arises at an intersection of specific interplays. … Instead of aligning desire with fantasy and illusion[,] … Deleuze explains desire as what produces the real, creating connections, relations, and alignments. … It is the primary producer of reality and all relations within it. (Roy, 2003, pp. 42–43)

From the desires in assemblage with the city, flower, wall, perfume, and desire itself, emerged a genesis of lived experiences in a wider framework of collaborative inquiry with the world—a relationship among individuals, objects, place, traits, and society collectively relaxing contentedly into the moment of a moving and evolving interplay.

Child: Lifecycle of perfume: It goes into the body when it is light; it leaves the body and goes up into the sky.

Just as in the opening narrative—a tale of walking around like a 5-year-old—these children and their teacher travelled nomadically while fully attuned to the traits of the city-space, traits such as a wall where there had been no wall before and the significance of a flower growing in pavement. A child says, "That plant wants to tell us it has perfume," and the children and their teacher embrace mystery and wonder, and, in a pedagogy of desire, adopt an attitude of encounter with the wiggly world.

Child: Desire comes to us by shooting stars, come at night on parachutes; knock on door and then child gets desire.

SOME THINGS ARE JUST BEGGING FOR A GOOD QUEERING

> Nearly every hegemonic account of the genesis of the child is [grounded in] the assumption that the child in question will either grow up into an adult or arrest at an inhuman stage that was on its way to that inevitable and singularly desirable outcome before something went horribly wrong. In any case, no matter what the deviation, the only way to think a child is in comparison with the adult it is meant to eventually become. … Little kids grow, we say. "Up." Up into what? Adults, and nothing else. (Gill-Peterson, 2013, para. 1)

Children are defined by what they are not, and so are plumbed with a standard metric of development that "subtends the concept of the child, the adult, the human itself … [and] is the reason why it seems nearly impossible to describe the child 'on its own terms'" (Gill-Peterson, 2013, para. 1). Rather, children are deemed incomplete and so, are thoroughly marginalized by their identity as child. Further inflating their marginalized state is the dominant perception of children as not only incomplete, but deficient. Hart (2006) noted that "children have been generally seen as developmentally immature, without sufficient intellectual growth to manifest anything that might be understood as meaningfully reflective and/or spiritual" (p. 163). Vaunted developmental theorists such as Kohlberg (e.g., 1981) and Piaget (e.g., 1932) place children at the lowest levels of moral and behavioral development, while Wilber (1996, cited in Hart, 2006) has described children's ways of thinking and feeling as "instinctual, impulsive, libidinous, id-ish, animal and apelike," likely guided by Piaget's 1968 stage model in which children are viewed as "largely incapable of meaningful reflection" (p. 163). Diminished by their identity category (child) as innocent, incompetent, incomplete, immature, and less-than, children's desires are seen as lack and their needs as inconveniences.

Johannes (2015) alludes to Deleuze and Guattari's explanation of the "powerful, overarching arborescent [as opposed to rhizomatic] cultural forces that can so quickly and decisively damage any of us, but particularly those who were less likely to be heard in the first place" (para. 8)—those such as children. Considering the vast quantities of research built on and building on the power of the cultural narrative that seeks to solve the problem of children, it is no wonder that "the emphasis on children's social incompetence and irresponsibility means … that children need to be brought to a state of sociability through the continued surveillance and control of a *benevolent* authority" (Thomas, 2002, p. 91, italics added).

Queering DAP

> Millennia ago, some genius discovered that such wiggles as fish and rabbits could be caught in nets. Much later, some other genius thought of catching the world in a net. By itself, the world goes something like [wiggle]. … The net has "cut" the big wiggle into little wiggles, all contained in squares of the same

> size. Order has been imposed on chaos. We can now say that the wiggle goes so many squares to the left, so many to the right, so many up, or so many down, and at last we have its number. … The net has thus become one of the presiding images of human thought. But it is always an image, and just as no one can use the equator to tie up a package, the real wiggly world slips like water through our imaginary nets. (Watts, 1972, pp. 52–54)

In the field of early childhood education, position statements and the treatise *Developmentally Appropriate Practice in Early Childhood Programs Serving Children From Birth Through Age 8* (DAP) (Copple & Bredekamp, 2009), published by the National Association for the Education of Young Children (NAEYC), serve as "benevolent" guidelines of biblical proportions. These documents operate to privilege certain claims as valid truth-knowledges about young children and to delineate ways of teaching and learning and being that are, in Tamboukou's (2008) words, "hierarchical, rule-intensive, strictly bounded and confining"—thus restricting teachers to follow deeply striated developmental norms and expectations, discouraging them from following the nomadic paths that walking with children might suggest.

> We know a great deal about how children develop and learn at various ages and what approaches and conditions tend to work best for them. (Copple & Bredekamp, 2009, p. 33)

Look straight ahead and act like you know where you're going.

> Knowledge of how children within a given age span typically develop and learn provides a general framework to guide teachers in preparing the learning environment, considering curriculum, designing learning experiences, and teaching and interacting with children. (NAEYC, 2009, p. 11)

Look straight ahead and act like you know where you're going.

> All young humans must … [learn] to regulate their emotions, behaviors, and attention. … Self-regulation means developing the capacity to manage strong emotions and keep one's attention focused. Throughout the early years, adults play significant roles in helping children learn to self-regulate. (NAEYC, 2009, p. 12)

Look straight ahead and act like you know where you're going.

In their training, early childhood education teachers are inducted into a DAP state of mind and are lulled into compliance by the security of the mandated and the known. In seeking the comfort of complacency, the teachers respond as instruments of the State—and the State is afraid and paranoid. As Sedgwick (1997) forewarns, "The first imperative of paranoia is there must be no bad surprises, and indeed, the aversion to surprise seems to be what cements the intimacy between paranoia and knowledge" (p. 9). Zembylas (2007) frames the risks that may cause paranoia to escalate—risks that are innate in the desiring assemblages that are children; he notes that

> desire presumes an invitation to forbidden terrains; clearly, such an invitation comes not only with possible pleasures but also with unavoidable risks. When desires challenge … the boundaries of what is sayable or visible … the fragility of these boundaries is exposed. (p. 333)

Gilbert and Gray (this volume, Chapter 4) hint at the bad surprises and bad terrains inherent in adults' relationships with children whose challenging thinking and doing "represent wildness, the frightening feral-thinking child."

Teachers, School, and the State soothe their paranoid fears of such an exposure through knowing what is to come, through naming and defining the terrifying and grotesque desires of children, and through having in place "best practices" to guide their way—"because there must be no bad surprises, and because learning of the possibility of a bad surprise would itself constitute a bad surprise, paranoia requires that bad news already be known" (Sedgwick, 1997, pp. 9–10). DAP promises protection from bad surprises. It already shares the bad news about the children one must control and change—and it delineates the binaries and boundaries that keep us apart and in control.

Developmental theories and school(ed) practices construct images of children in rigid binaries clearly distinguishing the fully formed adult from the incomplete child: big/small, smart/ignorant, powerful/weak, mature/immature, moral/pre-moral, teacher/student, important/cute, ready/not ready, capable/incompetent, in control/out of control, independent/dependent, calm/overactive, rational/irrational, good/bad, informed/naïve, wise/innocent, motivated/needs motivating, manager/managed, and so on. And within those binaries of adult and child, there lie degrees of what is perceived as immaturity and incompetence. A first-grader will be told when acting in ways deemed "childish," *Do you want to go back to kindergarten?* as an admonishment that suggests younger is lesser. Lacking.

> The features of lack are quite prevalent within the practice of early childhood education. … Within this setting, the desiring-repression is evident when we see what children do[;] reducing their desire to lack … these constructions and definitions repress and tame children's desires into already defined schemas of development. … Children need something because of their lack of it, or they need to get rid of something (over-activity, too much imagination, etc.) or their particular desire needs to be redirected so as to attain a specific goal or standard. (Olsson, 2009, p. 143)

We see evidence of the role that binary perceptions play in early childhood classrooms in coercive structures and power plays of bribery, punishments, and control, and in narrow and myopic interpretations of curriculum that are gross underestimations of children's intellect and purpose. Wallin (2013) employs a Zarathustrian pedagogy to explain School's commitment to shrinking and narrowing the child's experience with the world, noting that the "child's fabulations [such as "child-becoming-lion-becoming camel"] become overcome by a rational appeal to what is the production of a moribund life" (p. 197). The following story of kindergarten children negotiating the sequences of a midmorning science activity illustrates essentializing notions of children as needing to be managed and coerced into engaging in mundane and unsophisticated exercises in science and math—exercises that, like most School activities, diminish children's relationships with scientific and mathematical ways of seeing and being.

Spiders and Eyeballs

In a kindergarten classroom, the clock reads 10:20. Returning from a scheduled restroom visit, 24 energetic bodies fill the room like air molecules being pumped into a tire. The children quickly flow to every corner … touching, moving, laughing … the room is active, alive, smooth. But not for long because, as designated on the colorful poster clearly visible on the front board, it is 10:23 and time for (School) science—SCIENCE: 10:23–10:43.

Scott observes as, with one command, the teacher sends the children to their seats placed in straight rows and facing the same direction, and, without discussion, she tosses colorful plastic bracelets to those students who are sitting quietly and attentively, face front and hands unoccupied. (*Look straight ahead and act like you know where you're going.*) On this day students are directed to "integrate" science and math by identifying eye color and doing a graphing activity. The teacher circulates with a single mirror, telling the students to look in the mirror, name his/her eye color, and then use colored pencils to color in a diagram of an eye preprinted on a worksheet.

The coercive striating power of the plastic bracelets has quickly worn off and many of the students become restless. But not Alex and not Stephan. Alex is completely enthralled with this easily completed and predictable activity—as a good school-child, she has found her way into the comfort of knowing and has learned well the self-control and external motivation that make School so easy to do. She does what she is told when she is told and happily follows the straight-ahead path, knowing where she is going: wherever her teacher and School tell her. Alex has found a safe harbor, comfortable and secure and enjoying a fixed sense of who she is—a good student, a good girl—as her storied journeys arc toward the safely familiar (Broadway, 2011; Carlson, 2002). Today, Alex is assisting in the activity by acting as a mirror for students around her. As she looks into the eyes of one of her peers, she says, "You have blue-green eyes. Here, take a blue pencil and a green pencil so you can color in your worksheet." She moves easily from desk to desk repeating the process and distributing the appropriate colored pencils to several more of her classmates.

Stephan is also apparently engaged in the task at hand: head bent to paper with his pencil moving across the page. However, on closer inspection, Scott notices that he is not coloring an eye already printed on paper; he has flipped his worksheet over to the good side (the blank side) and, oblivious to the noise of the classroom, has begun to draw on the paper. Paper that desires to be drawn upon. As the teacher compulsively attempts to regain appearances of attention, order, and control of the restless children with threats ("no cookies later") or bribes ("Who wants a bracelet?"), Stephan quietly leaves his seat and walks toward Scott. With an air of confidence and pride, he displays his worksheet: He has drawn a fuzzy spider with not one but four eyes. Stephan has wiggled away from the straight-ahead curricular path and so has queered the curricular assemblage of paper + pencil + eyes to include something he desires to share—something that he loves. Drawings desire to be seen.

As Stephan explains the details of his drawing, the teacher notices. She comes over and a smile crosses her face. "Stephan is very talented," she says to Scott. Stephan gives her a hug. In this moment, the teacher has crossed over—briefly traveling the nomadic path taken by a child who is capable of composing a lived-curriculum: "a smooth space between curriculum-as-plan and the virtuality of a life[;] … [this space is] encountered in the mode of improvisation [as] the immanent flows of classroom life deform and assemble in singular ways" (Wallin, 2013, pp. 201, 202). In this brief, queer moment, the teacher has taken on the hard work of relaxing her attachments to what she thinks she knows and desires (Fifield & Letts, 2014) about children in the context of moribund classroom tasks.

Queer theory requires us "to interpret the minutiae of differences among persons, not merely among categories of persons" (Sumara & Davis, 1998, p. 198). In the tale of spiders and eyeballs, we see children and their teacher in a kindergarten classroom. We see hard workers and noncompliants and restless and talented and coercive and brave and proud and, and, and, and. In all tales set in all classrooms of children and teachers, Deleuze and Guattari's (1987/2004) "AND" logic applies: "think with AND instead of thinking IS, instead of thinking for IS" (p. 57). Also useful is the Deleuzian call for "'a thousand tiny sexes' rather than the binary categories of homosexual and heterosexual" (Lather, 2006, p. 43). We suggest that teachers might view children and their actions in thousands of tiny qualities rather than the straight and narrow binary categories insisted upon by developmentalism and DAP. In this way, one is compelled to consider multiple ways to be human beings together, because "such nomadic rather than sedimentary conjunctions produce fluid subjects, ambivalent and polyvalent, open to change, continually being made, unmade and remade" (Lather, 2006, p. 43).

It would be easy to place Alex and Stephan in suspiciously comfortable categories of identity. On this day—and maybe even most days—Alex chose to embrace the normalizing practices of School, sliding safely into the deep striations of the institutions, while Stephan quietly resisted those expectations, nomadically improvising a smooth space for himself and his spider. But queering makes slippery and smooths the way, resisting immovable identity categories and championing "those who refuse to be defined … by the (moral) codes of behaviour and identification set down by the dominant [State and striated] society" (Giffney, 2009, p. 3). So Alex is more than the good girl, and Stephan is more than the inattentive boy—they are those things AND all that they desire. In queering DAP, we love that there will be "bad surprises," as we look "for how children's desires deploy themselves away from the institution's repression of desire" (Olsson, 2009, p. 147). Queering honors children as complete and real and fully human beings—albeit human beings who often do human being differently than older beings tend to do.

Queering STEM-Talk

> If a fool throws a stone into water," goes an old saying, "even a hundred sages can't bring it back." ... But an ingenuous child, who still thinks in pictures, might ask, "Isn't the world full of stones—so why should a hundred clever people try so hard to get back this one? Why don't they look around? If they do, they might find all kinds of new treasures they can't see because they are so busy searching in vain in the water. (Miller, 1981, p. xvii)

Like the old saying in Miller's quote, authorities in fields of STEM education generate narratives that emerge as undisputed truths. These truth narratives then serve as foundations for thoughts and actions that reduce the wonder of children's science, queered and nomadic, to a series of predetermined procedures and normalized judgments—a pursuit of lost stones. Where children might see a world full of stones, adults—having learned self-control and denial of desires; having agreed to forego childish pleasure; having accepted the tasks and constraints that one may hate, suppressing their emotions and desires; having been forced to settle down, to behave properly in proper time and space; having abandoned play and learning to work in a compulsive manner; and having accepted without question the discipline of externally imposed rules—feel compelled to follow strictly the sage advice that might send them fruitlessly seeking a single stone that has been named Truth (or Best Practice, or High Quality, or ...).

Too often, narrow interpretations of STEM educational reform keep interested parties looking straight ahead, focused on the end products, remaining in safe proximity to current practices and in search of the elusive lost stone. Should these seekers ask a child, they might learn ways—such as Willie's exploriments, and the school roof as a true-to-life black box experiment—that STEM is traveled nomadically in a "world full of stones." When a sedentary STEM vision becomes a "central conduit for providing information, communication, and resources focused on the enhancement of STEM education" (American Association of Colleges & Universities [AACU], n.d., para. 2), it perpetuates current standardizing practices that normalize science-related learning as truth messages that signify *look straight ahead and act like you know where you're going.*

Just as the proponents of DAP have calculated children and devised the appropriate and research-based means for controlling the nomadic child and her relationship with the world, so have a range of State apparati sought means of bringing the nomadic nature of science in line by binding the inquiry and discoveries in the natural sciences to the more pragmatically constructed worlds of engineering and technology. Seeking certainty and predictable ends, agencies of the State determine STEM standards that dictate what all children should know in every grade level and how they should know it, translating the designated knowledge into what is needed to pass the class or achieve higher test scores. STEM has become another cloak to mask long-standing traditional didactic practices. The STEM vision to "help" students—similar to DAP's vision—has become a benevolently simplistic view of what can and should be known regarding science. The rhetoric of "learner-centered STEM … education through evidenced-based practices" (AACU, n.d., para. 1) lacks the voice of the learners, especially children, and conveys the already determined "known": what can be known, scripted, and anticipated in advance—no bad surprises here. This forced tunnel vision from the first step to the specified end has created, or simply perpetuated and maintained, a rigid progression on an entrenched path already determined—safe, efficient, and known. It simply doesn't allow anyone to look side to side, and so, risks missing all the surprises inherent in explorations of science—and math, technologies, and engineering, too.

According to Johannes (2015), "a nomadology approach can honour the less powerful" (para. 8). The less powerful are obviously the children compelled to classrooms, but are also teachers. As State-sanctioned purveyors of content, teachers have become charged with pressing predetermined content knowledge on to deficient students. Uncorrupted inquiry can be a pretty scary path to take for someone who is operating in cahoots with the paranoid State. When a teacher's sense of self, belonging, safety, and control are "deeply wrapped up in [the] ability to know … then, faith in ignorance, following children's lead, and facing the world in all of its beauty and ugliness … can provoke, I think, a real crises of faith and identity" (G. Pond, personal communication with Leafgren, July 28, 2016).

Teacher paranoia and powerlessness are perpetuated by the reality that, in contrast to Deleuze's explanation of desire as "generative and seeking" (Jackson & Mazzei, 2012, p. 91), much of STEM reform, in its pursuit to become a "foremost leader" (AACU, n.d., para. 2), adopts the traditional meaning of desire as something it wishes to acquire, thus leaning forward into a desire that is lack. In the straight-and-striated STEM-talk one hears from AACU and from many presenters at STEM education conferences, the world is known, with no room for bad surprises or exploriments that might gum up the mechanistic mental flow and so challenge our routine way of thinking about science education per STEM. In straight STEM-talk, learner growth is oppressively and patriarchally measured by test scores and increases in passage rates. In school(ed) STEM-talk, inquiry and wonder are on the books, but are they in the safely tamed walks that teachers and children take together? If the destination is determined and the route already plotted, we wonder, where is the wonder?

Disrupting STEM through a Deleuzian embrace of desire creates rumble strips that cause one to pay attention to things one would normally cruise right past without conscious notice—things one begins to wonder about. Children's walks through the city, with their curious attention to unexpected noises and their willingness to embrace surprise and engage in the joy of not-knowing, are the embodiment of a queer and nomadic STEM. We suggest that in queer(ed) STEM-talk, learners as human beings live and desire in relation with the world, with bodies and objects, happenings and events, signs and utterances—all moving fluidly, joyfully, smoothly in, out, with, and through the concepts and surprises inherent in engineering, science, mathematics and technologies.

A WIGGLY CONCLUSION

A colleague was once called to her first-grade child's school to meet with the child's teacher. The teacher had "grave concerns" about the colleague's son: He was "too wiggly." The teacher wanted him to settle down and straighten up. She wondered to the boy's mother how he could possibly learn and get his work done with all that wiggling going on. The boy's teacher could not bear the bad surprises, she could not bear the wiggly, and she just wanted things to be the way they are supposed to be in School. We think Alan Watts has an explanation:

> Nature is wiggly. Everything wiggles. The outlines of the hills, the shapes of the trees, the way the wind brushes the grass, the clouds, the tracks of streams. It all wiggles. And for some reason or other, we find wiggly things very difficult to keep track of. And you know, we say to people, *Keep still so that I can see you. Keep still for the camera.* And we say, *Well, let's get things straightened out. Let's get this ironed out. Let's get this all squared away*. And then somehow we think we understand things when we have translated them into terms of straight lines and squares. (Watts & Putterman, 1971a, video 1:45–2:42, italics added)

We suggest that straightening things out only makes them straight.

REFERENCES

AACU (Association of American Colleges & Universities). (n.d.). *PKAL regional network: Ohio (OH-PKAL)* [Website]. Retrieved from https://www.aacu.org/pkal/regional/ohio

Aldea, E. (2014, September 10). *Nomads and migrants: Deleuze, Braidotti and the European Union in 2014* [Online essay]. Retrieved from https://www.opendemocracy.net/can-europe-make-it/eva-aldea/nomads-and-migrants-deleuze-braidotti-and-european-union-in-2014

Beaulieu, A. (2009). Gilles Deleuze's politics: From Marxism to the missing people. In C. Boundas (Ed.), *Gilles Deleuze: The intensive reduction* (pp. 204–215). New York, NY: Continuum.

Bond, D. (2007). *Excavating a Deleuzo-Guattarian place concept.* Paper presented at the Constructing Knowledges Conference, Rutgers University, New Brunswick, NJ.

Britzman, D. (1995). Is there a queer pedagogy? Or, stop reading straight. *Educational Theory, 45*(2), 151–165.

Broadway, F. (2011). Queer (v.) queer (v.): Biology as curriculum, pedagogy, and being albeit queer (v.). *Cultural Studies of Science Education, 6*, 293–304. doi:10.1007/s11422-011-9325-7

Cardy, T. (2011, November). *Desiderio wish: The experience of the covered market.* Paper presented by at the International Study Group, International Centre Loris Malaguzzi, Reggio Emilia, Italy.

Carlson, D. (2002). *Leaving safe harbors: Toward a new progressivism in American education and public life*. New York, NY: Taylor & Francis.

Copple, C., & Bredekamp, S. (Eds.). (2009). *Developmentally appropriate practice in early childhood programs serving children from birth through age 8* (3rd ed.). Washington, DC: National Association for the Education of Young Children.

Deleuze, G., & Guattari, F. (1987/2004). *A thousand plateaus: Capitalism and schizophrenia* (B. Massumi, Ed. & Trans.). London: Continuum.

Dewey, J. (1938). *Experience and education*. Toronto: Collier-MacMillan Canada Ltd.

Fifield, S., & Letts, W. (2014). [Re]considering queer theories and science education. *Cultural Studies of Science Education, 9*, 393–407. doi:10.1007/s11422-013-9509-4

Gao, J. (2013). Deleuze's conception of desire. *Deleuze Studies, 7*(3), 406–420. doi:10.3366/dls.2013.0120

Giffney, N. (2009). Introduction: The 'q' word. In N. Giffney & M. O'Rourke (Eds.), *The Ashgate research companion to queer theory* (pp. 1–13). Surrey: Ashgate Publishing.

Gill-Peterson, J. (2013, April 29). *Childhood blocks: Deleuze and Guattari's infant affects*. Retrieved from http://juliangillpeterson.wordpress.com/2013/04/29/childhood-blocks-deleuze-and-guattaris-infant-affects/

Hart, T. (2006). Spiritual experiences and capacities of children and youth. In E. Roehlkepartain, P. Ebstyne-King, L. Wagener, & P. Benson (Eds.), *The handbook of spiritual development in childhood and adolescence* (pp. 163–177). Thousand Oaks, CA: Sage Publications.

Jackson, A., & Mazzei, L. (2012). *Thinking with theory in qualitative research: Viewing data across multiple perspectives* (1st ed.). New York, NY: Routledge.

Johannes, A. (2015, May 20). *So much to be uncertain about ... #Rhizo15* [Web log post]. Retrieved from http://imagineacircle.com/2015/05/20/so-much-to-be-uncertain-about-rhizo15/

Kesson, K. (2005). On bumblebees and Pleiades: Schooling and the great journey of the soul. *Journal of Curriculum and Pedagogy, 2*(2), 42–44.

Kohlberg, L. (1981). *The philosophy of moral development: Moral stages and the idea of justice*. New York, NY: Harper & Row.

Lather, P. (2006). Paradigm proliferation as a good thing to think with: Teaching research in education as a wild profusion. *International Journal of Qualitative Studies in Education, 19*(1), 35–57.

Leafgren, S. (2009). *Reuben's fall: A rhizomatic analysis of disobedience in kindergarten*. Walnut Creek, CA: Left Coast Press.

Leafgren, S. (2013). This is not a pencil: A Deleuzoguattarian re-imagining of the classroom milieu. *Global Studies of Childhood, 3*(3), 276–288.

Miller, A. (1981). *The drama of the gifted child*. New York, NY: Basic Books.

NAEYC (National Association for the Education of Young Children). (2009). *Developmentally appropriate practice in early childhood programs serving children from birth through age 8: A position statement of the National Association for the Education of Young Children*. Retrieved from http://www.naeyc.org/files/naeyc/file/positions/PSDAP.pdf

Olsson, L. (2009). *Movement and experimentation in young children's learning: Deleuze and Guattari in early childhood education*. New York, NY: Routledge.

Piaget, J. (1932). *The moral judgement of the child*. London: Routledge and Kegan

Ringrose, J. (2011). Beyond discourse? Using Deleuze and Guattari's schizoanalysis to explore affective assemblages, heterosexually striated space, and lines of flight online and at school. *Educational Philosophy and Theory, 43*(6), 598–618.

Roy, K. (2003). *Teachers in nomadic spaces: Deleuze and curriculum*. New York, NY: Peter Lang.

Sedgwick, E. (1997). *Novel gazing: Queer readings in fiction*. Durham, NC: Duke University Press.

Sumara, D., & Davis, B. (1998). Telling tales of surprise. In W. Pinar (Ed.), *Queer theory in education* (pp. 197–219). Mahwah, NJ: Lawrence Erlbaum Associates.

Tally, R. (2010). Nomadography: The 'early' Deleuze and the history of philosophy. *Journal of Philosophy: A Cross-Disciplinary Inquiry, 5*(11), 15–24.

Tamboukou, M. (2008). Machinic assemblages: Women, art education and space. *Discourse, 29*(3), 359–375.

Thomas, S. (2002). The silence of the lambs: Construction of childhood in the public sphere. *Journal of Curriculum Theorizing, 18*(3), 91–104.

Wallin, J. (2013). Morphologies for a pedagogical life. In I. Semetsky & D. Masny (Eds.), *Deleuze and education* (pp. 196–214). Edinburgh: Edinburgh University Press.
Watts, A. (1972). *The book: On the taboo of against knowing who you are*. New York, NY: Knopf Doubleday Publishing Group.
Watts, A. (Writer), & Putterman, Z. (Director). (1971a). *Alan Watts: Conversation with myself: Part 1* [Video]. Retrieved from https://www.youtube.com/watch?v=8aufuwMiKmE
Watts, A. (Writer), & Putterman, Z. (Director). (1971b). *Alan Watts: Conversation with myself: Part 3* [Video]. Retrieved from https://www.youtube.com/watch?v=3RcjATFcbq4
Zembylas, M. (2007). Risks and pleasures: A Deleuzoguattarian pedagogy of desire in education. *British Educational Research Journal, 33*(3), 331–347. doi:10.1080/01411920701243602

ABOUT THE AUTHORS

Sheri Leafgren is an associate professor of teacher education at Miami University, in Oxford, Ohio (USA). From her many years of teaching and learning from young children, she is particularly interested in the spiritual and moral wisdom of children, and how they find space to enact their moral and spiritual selves while swaddled tightly by the rules, procedures, and surveillance of the schoolroom. Of late, she has been expanding this inquiry to examine the spiritual and moral wisdom of early-career teachers similarly constrained by State rules, striated spaces, and surveillance. Sheri would welcome hearing from you at leafgrs@miamioh.edu

Scott Sander is a clinical faculty member in the Department of Teacher Education at Miami University in Oxford, Ohio (USA). His research interests include science teacher education as it intersects with social justice, diversity, and inclusion. His work in large lecture courses looks to challenge traditional notions of science teaching and learning as he positions students as scientists/scholars who critically analyze both physical and social phenomena. You can reach Scott at sandersa@miamioh.edu

ANNA MACDERMUT AND ADRIAN ZONGRONE

12. INVITING THE MESS

A Children's Museum's Transgressive Tactics for Unleashing Play

INTRODUCTION

> Perhaps it is like a flash of lightning in the night which, from the beginning of time, gives a dense and black intensity to the night it denies, which lights up the night from the inside, from top to bottom, and yet owes to the dark the stark clarity of its manifestation, its harrowing and poised singularity; the flash loses itself in this space it marks with its sovereignty and becomes silent now that it has given a name to obscurity. (Foucault, 1977, p. 35)

The dialogue surrounding science education is derived primarily from assumptions, ideas, and insights regarding *formal* schooling. While the particulars of this experience may vary, certain elements remain relatively consistent: there is a classroom of some sort, that classroom is typically populated by one or more instructors as well as students, and those students are likely under some sort of obligation to be present. We speak, instead, of our own experiences with STEM education contextualized through the lens of a children's museum, where we served together as educators. We argue that the children's museum, as a distinct agent of teaching and learning, has a unique opportunity and moral imperative to transgress common definitions of education and help to shift the paradigms regarding science, student, teacher, child, desire, play, learning, and more.

The most prolific form of STEM learning happens both within and outside of educational institutions as *informal* education: the lessons one learns when touching a hot stove, dropping a pebble in a lake, or testing the structural integrity of a tree branch. Within and beyond the confines of the children's museum, museum staff are likely to both encourage this kind of informal learning as well as guide *nonformal* educational experiences. Nonformal education typically occurs in the absence of any grading or graduation and may follow less-formal curricula. It may happen in a museum, a library, a community center, or a school classroom. While the lines that separate informal, nonformal, and formal education are porous and nebulous, an essential aspect of the children's museum is that it is a space relatively free of formal education.

We do not intend to upend or deny agency to the plurality of educational models that one might describe as "formal," but we do demand that the queering of STEM

DOI:10.1163/9789004331068_014

education cannot occur solely in the classroom. Indeed, a STEM education that seeks to be queer ought to be "at odds with the normal, the legitimate, the dominant" (Halperin, 1995, p. 62). For the children's museum, that "positionality" (p. 62) may come from a proliferation of both informal and nonformal frameworks that bash back at the standing moral codes that surround much of formal education, uncovering the rich possibilities that have been precluded and revealing ways of embracing the unthinkable by teaching the unteachable. It is within this realm and with such tools that we seek to queer our own contemplation, comprehension, and dissemination of STEM curricula.

TRANSGRESSION

Likely the most consistent epistemological underpinning of the children's museum is play. The contemporary children's museum operationalizes play as an informal educational approach, an experiential methodology of learning that allows desires to be fulfilled and imaginations to take control of engaging investigations. While play remains a primary heuristic framework, in semistructured educational programs, inquiry- and project-based learning approaches may take precedence. These nonformal techniques imbue young people with authority over their own STEM lessons so that they may rearrange concepts and create unique orders and structures of importance. At their most elegant, these approaches expand beyond knowledge building and into narratives that contextualize concepts in a sociocultural framework, giving rise to questions, questions, and more questions that typically outnumber accompanying answers. As in their day-to-day experiences, children are presented with problems, are invited to devise solutions, and, under various conditions and constraints, construct paths to those solutions. A lesson in the engineering design process foregoes factoid acquisition and teaches problem solving as a cyclical, unending pathway to imperfect and shifting answers. Children are encouraged to generate fruitful questions and to respond to them.

The museum educator may present STEM concepts such that they beg to be broken down, investigated, and reconstituted, covered in small fingerprints. This creates a space for children to negotiate critical ideas with each other and instructors in facilitated experiences through a symbiotic honoring of children's agency within their education. These hands-on moments of exploration and discovery are inherently unique each time they occur due to their cocreation by unique hands and minds, and this cultivates a joyful myriad of heterogeneous knowledges that prevent scientific thought processes from being collapsed into singular fact.

These modes of education can be challenging and messy for all parties involved. Higher-order improvisational skills are required of both facilitator and participant to navigate such dynamic learning spaces. However, we feel that a joyful myriad of understandings can best be achieved when students are presented with ideas that exist beyond the confines of academic standards and may transgress community standards as well. Importantly, this process may be facilitated both within the physical bounds of the museum as well as within other community settings through outreach experiences or pop-up exhibits. One of us (Adrian) once delivered such a program about the human immune system at a local middle school.

> This was part of a professional development opportunity for the students' school teacher, wherein she stepped out of the role of instructor and became classroom voyeur. I came to present a science lesson so that she could observe and analyze how her students engaged during a nonformal educational program, and then discuss her lessons-learned with an educational consultant later in the week. I brought a program that integrated South Carolina's state science standards and health standards for seventh grade. The program explored the human immune system by having students perform a simulated HIV test. Included in the lesson were vials of simulated blood samples that students could "test" for HIV. Each vial also came with background information that described why the imagined individual felt they may have been exposed to HIV. Midway through, school administrators quickly collected all handouts that had been passed to students. An official pulled me aside to inform me that any discussion of sexuality or mention of the word *semen* was inappropriate for science class; a detailed discussion on HIV transmission should only occur in a strictly regulated health class. The principal had been notified, and I was nearly asked to leave. In the end, I was allowed to continue the lesson on the human immune system by replacing HIV with Lyme disease.

We were not present to witness the professional development discussions on classroom engagement that took place later that week, so we can only imagine what wisdom this teacher might have gleaned from observing her students as administrators collected the profiles of imagined individuals seeking an HIV test. At the beginning of class, a dismayed young student did not know what to make of the narratives on sexuality and drug use that were in her hands. The teacher coolly assured her that this lesson was a reflection of "real life." Several minutes later, adults filed into the room to collect and hide these reflections of real life, and the teacher was given the opportunity to observe how both her students and her colleagues engaged with the classroom's guest educator. Did she register the confusion and fear that filled her classroom from child and adult alike?

> As an administrative official privately explained to me the consequences I would face if I continued teaching about HIV, I noticed the HIV/AIDS red ribbon pin on his school lanyard. He assured me that he understood where I was coming from. But I registered desperation in his eyes as he swiftly responded to this disruption to policy and curriculum that he had been trained to neutralize.

To be clear, this visit was orchestrated as transgression—that which Foucault (1977) describes as an affirmation of the limit and the limitless:

> Transgression contains nothing negative, but affirms limited being—affirms the limitlessness into which it leaps as it opens this zone to existence for the first time. But correspondingly, this affirmation contains nothing positive: no content can bind it, since, by definition, no limit can possibly restrict it. Perhaps it is simply an affirmation of division; but only insofar as division is not understood to be a cutting gesture, or the establishment of a separation of the measuring of a distance, only retaining that in it which may designate the existence of difference. (pp. 35–36)

The merit found in transgression, then, does not reside in that which lies beyond limit, but in the crossing itself: the exposure to thoughts and perceptions that exist outside the borders of a seventh-grade biology unit, and the affirmation that such a border exists. Indeed, as administrators collected the imagined profiles of individuals seeking an HIV test, they reinforced the weight of that limit between conceptually learning about HIV and exploring its greater context and meaning. In this sense, the children's museum draws its transgressive power from that learning space it might seek to transgress: the school classroom. Within that normative space, the role of the student is structured on a daily basis by the presence of familiar faces and norms operating in a repetitious-though-changing fashion.

The students' acquiescence to this role creates a static, limited space that they continually occupy, which we will call *student-being*. The unfamiliar face of a museum educator invites a substantial shift in a school classroom's practices and social hierarchies. The limit of student-being is interrupted, opening doorways to new paths of knowing. The museum educator's job mandates that they also interrupt the limit of *teacher-being* by inviting students to disrupt the teacher/student duality with pointed and poignant questions: How would you explain gravity? How has DNA affected your life? It is empowering and unnerving for seated students to articulate experiences, insights, and ideas as an essential element of the educational construct, a sharp contrast to the educational method of folding demarcated facts and theories into an individual's growing body of knowledge. What use is the cell unless it is understood to be a different and sometimes useful way of contemplating organisms that one has always known? What use is an immune system unless it is presented within the messy sociopolitical contexts of vaccinations, Ebola, and AIDS? Affirming the limits of student-being, many students clam up when presented with the opportunity to speak about their own experiences and explore a limitless space of uncertainty and amateur guess-work.

We are fully aware that few museum educators are bound by any formal professional mandates to engage in these transgressions. Like any other wholesome institution with some authority over young minds, our own unique set of limits is imposed by protective parents, consumer-driven donors, and community-minded board members, all guided by their own standards of respectability. Rather, the mandate to transgress can be understood as an ethical necessity that arises from teaching in an educational institution that can operate beyond government-regulated academic standards—a necessity to call attention to the boundaries of student-being and teacher-being. However, the museum's aim cannot be to deride or unseat K–12 classroom science education as a useful and beneficial institution. This would be a fruitless endeavor that would sever mutually beneficial relationships between schools and museums, just as the transgression and limit require one another: "The limit and transgression depend on each other for whatever density of being they possess" (Foucault, 1977, p. 34). Academic standards, like all limits, were created because they are crossable, because there are taboo unknown desires haunting the edges of the carefully constructed classroom experience. So we are drawn to and aroused by those limits—those academic standards that delineate what is and is not appropriate, what is and is not of value, and what does and does not have meaning. From the crossing of those limits, we engage in the transformative power of play, a recreational science. In a classroom of 30 students, pleasure and play may promote an unwieldy degree of imprecision and chaos. To engage with this chaos is both to confirm the border that lies between knowing and the unthinkable, and to puncture that border. How easy it is to criticize the joyful student as a disruption to learning when unique and uncharted connections are made between science, art, culture, family, drugs, church, literature, sex, and so forth.

THE RHIZOME

The transgressive tactics that children's museums are well-suited to utilize disrupt not only the external limits of the learning experience, but also its internal organizing processes and the power dynamics betwixt participants who uphold this structure. To illustrate this idea, we want to explore one of Gilles Deleuze and Felix Guattari's (1987) signature propositions—the rhizome. They write:

> Principles of connection and heterogeneity: any point of a rhizome can be connected to anything other, and must be. This is very different from the tree or root, which plots a point, fixes an order. … A rhizome ceaselessly establishes connections between semiotic chains, organizations of power, and circumstances relative to the arts, sciences, and social struggles. (p. 7)

They conceptualize a rhizome as a type of underground stem (e.g., those of turmeric and ginger plants) that extends shoots and roots in multifarious directions, appendages that can and often do make connections between themselves. Deleuze and Guattari contrast the image of the rhizome with the hierarchical structure of a tree. Like a teacher might scaffold sequential lessons, the tree "fixes an order": it determines a ranked structure of relations between different parts of the organism. The rhizome allows for differing types of relations between similarly and heterogeneously comprised elements.

We argue that the free-flowing exhibits, open-ended discussions, adaptive learning experiments, and cross-disciplinary style espoused by many children's museums gain much of their transgressive power through their rhizomatic movements. Such methods allow for connections to spring forth in a nonlinear, unorganized fashion. No single thought or idea creates the rest; no conclusion is an ending; no person is the sole possessor or producer of knowledge.

Deleuze and Guattari further their discussion of rhizomes by talking about the relations between organisms in symbiotic relationships, which they describe as rhizomatic. They focus on the example of orchids and wasps, who mimic each other's physicality in their mutually supported survival: The wasps disseminate the orchids' pollen in their movements, and the orchids supply food for the wasps. However, Deleuze and Guattari see beyond these utilitarian movements, saying:

> At the same time, something else entirely is going on: not imitation at all but a capture of code, surplus value of code, an increase in valence, a veritable becoming, a becoming-wasp of the orchid and a becoming-orchid of the wasp. (p. 10)

By recontextualizing the term *valence*, the quantifiable power of elements to combine with one another, Deleuze and Guattari provide critical insight into a powerful, generative process that flows without beginning or end. Much as the orchid and the wasp reflexively territorialize one another in a dance of reciprocity, transgressive educational tactics seek to empower children with this increased ability to combine ideas, selves, knowledges, exercises, experiments, lessons, and more through such intermingling and through awareness of these fusions. New relations arise out of interconnectivity rather than out of ascent or descent along a fixed path. When a child uses pipettes to mix solutions of red, yellow, and blue into colors of the rainbow, fine motor skills and lab technology are enveloped into a lesson on primary colors and mixing solutions—and vice versa. Deleuze and Guattari go on to elaborate upon this distinction from tree-like systems of relation: "Evolutionary schemas would no longer follow models of arborescent descent going from the least to the most differentiated, but instead a rhizome operating immediately in the heterogeneous and jumping from one already differentiated line to another" (p. 10). A rhizomatic understanding allows us to see relationships among entities that are presupposed to be dissimilar, allowing connectivity to arise as desired rather than through a path determined by homogeneous elements.

This idea for remodeling our idea-systems from arborescent to rhizomatic can provide a unique perspective regarding how children gain knowledge and the ability to synthesize information. Within the traditional classroom, knowledge and the student's capacity to utilize it are measured in relation to predetermined knowledge structures called educational standards. These standards are governed by educational administrators who establish the greater and lesser importance of particular pieces of knowledge. Standards are arranged in an arborescent structure in which ideas are introduced, evolve, and connect in preconceived ways along prescribed pathways over many years of tutelage. We do not intend to ignore or discredit the reality that many teachers, schools, and students make connections that disrupt this sequence, jumping among disciplines and levels of knowledge. We say this to draw attention to the relations that are often, though not always or necessarily, precluded by such a model. Although one idea may lead to other ideas, one idea is not inherently the root of other ideas. The progression from one idea to another along a fixed path from standard to standard creates a tree-like trail that frequently establishes certain concepts as roots, trunks, limbs, branches, etc.

In a sense, educational standards and exams mirror Heisenberg's principles of uncertainty: The more precisely we observe, measure, and document the position of our young learners, the more we derail their momentum. As frequent, standardized, state-mandated grading practices are used to pinpoint a student's position along a linear, vertical path, their learning is accelerated and decelerated under false pretense. Students who succeed under these conditions may do so with little effort, much praise, and greater access to both academic enrichment and career opportunity. Meanwhile, those who do not perform as well may be marginalized in remedial courses, led down a path less privileged, carrying the weight of grades that may literally demarcate them as failures.

A rhizomatic structure of learning allows students to move freely in multiple dimensions, out and around in addition to up and down, shifting among ideas and concepts. This framework makes measurements of ascent obsolete and invokes new methods to define and promote success. A rhizomatic understanding also eschews troubling binary relationships between student and instructor by resisting normalizing, linear, arborescent models that, as Silin (2014) points out, have "identified some people—notably children, queers, and women—as undeveloped, lacking in 'adult' attributes" (p. 57). One position of identity need no longer serve as the core or root from which others deviate. These dualities of child/adult and student/teacher are arborescent, intrinsically asserting a growth process of linear ascension, and may promote stifling adultism and self-righteous instructors.

When the student's path to understanding is less predictable, the instructor's path to facilitating is less commanding and more supportive. As a result, the student wields much more influence over their learning process. Such collaborative approaches transgress ideas of what children are and might be, overcoming inherent power discrepancies by eroding the duality of adult/child. In the children's museum, guests assume various roles as they move through lessons or different exhibits, becoming-adult and becoming-child, becoming-student and becoming-teacher, shifting among and beyond these dichotomies. A child who teaches an adult, queers the prevalent moral infrastructure that restricts students to desks and passive roles.

The modalities within the children's museum of exploration, discovery, and imagination allow children, adults, teachers, students, hosts, visitors, and others to rhizomatically learn. This reaching outwards in many directions allows them to establish connections in unexpected locales, creating a rich spatial expansion. These uncharted, unplanned associations and relations exceed the cultural expectations and norms that surround and restrict childhood, sparking the pleasure of learning. Seeing children and adults as shoots within a rhizomatic space grants each the autonomy to encounter the other with respect, curiosity, and engagement.

LYSIS

The rhizome, in constantly seeking to forge new and unexplored connections, represents a kind of underground stem of latent and fulfilled desires. As a model for learning, it also brings to mind a popular museum activity in which guests are invited to extract, observe, and engage with the genetic material of cells. During the process of extraction, there is a certain necessary step before the DNA precipitates: lysis. We mix a soapy saline solution with a sample of cells to dissolve the outer limit of each cell—its membrane—while leaving the internal organelles intact. These corporeal components comingle and shift within the lysate, a conglomerate of cellular structures no longer bound by porous borders. Among this population of cells, the importance and order of connections afforded by bounded positions is disrupted, and an ever-changing disarray of deprioritized components emerges. Beyond the confines of inclusion and arboreal structures lies a similar loosing, an unbinding of limits and delineations—a heterogeneous, ever-expanding space where tolerance and intolerance are replaced by intra-action (Barad, 1991) among truly challenged and hopefully generous agents. How do we nurture such deprioritized agents? What sort of pedagogical tools serve to puncture the tenuous bubbles of childhood identity, loosing the desires, ideas, and intelligence within them?

We have discussed the power of heuristic learning approaches executed outside of standardized curricula and classrooms. But perhaps the space and pedagogical tool that is most characteristic of museums is the exhibit—a largely static yet playful and educational zone. Self-directed exploration, thought, and play allow childhood learning to begin to shirk curricula imposed by proctors of the state. Open layouts allow exhibits to blend and blur into one/an/other and expose guests of any age to multiple disciplines and to undisciplined learning.

Often, the borders between exhibits encroach on one another. For instance, to teach about agriculture, peach trees, with orange balls that signify peaches, are built into the walls of the museum where we worked. Tunnels in these trees let children pick and collect the peaches in bags. Nearby there is a construction site with a massive crane children use to lift yellow balls—which signify dirt or other debris—and move them to the loading zone. Staff who spend time moving peaches back to the tree and debris back to the construction site know how common it is for young visitors to effortlessly move from exhibit to exhibit, synthesizing similar concepts presented as separate. Through this sort of play, we witness the becoming-construction-crane of the peach tree and the becoming-peach-tree of the construction crane—a lysing of categorization grounded in play that reveals an inter- and intra-relatedness (Barad, 1991) and coexisting generative power perhaps previously unseen.

Yet, by virtue of their age, children are typically afforded little agency over their own explorations and desires. There is a tension between the desires of children and the desires of adults, even when they seem to align. When the becoming-construction-crane of the peach tree beckons children toward a new type of investigation, the critique, “That’s not how you play with that!” may halt these lysed explorations.

Museum staff ask that children remain with their grown-ups, and these grown-ups may easily utilize their mandated authority to impose certain orders and structures, ossifying a time of play into a rigid pantomime. We don't want adults to sit idly on a bench while their children meander through the museum—that would violate not only the norm of adult supervision, but the museum directive to learn and play together. Unfortunately, we have seen caretakers scold and direct young visitors' exploration, quashing their capacity to wonder and wander (see Gilbert & Gray, this volume). There are caretakers who oversee a child's every action to create a visit that the adult finds desirable. In a recent example, an adult directed their children into an art activity, dictated the details of how they ought to compile their collage, and, before long, declared, "Hurry up, I'm ready to go see more of the museum." Still other adults might put effort into allowing children to lead explorations, but passively still exert influence by engaging with some activities more than others. The museum visit is directed by adults before it even begins, as the choice to go to the museum lies with the caregiver who pays and provides transportation, whether in response to a child's request or not.

This tension between structure and lysate exists as an undercurrent throughout the museum. It exists in a child escaping their chaperone. It exists in the area meticulously designed for use as a grocery store that also bears witness to scenes of engineering, construction, and demolition. It exists in the child who continues how they started playing despite an adult's claim that they are "doing it wrong." It exists in the yellow balls created to signify debris in a construction zone that become projectiles in a game of dodgeball. It exists in experiential educational programming, where educators balance instruction and comprehension with critical, playful exploration. These moments of tension help to confirm that the limitlessness Foucault describes is not a sustainable state of being. Rather, we find that a transgressive mode of education exists in such momentary border crossings—and, by extension, in the borders themselves. For those of us who seek to queer STEM education, complicating and differentiating its aims and methods, these moments do more than expand the possibilities of STEM education. Traversing these boundaries sustains us.

IRREVERENCE

The centerpiece of the museum where we worked is an exhibit that often elicits the strongest immediate reaction from our visitors, both children and "their grown-ups": a 40-foot-tall, 17.5-ton, 10-year-old boy named Eddie that is so big and self-contained that it has its own building code and sprinkler system. It was built before the rest of the museum which was then built around it. Guests can climb inside the exhibit to explore many of the human body's internal organs. This exhibit holds the record for being the world's largest child. To encounter something as conventional and vital as the human body in such a drastically oversized and simplified form is, for young and old alike, uncanny. It creates horror and revulsion in some, stirs confusion and ambivalence in others, yet draws forth wonder and attraction in others. A common day working the front desk of the museum is punctuated by children cowering and crying in reaction to this exhibit, and others running towards it with grown-ups in tow calling them back to be formally admitted to the museum. It features the most familiar subject matter of all the exhibits and yields a multiplicity of interpretations and reactions.

At the top, an imposing dark chamber emits intermittent buzzing and zapping sounds. Bright flashes of light momentarily illuminate pink and white patterns on the ceiling and walls, revealing, in bursts of clarity, the structures of brain and bone inside Eddie's skull. This strobe and speaker system create an interplay of sound and light that illustrate electrical signals traveling across the synapses of a young child's mind—a veritable electrical storm, a sporadic and unpredictable string of lightning bolts directing thought, action, and identity. Just as Foucault's lightning bolt declares the night in denying it, moments of childhood impunity also help to clarify and shore up borders. A child throwing, rolling, dropping, squeezing, and crushing museum artifacts may glean a greater understanding of their physical properties as well as a lecture on order and cleanliness. Dressing in a story-time fairy or knight costume may yield unsolicited lessons on gender roles and expectations. In espousing the tools we have outlined in this essay, a rhythm of undulating obediences and disobediences is immanent. In fact, the potency of these tools depends largely upon rhizomatic connections forged by childhood transgressions. The questionable cliché that all children are naturally curious scientists may actually refer to an innate irreverence for limits that have not yet been fully grasped. Even as we cross such borders, knowingly or unknowingly, we are pulled back inward by the entrenchment of norms we have defied—but if we return, we do so with new ideas. With the stark clarity of each transgression's manifestation, and the flashes of light it sheds, we gain a vision that is perhaps uncanny, profound, absurd, fun, unreasonable. We also gain a vision of more boundaries and, in turn, more opportunities to devise their undoings.

REFERENCES

Barad, K. (1991). Agential realism: Feminist interventions in understanding scientific practices. In M. Biagioli (Ed.), *The science studies reader* (pp. 1–11). New York, NY: Routledge.

Deleuze, G., & Guattari, F. (1987). *A thousand plateaus: Capitalism and schizophrenia* (B. Massumi, Trans.). Minneapolis, MN: University of Minnesota Press. (Original work published 1980)

Foucault, M. (1977). A preface to transgression. In D. F. Bouchard (Ed. & Trans.), *Language, counter-memory, practice: Selected essays and interviews* (pp. 29–52). Ithaca, NY: Cornell University Press.

Halperin, D. (1995). *Saint Foucault: Towards a gay hagiography*. New York, NY: Oxford University Press.

Silin, J. (2014). *Reconceptualizing early childhood care and education: Critical questions, new imaginaries and social activism*. New York, NY: Peter Lang.

ABOUT THE AUTHORS

Anna MacDermut is a native of Columbia, South Carolina (USA) and a proud alumna of Mount Holyoke College, where she earned her BA in gender studies and anthropology. She enjoyed exploring modes of educating while at EdVenture Children's Museum in Columbia, South Carolina. She is currently a controller for the local Episcopal diocese. While navigating accounting by day, she enjoys creative writing and bringing queer theory into everyday experiences. Research interests include the intersections of childhood, safety, and technology in relation to the discrepancy between technology's rate of change and that of standing legal systems. Anna would be delighted to hear from you at anna.macdermut@alumnae.mtholyoke.edu

Adrian Zongrone is a research associate with GLSEN in New York City (USA), examining school climate for LGBTQ K–12 students. He is also a recent graduate of New York University's College of Global Public Health, though his career in education began at EdVenture Children's Museum in Columbia, South Carolina. Through his experiences there, as well as at the Brooklyn Children's Museum, he has critically examined and experimented with how nonformal institutions of learning engage with their queer audiences. He can be reached at adrian.zongrone@gmail.com

MICHAEL J. REISS

13. THINKING LIKE A FOX

Queering the Science Classroom When Teaching about Sex and Sexuality

The fragment attributed to the ancient Greek philosopher Archilochus—πόλλ’ οἶδ’ ἀλώπηξ, ἀλλ’ ἐχῖνος ἓν μέγα (the fox knows many things, but the hedgehog knows one big thing)—is well known. Isaiah Berlin (1953) developed this fragment into an argument in an essay in which he divides intellectuals and authors into two categories: hedgehogs, such as Plato, Dostoevsky, and Nietzsche, who see the world through a single lens; and foxes, such as Aristotle, Shakespeare, and Goethe, who draw on a range of ways of understanding the world.

My own intellectual journey has been from the position of a hedgehog—my doctorate was in applied mathematics, and my first academic book consisted of exploring the implications of a single equation ($dM/dt = k_2M^b - k_1M^a$) for the growth and reproduction of organisms (Reiss, 1989)—to one of a fox, largely on the grounds that education is a complex field rather than a single discipline, and it is therefore rare that there is a single, widely agreed framework within which people agree that educational questions can be asked and answers proposed.

In this chapter I therefore, as befits a fox, look at a single question from a number of viewpoints. Encouragingly, my conclusion will be that we obtain not dissimilar answers whichever of these viewpoints we adopt. This therefore suggests that a way forward can be proposed, a way forward that can, by a sort of epistemological triangulation, be more robustly defended than would be the case if I had behaved as a hedgehog and, to mix metaphors, put all my eggs into one basket (both hedgehogs and foxes do eat eggs).

Of course, too great a convergence would be suspicious—just as we would be suspicious if the readings of the height of a mountain were too similar, which would suggest that our measurements were not independent or even that they had been "fixed"; queer theory is obviously suspicious of things that are fixed! And so should science be too. Scientific advancements are always on the edge and messy; the temptation is to strive too hard to make sense of them either at the time, for example by omitting data that "don't fit," or in retrospect, as Foucault (1969/1972) argues in his critique of the way that neat histories are produced.

The question I will address is How should we teach about sex and sexuality in school science? The topics of sex and sexuality are, of course, only a small part of the typical school science curriculum. Nevertheless, it is obvious, given the historical

 | DOI:10.1163/9789004331068_015

roots of queer theory, its ongoing interest in sexualities, and the continued need for it to tackle heteronormativity (Holland, Ramazanoglu, Sharpe, & Thomson, 1998), that queer approaches to science education will have something to say about sex and sexuality, though it is, again, obvious that queer approaches to science education are not restricted to such topics (as other chapters in this book indicate). As Giffney (2009) points out, "*queer* loosely describes a diverse, often conflicting set of interdisciplinary approaches to desire, subjectivity, identity, relationality, ethics, and norms" (p. 2, italics added). However, what is not so obvious about how we should teach about sex and sexuality is that, as I shall argue, one reaches similar answers whichever of the following frameworks one adopts:

- The science curriculum should promote human flourishing.
- The science curriculum should enable students to acquire powerful knowledge.
- The science curriculum should advance social justice.

Important as out-of-school settings are, I focus on schools for reasons of both pragmatism and equity: some 95% of each cohort of students between the ages of about 5 and about 16 attend schools in many countries (others are home-schooled, truant, absent because of bullying, unwell, or in various nonschool institutions). Schools are therefore the easiest route by which educators can reach the overwhelming majority of young people in this age group; other routes (e.g. museums and online) while important do not have the coverage of each cohort that schools do. At the same time, education in out-of-school settings (e.g., http://www.wellcomecollection.org/exhibitions/institute-sexology) can play an important role in disturbing the conventional school curriculum, and schools can, and should, avail themselves of such opportunities.

THE SCIENCE CURRICULUM SHOULD PROMOTE HUMAN FLOURISHING

There are a number of ways in which curricula can be developed. National curricula, especially at high school level, typically start with a list of subjects: mathematics, the predominant language, science, and so on. Aside from minor skirmishes (for example, over citizenship, drama, geography, and dance) they take for granted a dozen or so discrete school subjects and the knowledge they entail. There is a general implicit presumption that agreement exists as to the purpose of school education, without these purposes being spelt out and examined in any depth, and that the way to construct a curriculum is to start with familiar subjects.

An alternative to starting with subjects is to start further back, with aims (Reiss & White, 2013). Education has had diverse aims over the years. As Harris (1999) has put it:

> In the very first lecture of every course I give, I stress that 'education' is a changing, contested and often highly personalized, historically and politically constructed concept. To illustrate this I read a few dictionary definitions of

> 'education', as well as a selected set of stated 'aims of education'. When students hear that D. H. Lawrence claimed education should aim to 'lead out the individual nature in every man and woman to its true fullness', that for Rousseau the aim of education was 'to come into accord with teaching of nature', that R. M. Hutchins saw the aim of education as 'cultivation of the intellect', that A. S. Neill believed the aim of education should be to 'make people happier, more secure, less neurotic, less prejudiced', and that John Locke claimed 'education must aim at virtue and teach man to deny his desires, inclinations and appetite, and follow as reason directs'; hopefully the penny has dropped. (p. 1)

Nevertheless, within even the examples that Harris cites, chosen to represent their diversity, we can see two broad groupings. First, those where the intention is to develop the individual for her/his own benefit; secondly, where the intention is to develop individuals so that they collectively contribute to making the world a better place. John White and I (Reiss & White, 2013) contend that there should be two fundamental aims of school education, namely to enable each learner to lead a life that is personally flourishing and to contribute to building up societies (whether at the local or global level) that enable others to do so too.

The idea that humans should lead flourishing lives is among the longest established of ethical principles, one that is emphasized by Aristotle in his *Nicomachean Ethics*. There are many analyses as to what precisely constitutes a flourishing life. A hedonist sees it in terms of maximizing pleasurable feelings and minimizing painful ones. Related to this, a person may wish to maximize their wealth, fame, consumption or, more generally, to satisfy their principal desires, whatever these may be. Admittedly, there are difficulties with all these accounts (White, 2011). For example, a problem with desire satisfaction is that it allows ways of life that virtually all of us would deny were flourishing, a life mainly devoted to drinking very large amounts of alcohol, for instance.

A life filled with whole-hearted and successful involvement in more worthwhile pursuits—significant relationships, meaningful work, and other activities we find satisfying—is on a different plane. Virtually all of us would rate it fulfilling. At the same time, nearly all of us in a modern society presume it is largely up to each of us to choose the mix of relationships and activities that best suits us (certain family obligations are generally excepted from this generalization, though less than in the past). A central aim of the school should therefore be to prepare students for lives of autonomous, whole-hearted, and successful engagement in worthwhile relationships, activities, and experiences.

Such an aim is facilitated by a much richer, less binary account of sex and gender than is usually presented in science lessons. I shall have more to say about this in the rest of this chapter, but the key point here is that by queering the way we teach about sex and sexualities in school science we can provide a curriculum and a pedagogy more appropriate for *all* students. This claim may appear counterintuitive. While queer approaches may evidently help LGBTQI (lesbian, gay, bisexual, transgender, queer, intersex) youth, aren't they counterproductive for dyed-in-the-wool heteronormative or heterosexist youths? Aware that I may be accused of a Panglossian tendency, I think not. My argument is that the truth(s) will out and queer approaches to sex and sexuality are closer to these truths than the conventional, outdated fodder all too typically served up in bland science textbooks and standardized approaches to teaching human reproduction. Queer teaching is simply better science teaching—a point made by Susanne Luhmann (1998) and Jay Lemke (2011), with Lemke arguing that at least one meaning of queer approaches is to live up to science's own professed epistemic virtues better than science education usually does.

THE SCIENCE CURRICULUM SHOULD ENABLE STUDENTS TO ACQUIRE POWERFUL KNOWLEDGE

While historians tell us that what scientists study changes over time, there are reasonable consistencies. Science is concerned with the natural world and with certain elements of the manufactured world—so that, for example, the laws of gravity apply as much to artificial satellites as they do to planets and stars. Furthermore, as important as the scope of science is the way that science is undertaken. Fundamentally, science is about the building up of knowledge that is derived *empirically* and gives rise to *testable* models of the world.

This means, for instance, that scientific knowledge is distinct from mathematical knowledge (though there can be overlaps, as is the case with applied mathematics and theoretical physics). Mathematical knowledge about, for example, the characteristics of polynomial equations is not obtained by setting off into the field or the laboratory to isolate examples of polynomial equations that can be purified and then investigated. Equally, scientific knowledge about, for example, the causative agents of infectious diseases is not obtained by sitting in arm chairs attempting to find the solutions to the problems that worried Pasteur and Koch.

Similarly, scientific knowledge can be distinguished from aesthetic knowledge, from literary knowledge, from normative knowledge, and from a number of other knowledges. To say a bit more about just one of these, namely normative knowledge, science is fundamentally concerned with how things are rather than with how they should be. So, there is a science of gunpowder and in vitro fertilization without science telling us whether warfare and test-tube births are good or bad (Reiss, 2015). Of course, science plays a role in telling us whether such things are good or bad—any utilitarian approach to ethics relies on a detailed, accurate, and thorough knowledge of the consequences of human actions; but there are aspects of moral philosophy that cannot be reduced to science, as has been appreciated at least since David Hume made his famous is/ought distinction.

At the same time, to return to the example of in vitro fertilization, the science around this reproductive technology is made possible in part by all sorts of decisions and resource allocations that define it as a good and worthy thing to know about.

Science serves as a culturally validating force—because it is so in science lends credence and credibility to phenomena, making them seem *right*. Furthermore, the existence of new technologies, informed by science, alters the way we see ourselves and so affects our hopes and our behaviors. The promises of in vitro fertilization can profoundly affect human aspirations.

Scientific knowledge is therefore powerful and pervasive, but it is not all powerful. To think otherwise is to be guilty of scientism. The argument that schooling should enable learners to obtain the sort of powerful knowledge that they would otherwise be unlikely to acquire (e.g., from their families or from listening to television) has received a sustained defense in recent years from the writings of Michael Young (e.g., 2008). My point is that queer approaches to scientific knowledge, precisely by troubling what otherwise would remain unwarranted presumptions about how humans exist and behave, provide access to more powerful knowledge that would otherwise be the case.

Consider how sex is typically presented in school science classrooms. School biology typically examines issues of sex through the lens of reproduction (Reiss, 2007a). This immediately tends to presume heterosexuality. Biology is all too often assumed to be a neutral subject, so that many biology teachers in schools continue to teach gender and sex as unquestioned fact. In particular, differences between females and males are often presented as clear-cut and inevitable, and the study of school biology textbooks shows that they are often sexist and typically ignore lesbian and gay issues (Bazzul & Sykes, 2011; Reiss, 1998; Snyder & Broadway, 2004). For example, biology textbooks in England for 14- to 16-year-olds often omit all mention of the clitoris and, when they do refer to it, frequently talk of it in a belittling way as the female's equivalent of a penis. Males are rendered visible, females less so; and the female exists by virtue of comparison with the male. When the possibility of being gay or lesbian is addressed (the furthest that school biology textbooks ever get from heteronormativity), the impression is generally given that this is a sort of second-best option which the reader may well grow out of.

However, closer examination of sex in human biology provides plenty of space for critical reflection and allows for a richer understanding of what it is to be a sexual person. Emily Martin (1991) has shown that while menstruation is viewed in scientific textbooks as a failure (you should have got pregnant), sperm maturation is viewed as a wonderful achievement in which countless millions of sperm are manufactured each day. Furthermore, sperm are viewed as active and streamlined, whereas the egg is large and passive and just drifts along or sits there waiting. It was back in 1948 that Ruth Herschberger argued that the female reproductive organs (it is difficult in the scientific discourse around sex to avoid referring to reproduction) are viewed as somehow being less autonomous than those of the male. The way the egg is portrayed in science textbooks has been likened to that of the fairy tale *Sleeping Beauty*, in which a dormant, virginal bride awaits a man's magic kiss. However, for well over a decade, biologists have seen both egg and sperm as active partners. Just as sperm seek out the egg, so the vagina discriminates between sperm, and the egg seeks out sperm to catch. Nevertheless, as Martin points out, even when acknowledged, such biological equality is still generally described in a language that gives precedence to the sperm. When the egg is presented in an active role, the image is one of a dangerous aggressor, "rather like a spider lying in wait in her web" (Martin, 1991, p. 498).

Social historical research on sex hormones has also shown how the way that such hormones are presented in textbooks and scientific papers gives messages that go well beyond what the data indicate. Despite the fact that it has been known since the 1920s that each sex contains the "other's" hormone—so males contain estrogen and females testosterone—school textbooks typically ignore both this fact and the close chemical similarity between estrogen and testosterone (Roberts, 2002). Indeed, a different reading of the data to that usually presented in school textbooks—but one more in line with the scientific evidence about the working of sex hormones—is that conventional understandings of femaleness and maleness lie on a continuum. Such a model of the consequences of the actions of the sex hormones became common among endocrinologists in the 1940s.

While this model can lead to an essentialist understanding of sexuality and sexual orientation—and it correlated with a rise in the number of studies of the presumed femininity of gay men (Oudshoorn, 1994)—it can also be seen as allowing a far more fluid understanding of sexuality, accommodating, for example, some forms of intersexuality (cf. Callahan, 2009). The principle of intersexuality dates largely from Magnus Hirschfeld's pioneering work in the first three decades of the 20th century on sexual difference. By rejecting the discrete categories of male and female, arguing instead that each of us is on a continuum, Hirschfeld radically deconstructed the sexual binary (Bauer, 2003). One of the things good teaching can help students to appreciate is the way in which males and females are pressured, respectively, to perform maleness and femaleness, discourses that are structured largely in opposition to each other (cf. Butler, 1990; Martino & Pallotta-Chiarolli, 2005). My experience of teaching a nonbinary version of human sex to school students is that many of them find it fascinating; it can help them to see the world, and themselves, in a new light. Indeed, enabling students to see classifications that relate to gender, sex, and sexuality more fluidly has the potential to lead students to question, even disrupt, other rigid typologies, facilitating the beginnings of an intersectional analysis.

Biological indicators of sexual orientation have long been sought and continue to fascinate commentators, attracting some and worrying others in the gay and lesbian communities. Hardly a month goes by without a report of some such biological indicator. Precisely which indicator is flavor of the month (a hormone, a gene, parental upbringing, relative finger length, etc.) may tell us more about research fashions and the power of statistical analysis than much else. Around the middle of the 20th century, hormones were widely thought to be all powerful and responsible for our sexuality. Towards the end of the 20th century the focus shifted to genes. Genes are responsible for the chemicals, including hormones, made in the body, and a reductionist perspective sees them as determining not just sex and sexuality but almost all of what it is to be ourselves.

Much of the literature about the "causes" of sexuality concentrates on gayness, though Lynda Birke, a biologist as well as a feminist and a lesbian, provides a valuable review about lesbianism and notes that she has "spent much time and energy refuting the allegations that any social categories (of gender, race or sexuality) are fixed by biology" (Birke, 1997, p. 58). However, as Birke points out, there are, of course, several reasons for hesitating in rejecting entirely biological notions of sexual orientation. For one thing, some have used such notions politically to argue for gay rights (though this approach is hotly contested—see Schüklenk & Brookey, 1998); more prosaically, it may well yet turn out that there are biological bases to at least some people's sexuality.

All of which leads us on to how might biology be taught better in schools? Much biology teaching is focused around the use of textbooks, yet "teachers can read subtextually and resistantly and can help their students to do likewise. Too rarely are students encouraged to critique their science textbooks; too often are textbooks used as if they contained only unquestionable truths" (Reiss, 1998, p. 148). This is a simple message but one that provides a teacher—and her/his students—with a powerful tool, for it avoids buying into the general assumption of teacher as the expert repository of facts, instead sitting more comfortably with critical and emancipatory understandings of education. This can be more satisfying for teacher and students alike, and fits well with an information society which provides students with plenty of opportunities to obtain many of the facts they want/need to know at the right pace for themselves.

A fuller account of what biology teachers can do is provided by Anne-Marie Scholer (2002) in her description of her teaching program for a two-semester intermediate-level college course in anatomy and physiology, required for first-year students in nursing, athletic training, and physical therapy majors. Scholer begins with the idea that male/female is not a dichotomy. Here she draws on the various causes of indeterminate sex in humans, the sex hormone story outlined above, the existence of breast cancer in men, and transgender. As she says,

> While the foregoing material is undoubtedly familiar to individuals in the fields of sexuality education, it is quite new to most of my students and peers. I have found such examples to work well in class, creating vocal displays of cognitive dissonance. (Scholer, 2002, p. 78)

As every teacher knows, cognitive dissonance, if well handled, can be a powerful incentive to learning. (Handled badly it can merely reinforce prejudice or be rejected as confusing.) Scholer goes on to challenge the prevailing stereotype of eggs as passive objects, to discuss how sex is not just anatomy and hormones, to avoid heteronormativity, and generally to "create an inclusive environment in my classroom" (p. 82).

There may be some who think that school and college biology is not an important battle ground, even that to fight discrimination and injustice on this front is to risk allowing the discourse to be predetermined by the other side. However, as Mariamne Whatley pointed out in a chapter in Debbie Epstein's and James Sears's *A Dangerous Knowing*, "Using science to attack comprehensive sexuality education and to support abstinence-only education is one strategy being used currently" (Whatley, 1999, p. 238).

Similarly, Will Letts (2001) has explored how school science structures and is structured by norms of heterosexual masculinity. Letts's work is particularly valuable as he focuses on classroom examples of primary school science—when some might assume that at this age science is fairly neutral (see the work of the No Outsiders project in DePalma & Atkinson, 2009). He argues that science, including school science, functions as a grand narrative that seduces students and teachers; he concludes:

> As a plan of action, I advocate that school science becomes an active and generative site for critical science literacy. The words 'science literacy' in this phrase are intended quite differently than popular utterances of them have come to mean. 'Science literacy' does not simply mean an intake and consumption of science texts and 'facts', either purposefully or through acts of seduction. I am using critical science literacy to denote something akin to critical media literacy. (p. 270)

A school science classroom for critical science literacy, at any age, would be one in which the traditional virtues of science—its open-mindedness and refusal to accept tradition on trust—were more widely (and reflexively) applied. It would allow young people to think about themselves and their sexuality more meaningfully. It would help those uncomfortable with traditional descriptions of masculinity and femininity to realize that they are not alone in their rejection of such simple dichotomies. All this can be achieved without harming those students who are comfortable with such conventional descriptions. Sadly, such classrooms are still rare. In the long run, such teaching, idealistic though this may sound to some, would contribute to making the world a better place both overall and for the many individuals who otherwise feel or find that they don't fit. (But note, Britzman, 1995, points out that attempts to improve things for those who are excluded can often get it wrong, as when the inclusion of gay and lesbian studies in the curriculum causes the very exclusions it is meant to cure.)

THE SCIENCE CURRICULUM SHOULD ADVANCE SOCIAL JUSTICE

The first section of this chapter argued that schooling should enable human flourishing, both individually and collectively. The second section argued that schooling should give students access to powerful knowledge to which they otherwise would lack access, and then applied this dictum to teaching about sex and sexuality in the science classroom. This section begins from an oft-asked question, namely what the aims of school science education should be, and explores the implications of possible answers.

I have argued for a number of years that science education as currently undertaken in schools is generally too narrow in its conceptualization, its aims, its curriculum, its teaching, and its assessment, and that this is a major reason why it fails to engage many young people (Reiss, 1993).

Before designing a school science curriculum, one needs to determine its aims. Many aims for school science education have been proposed (e.g., Reiss, 2007b), though these are often implicit. A frequent aim of science courses has been to provide a preparatory education for the small proportion of individuals who will become future scientists (in the commonly understood sense as employed professionals). This aim has been widely critiqued on democratic grounds (e.g., Millar & Osborne, 1998). After all, what of the great majority of school students who will not become such scientists?

Another aim is to enable "scientific literacy." Although there has been a long-running debate as to the meaning of the term (e.g., Miller, 1983), generally, scientific literacy is seen as a vehicle to help tomorrow's adults to understand scientific issues (Gräber & Bolte, 1997). The basic notion is that science education should aim to enhance understanding of key ideas about the nature and practice of science as well as some of the central conclusions reached by science. Perhaps to be included within this category is the argument that to be an educated person in the 21st century is to understand something of science (e.g., Shamos, 1995). This is the "science as culture" argument; that science is as worth studying *in itself* as are, for example, literature and the arts.

A further aim is that many science courses hope that, as a result of what is learnt, students both now and in the future as adults will be able to gain *practical* benefit from it. At its most straightforward this might be by entering paid employment that draws on what they have learnt in science. Although, as noted above, most students do not enter such careers, they too may still benefit individually from their school science. For example, in most science courses, in countries round the world, it has long been accepted that one of the justifications for the inclusion of certain topics is that knowledge and understanding of them can promote human health. Such topics may include infectious diseases, diet, reproduction and contraception, exercise, and the use of drugs (including smoking and alcohol).

Another, more mundane, way in which school science might help individual advancement is by providing what I have termed *science education for consumerism* (Reiss, 2007b). This is the hope that school science education might, for example, help us choose the most appropriate technological goods (Is it worth my paying *x*% more for a washing machine that uses *y*% less hot water?) or make broad decisions, for instance about climate change, on narrow criteria (Where should I live so as to minimize problems with rising sea levels or extreme weather events?). This is a subset of the more general and long-established argument that science education should be for public understanding (American Association for the Advancement of Science, 1990; Millar, 1996).

A further aim of school science education is that it should be for citizenship (Jenkins, 1999). A "weak" version of this consists of learning what a democracy is and the place that science plays in being an engaged citizen. A "strong" version entails using such knowledge to bring about desirable change. This is closely allied to claims that the aim of school science education should be to effect social justice or sociopolitical action (e.g., Calabrese Barton, 2001; Carter, 2005; Hodson, 2009). Angela Calabrese Barton draws on feminist approaches to show that many of the students with whom she and her colleagues work, whilst seen in school as poor attainers in science, are actually perfectly capable of high-quality science work provided they are given real choice in the science they work at.

It is evident that there are currently diverse aims for school science education. That science education should help advance social justice seems an aim that is worth arguing for. Science is so powerful a discourse that it has too often been used to marginalize and oppress others, those who do not fit into the mainstream, the majority. Stephen J. Gould (1981) pointed out how supposedly objective measures undertaken by the most reputable scientists were used by them and others to reinforce gender, racial, and class stereotypes.

It can be difficult for any of us, scientists or not, to ensure that our work is not overly influenced by the cultural presumptions of the society in which we live—structures are important. One approach that can be useful is that advocated by John Rawls (1971) in his *A Theory of Justice*. Rawls argued that we should live as if behind a veil of ignorance, i.e. that societal decisions should be made as if each of us did not know our own circumstances. The great advantage of this imaginative approach is that it puts us in the shoes of others and can help to make discrimination against minorities less likely. Rawls goes on to argue that matters should be arranged so that (a) each person has the most extensive liberty compatible with the liberty of others and (b) social and economic inequalities benefit the least advantaged members of society.

Rawls's arguments not only help us to consider how science might be more justly presented; they also have implications for such repeatedly contested matters as same-sex couples at high school proms and cases of students being omitted from yearbooks because of pregnancy or what a school deems to be inappropriate dress (the word has a pleasing ambiguity), e.g. young women not being allowed to wear a tux(edo) or a button-down shirt and tie.

CONCLUSIONS

Much of the early literature that addressed gay, lesbian, and queer issues in schooling (e.g., Garber, 1994; Unks, 1995) paid little or no attention to science. The more recent emphasis on queering school science (e.g., Fifield & Letts, 2014; this volume) is therefore to be welcomed. Around the world, a common argument in favor of science enjoying its privileged place in the curriculum is that a country needs large numbers of people to choose careers in science and related fields (e.g., engineering). This neoliberal argument is not to be ignored; human flourishing is helped (think of advances in medicine, transport, and communications) by scientific advances. Yet, there should be more to school science than its role in providing the next generation of scientists. In this chapter I argue that three different approaches to determining what might be in the science curriculum when teaching about sex and sexuality—a curriculum for human flourishing, a curriculum for powerful knowledge, and a curriculum for social justice—all conclude that conventional teaching about human sex and sexuality falls short.

Queer approaches to teaching these topics can aid human flourishing, enable students to gain more powerful knowledge, and address social injustices. Crucially, for those who believe that school science is an uncontested arena that should stick to solid and well-established science, it is increasingly the case that conventional teaching about human sex and sexuality is poor science. I am well aware that there will be some who think it overoptimistic to presume that science teaching can help when teaching about sex and sexuality, given how contentious these issues are in many countries (e.g., Irvine, 2002). However, science occupies a central space on the school timetable and may be able to achieve things more difficult to achieve through less powerful subjects.

Finally, what of those who maintain that the approach for which I have been arguing runs counter to how they see reality, for instance because of their understanding of the teaching of religion? Three main responses can be made. First, when religion is included in the school curriculum this does not mean that it should trump all other considerations (Reiss, 2012). Secondly, not all religions and not all interpretations of religion are conservative ones. Within Christianity, for example, a whole range of positions hold, from ones where scripture is read literally and noncritically to ones where scripture, while still held in high esteem, is read more critically (Reiss, 2014). Hedgehogs and foxes again: There is more than one way of seeing and being in the world. Thirdly, and perhaps more mundanely, there may be circumstances when even someone convinced of the various benefits of queering the science classroom may feel that it would be wise to avoid too much usage of the word *queer* (which can act as a red rag to a bull) and simply concentrate on the argument that all one is concerned with is the provision of good quality science education. This need not be seen as a "cop out"; just as teaching in biology about "the problem of the species" (hybrids, ring species, etc.) need not be taken as an attack on the notion of "kinds" in the book of Genesis, so a richer, more nuanced, and I would argue more accurate, teaching about sex and sexuality need not be read as an attack on religious faith.

REFERENCES

American Association for the Advancement of Science. (1990). *Science for all Americans: Project 2061*. New York, NY: Oxford University Press.

Bauer, J. E. (2003, October). *Magnus Hirschfeld's doctrine of sexual intermediaries and the transgender politics of identity.* Paper presented at the meeting of Past and Present of Radical Sexual Politics, Amsterdam. Retrieved from http://www.iisg.nl/womhist/radsexpol.html#papers
Bazzul, J., & Sykes, H. (2011). The secret identity of a biology textbook: Straight and naturally sexed. *Cultural Studies of Science Education, 6*, 265–286.
Berlin, I. (1953). *The hedgehog and the fox: An essay on Tolstoy's view of history*. London: Weidenfeld & Nicolson.
Birke, L. (1997). Born queer? Lesbians interrogate biology. In G. Griffin & S. Andermahr (Eds.), *Straight studies modified: Lesbian interventions in the academy* (pp. 57–70). London: Cassell.
Britzman, D. P. (1995). Is there a queer pedagogy? Or, stop reading straight. *Educational Theory, 45*, 151–165.
Butler, J. (1990). *Gender trouble: Feminism and the subversion of identity*. New York, NY: Routledge.
Calabrese Barton, A. (2001). Science education in urban settings: Seeking new ways of praxis through critical ethnography. *Journal of Research in Science Teaching, 38*, 899–917.
Callahan, G. N. (2009). *Between XX and XY: Intersexuality and the myth of two sexes*. Chicago, IL: Chicago Review Press.
Carter, L. (2005). Globalisation and science education: Rethinking science education reforms. *Journal of Research in Science Teaching, 42*, 561–580.
DePalma, R., & Atkinson, E. (Eds.). (2009). *Interrogating heteronormativity in primary schools: The No Outsiders project*. Stoke-on-Trent: Trentham Books.
Fifield, S., & Letts, W. (2014). [Re]considering queer theories and science education. *Cultural Studies of Science Education, 9*, 393–407.
Foucault, M. (1969/1972). *The archaeology of knowledge*. London: Tavistock.
Garber, L. (Ed.). (1994). *Tilting the tower: Lesbians teaching queer subjects*. New York, NY: Routledge.
Giffney, N. (2009). Introduction: the 'q' word. In N. Giffney & M. O'Rourke (Eds.), *The Ashgate research companion to queer theory* (pp. 1–13). Farnham: Ashgate.
Gould, S. J. (1981). *The mismeasure of man*. New York, NY: W. W. Norton.
Gräber, W., & Bolte, C. (Eds.). (1997). *Scientific literacy: An international symposium IPN 154*. Kiel: Institut für die Pädagogik der Naturwissenschaften an der Universitatät Kiel.
Harris, K. (1999). Aims! Whose aims? In R. Marples (Ed.), *The aims of education* (pp. 1–13). London: Routledge.
Hodson, D. (2009). *Teaching and learning about science: Language, theory, methods, history, traditions and values*. Rotterdam, The Netherlands: Sense Publishers.
Holland, J., Ramazanoglu, C., Sharpe, S., & Thomson, R. (1998). *The male in the head: Young people, heterosexuality and power*. London: Tufnell Press.
Irvine, J. M. (2002). *Talk about sex: The battles over sex education in the United States*. Berkeley, CA: University of California Press.
Jenkins, E. W. (1999). School science, citizenship and the public understanding of science. *International Journal of Science Education, 21*, 703–710.
Lemke, J. (2011). The secret identity of science education: Masculine and politically conservative? *Cultural Studies of Science Education, 6*, 287–292.
Letts, W. (2001). When science is strangely alluring: Interrogating the masculinist and heteronormative nature of primary school science. *Gender and Education, 13*, 261–274.
Luhmann, S. (1998). Queering/querying pedagogy? Or, pedagogy is a pretty queer thing. In W. Pinar (Ed.), *Queer theory in education* (pp. 141–155). Mahwah, NJ: Lawrence Erlbaum Associates.
Martin, E. (1991). The egg and the sperm: How science has constructed a romance based on stereotypical male-female roles. *Signs: Journal of Women in Culture and Society, 16*, 485–501.
Martino, W., & Pallotta-Chiarolli, M. (2005). *Being normal is the only way to be: Adolescent perspectives on gender and school*. Sydney: University of New South Wales Press.
Millar, R. (1996). Towards a science curriculum for public understanding. *School Science Review, 77*(280), 7–18.
Millar, R., & Osborne, J. (Eds.). (1998). *Beyond 2000: Science education for the future*. London: Nuffield Foundation.

Miller, J. (1983). Scientific literacy: A conceptual and empirical review. *Daedalus, 112*(2), 29–48.
Oudshoorn, N. (1994). *Beyond the natural body: An archeology of sex hormones*. London: Routledge.
Rawls, J. (1971). *A theory of justice*. Cambridge, MA: Harvard University Press.
Reiss, M. J. (1989). *The allometry of growth and reproduction*. Cambridge: Cambridge University Press.
Reiss, M. J. (1993). *Science education for a pluralist society*. Milton Keynes: Open University Press.
Reiss, M. J. (1998). The representation of human sexuality in some science textbooks for 14–16 year-olds. *Research in Science & Technological Education, 16*, 137–149.
Reiss, M. J. (2007a). Representing the world: Difference and science education. In M. Reiss, R. DePalma, & E. Atkinson (Eds.), *Marginality and difference in education and beyond* (pp. 61–72). Stoke-on-Trent: Trentham.
Reiss, M. J. (2007b). What should be the aim(s) of school science education? In D. Corrigan, J. Dillon, & R. Gunstone (Eds.), *The re-emergence of values in science education* (pp. 13–28). Rotterdam, The Netherlands: Sense Publishers.
Reiss, M. J. (2012). What should be the role of religion in science education and bioethics? In S. A. Øyen, T. Lund-Olsen, & N. S. Vaage (Eds.), *Sacred science? On science and its interrelations with religious worldviews* (pp. 127–139). Wageningen: Wageningen Academic Publishers.
Reiss, M. J. (2014). Sex education and science education in faith-based schools. In J. D. Chapman, S. McNamara, M. J. Reiss, & Y. Waghid (Eds.), *International handbook of learning, teaching and leading in faith-based schools* (pp. 261–276). Dordrecht: Springer.
Reiss, M. J. (2015). The nature of science. In R. Toplis (Ed.), *Learning to teach science in the secondary school: A companion to school experience* (4th ed., pp. 66–76). London: Routledge.
Reiss, M. J., & White, J. (2013). *An aims-based curriculum: The significance of human flourishing for schools*. London: IOE Press.
Roberts, C. (2002). 'A matter of embodied fact': Sex hormones and the history of bodies. *Feminist Theory, 3*, 7–26.
Scholer, A.-M. (2002). Sexuality in the science classroom: One teacher's methods in a college biology course. *Sex Education, 2*, 75–86.
Schüklenk, U., & Brookey, R. A. (1998). Biomedical research on sexual orientation: Researchers taking our chances in homophobic societies. *Journal of the Gay and Lesbian Medical Association, 2*(2), 79–84.
Shamos, M. H. (1995). *The myth of scientific literacy*. New Brunswick, NJ: Rutgers University Press.
Snyder, V. L., & Broadway, F. S. (2004). Queering high school biology textbooks. *Journal of Research in Science Teaching, 41*, 617–636.
Unks, G. (Ed.). (1995). *The gay teen: Educational practice and theory for lesbian, gay, and bisexual adolescents*. New York, NY: Routledge.
Whatley, M. H. (1999). The 'homosexual agenda' goes to school. In D. Epstein & J. T. Sears (Eds.), *A dangerous knowing: Sexuality, pedagogy and popular culture* (pp. 229–241). London: Cassell.
White, J. (2011). *Exploring well-being in schools: A guide to making children's lives more fulfilling*. London: Routledge.
Young, M. F. D. (2008). *Bring knowledge back in: From social constructivism to social realism in the sociology of education*. London: Routledge.

ABOUT THE AUTHOR

Michael J. Reiss is professor of science education at the UCL Institute of Education, University College London (UK), and a fellow of the Academy of Social Sciences. The former director of education at the Royal Society, he has written extensively about curricula, pedagogy, and assessment in science education and has directed a very large number of research, evaluation, and consultancy projects over the past 20 years funded by UK Research Councils, government departments, charities, and international agencies. He can be contacted at m.reiss@ucl.ac.uk

CECILIA RODÉHN

14. EXHIBITING DOCTORS AND NURSES

Queering Professional Education in a Medical Museum

OLD FRIENDS

Some old friends greet me as I enter the door at the Museum of Medicine in Uppsala. There are four of them: two white-skinned mannequins, a man and a woman, alluding to a married heterosexual couple; and a hunchback skeleton of a man who suffered from rickets. Further back there is also a brown-skinned mannequin depicting a doctor (Figure 14.1). As I enter the door, I also meet medical and nursing studies students from Uppsala University. The students will join guided tours as a mandatory part of their education, and I am here to study how the museum is used in their education. They seem unaware of my old friends, and it puzzles me that they are. My old friends point out the norm and the "other" in this museum, and by pointing these out they offer ways to discuss how gender, sexuality, and desire play roles in shaping nursing and medicine.

The curators I talk to reveal that the brown-skinned mannequin has been involved in several incidents connected to sexuality and desire. They tell me a story about how a teenage boy started making out with the mannequin and how they thought that this was a bit strange since there were so many female mannequins to choose from. One of the curators also tells me another story: "At one point when I offered a guided tour, there was a little boy that stopped in front of the mannequin, pointed and said, 'But, that's a girl.'" The curator answered, "No, it is a man." But the boy insisted and he made the curator investigate the mannequin's identity. After the guided tour, one of the curators undressed the mannequin. "He had breasts. Someone had covered it up by stuffing the midsection with paper to give him a more masculine look." After looking and establishing that the mannequin was physically female-bodied, the curators nevertheless continued to refer to the mannequin as he/him.

The stories may seem innocent, but they "reveal the regulatory mechanisms through which sexual difference is enforced" (Hird, 2000, p. 359). They also allude to harsh realities that many queer people experience. Calling someone out and insisting that a person should be of a different gender, sexually assaulting a person, or undressing a person as a way to establish and determine the "right sex" are part of sexualized violence against queer bodies (cf. Bremer, 2011). The stories, therefore, reveal what is thinkable in this museum, and they also reveal the vulnerable position of the other. In this museum there are several other *white-skinned* female-bodied

DOI:10.1163/9789004331068_016

Figure 14.1. The brown-skinned mannequin (photograph by Patrik Klingborg)

mannequins portraying male medical doctors, but they have never been sexually harassed, called out, or undressed. Sara Ahmed (2007) argues that the orientation of whiteness involves making bodies that are not white exposed and visible. Although Ahmed (2007) writes this in a critique of whiteness, I suggest that the exposure and visibility of brown-skinned, transgender bodies can mean something positive for this museum. The brown-skinned mannequin, in being visible and exposed, makes a queer mark in the room in a way that white-skinned female-bodied mannequins do not. He makes queerness present and in being exposed and visible he also challenges heteronorms.

The brown-skinned mannequin and my other old friends, nevertheless, point to the norm and the other in the museum, revealing that this museum, and many other museums, present science and knowledge constructed on heteronorms and around binary pairs such as man/woman, heterosexual/homosexual, and able/disabled. The students I meet seem to overlook this completely. It is not only the students who overlook how museums sort the world into orderly pairs of normal and other.

Researchers such as Justin Dillon (2012), John Falk and Lynn Dierking (1997, 2000), and Eva Davidsson and Anders Jakobsson (2012) seldom comment on this when discussing science education in museums. Instead they propose that museums can facilitate qualitative learning. This is, however, countered by researchers in critical heritage studies who propose that museums are political spaces that reflect and authorize the heteronormative, masculinized, and Eurocentric agendas of society (e.g., Golding, 2016; Macdonald, 2011; Rodéhn, 2016; Smith, 2006). Following these researchers it is clear that conducting higher education in museums is undoubtedly a complicated matter that needs further consideration. This chapter aims to contribute in this regard by investigating how the Museum of Medicine is used in the education of students studying nursing and medicine at Uppsala University. I want to find out what kind of higher education is made possible in this museum and what role sexuality and desire play when teaching in the museum. Is queer teaching possible in this museum?

Queer theory has been subject to limited discussion in museum and heritage research. It has been suggested that there is a need to create queer archives and collections, as well as to include "queer" in classification systems and provide interpretations from a queer perspective in exhibitions and pedagogic activities (e.g. Oram, 2011; Romesburg, 2014; Steorn, 2012; Tyburczy, 2012). A call for including the excluded is made, but Robert Mills (2008) and Steve Fifield and Will Letts (2014) challenge this and call it a neoliberal agenda working along heteronormative lines. They call for a fracturing of heteronormativity and a rethinking of science and museums at large. What Mills, and Fifield and Letts, suggest is a much larger and more complicated project than working to include those that have been excluded, and I agree with them. Queer is that which lies outside the norm; it is that which disturbs the order, and it offers a way to deconstruct categories and identity positions. It is, thus, not about including identities in, for instance, museums; its purpose is to stir, disturb, and fracture categories (Ahmed, 2006b; Ambjörnsson, 2006; Rosenberg, 2002). As Judith Butler (1993) describes it, queer is "a site of collective contestation" (p. 223) about rethinking heterosexuality, and a critique of heterosexual norms that cast queer experiences as the other (see Butler, 2006; Rich, 1980).

The term *queer* has also been extended to questioning how disabled bodies are understood (McRuer, 1997, 2013; Sandahl, 2003). Robert McRuer (2013) argues, drawing on queer scholar Adrienne Rich (1980), that society upholds an idea of *compulsory able-bodiedness* that produces disabled bodies as outside the natural order of things—as the other to the able body. Queer has also become a way to question ability and its intersections because it argues against the compulsion to observe norms of all kinds (McRuer, 1997; Sandahl, 2003). Queer, in the words of Fifield and Letts (2014), is thus about resisting and rethinking science and science teaching as a system of identity politics, or as Mills (2008) suggests, queer entails challenging the whole idea of the museum as a "meaning-making machine" (p. 46).

If queer is a challenging and fracturing of norms, then we need to understand how norms direct us in museums and direct teaching in museums. Ahmed's (2006a, 2006b) queer phenomenology and the concept of *orientation* assist in illuminating this. Orientation is Ahmed's (2006a, 2006b) way of theorizing sexual orientation and how the bodily, the social, and the spatial are intertwined. For Ahmed orientations are not only about starting points, that which is already known (heteronormativity), but also about how bodies move, and are moved, in space. Being oriented means being asked to follow *directions*, which is her way of explaining how bodies are put in normative lines (Ahmed, 2006b). Ahmed (2006b) suggests that the body is oriented from *lines*—norms—that we are forced to, or freely, align with. Yet, lines are not prescribed but recreated by being followed—they are performative. Following lines is, thus, a social investment in heteronormativity and a constant reproduction of the heteronorm. In this chapter I use Ahmed's concept to discuss how certain kinds of knowledge and histories are extended into the museum space, whereas others are not. I discuss how the exhibition space orients teaching, how volunteers and curators create lines of orientation when teaching, and how that teaching orients students.

In order to suggest a queer pedagogy, I appropriate Ahmed's concept of *disorientation*. She explains this as a situation that makes us aware of norms—aware that we are oriented in the first place. Ahmed explains disorientation as a failed orientation in terms of not following, or not conforming to, norms. Disorientation could be explained as becoming, or making things, oblique—making things queer and participating in queering the world. It is about making the strange familiar, about bringing to the front that which is in the background or slips away, and about disturbing the picture and opening up another angle of the world (Ahmed, 2006b).

I begin with a critical ethnography of exhibitions at the Museum of Medicine, move on to a critical discussion of teaching in the museum, and lastly offer ways to queer that teaching.

CURATING NORMS: A CRITICAL ETHNOGRAPHY

The Museum of Medicine is situated at Ulleråker, a former psychiatric hospital in the outskirts of Uppsala, in southeastern Sweden.[1] As this museum is used in the teaching of nursing studies students and medical students at Uppsala University, it is crucial to understand what kinds of teaching and learning are enabled and which are foreclosed.[2] Thus, I took a tour around the museum while doing thick descriptions of the exhibitions. This method involves documenting the exhibitions while being aware of my own social position (Geertz, 1994; Rooke, 2009).[3] I started the investigation with Dennis Sumara and Brent Davis's (1999) words in mind that "queer theory does not ask that pedagogy become sexualized" (p. 192). Instead, they stress that we should show how pedagogy is already sexualized, begin to see that it is heterosexualized, and start interrupting it. I focused on how heterosexual desires are fundamental in knowledge productions and in upholding and explaining different professions. Investigating and interrupting these foundations offers possibilities for suggesting queer knowledge productions.

I started my tour in a large, downstairs room stuffed with medical equipment (Figure 14.2). The overcrowded room focuses predominantly on the history of surgery

Figure 14.2. The Museum of Medicine in Uppsala (photograph by Patrik Klingborg)

as depicted through various objects, including ventilators, EKG machines, monitors, heart-lung machines, scalpels, other operating equipment such as retractors, and human remains in the form of different parts of skeletons. "You should have seen it a couple of years ago," a curator tells me. "You would not believe me, but there were even more things on display." The museum resembles a contemporary cabinet of curiosities; the number of objects makes me disoriented and I do not know where to start. Ahmed (2006b) would suggest that I am experiencing a failure to organize and to hold things in place. Although disoriented, I realize that the nonconformity of the exhibition is a benefit when teaching here as there is no predetermined path through the museum, which opens it for any kind of teaching. The museum is, in part, what Mills (2006, 2008) suggests a queer museum should be: a museum that rejects linear temporality and narratives. A queer museum according to him is a museum in anarchy, one that adopts the style of a scrapbook or collage, showing fragments, resisting grand narratives and categorical assertions.

Yet there are some lines of orientation in this exhibition that direct the understanding of different caring professions.[4] An example of such orientation is the heteronormative construction of gender as binary, complementary (but not equal) pairs (cf. Butler, 2006; Hirdman, 2007). This norm can be seen in how the histories of white male medical doctors occupy the largest space in exhibition. Moreover, the objects associated with white men stand in contrast to the presentations of (female) nurses. The history of midwives and nurses is told in corners of the large room and at the far back of the museum. They are literally on the edges, occupying a smaller space compared to the history of surgery. I suggest, following Ahmed (2006b), that the histories of white male medical doctors extend into the museum space and that the visitor is oriented in relation to this extension.

Figure 14.3. Representation of a medical doctor in the downstairs exhibition (photograph by Patrik Klingborg)

White male medical doctors are depicted in roles such as scientists and discoverers, and, in my view, they are described as saviors or heroes liberating the population from unwanted diseases and disabilities and, thus, connected to the idea of men's roles in controlling bodies. This enactment of masculinity, understood as a heterosexual trope connected to values such as intelligence and fortitude, is often associated with a domineering form of masculinity. R. W. Connell (1996) calls this "hegemonic masculinity" and she proposes that it assumes a dominant position over other men, especially gay men, and women. The hegemonic position of male medical doctors is important to discuss as it affects the medical profession today, especially considering that a recent study shows that 81% of the students (male and female) in the medical program at Uppsala University have experienced gender discrimination or gender and/or sexual harassment during their studies, and that these mostly occurred while gaining work experience in hospitals (Svensson, 2014).

In being both absent and present, sexuality and desire play important roles in how the caring professions are depicted and explained. Kristina Eriksson (2003), who interviewed medical doctors and deconstructed how they express gender relations, writes that they publicly depict their profession as nongendered, asexual, and unemotional. I suggest that this is clearly seen in the exhibition where surgeons engaging with medical equipment in a seemingly nonemotive way are often facing away, or moving away, from the patients (Figure 14.3). This depiction is no surprise considering that Connell (1996) argues that emotional restraint is a way to uphold successful masculinity in a heterosexual society. Eriksson (2003) goes on to show that medical doctors speak of nursing, in contrast to medicine, in personal, emotive, and gendered ways that present nurses as engaged with and caring for patients. Yvonne Hirdman (2007) comments that the male norm is often seen as ungendered whereas women are often constructed as gendered, sexualized, emotional, and lacking control of their desires.

This normative conceptualization of men and women can also be seen further back in the museum where pictures show nurses caring for patients. One display text refers to the nurse as "the surrogate mother." This is not surprising as nursing in Sweden since the 1800s has been constructed on the ideal of women's biological abilities as "natural" caregivers (Dufwa, 2010; Hirdman, 2007). The notion of women as natural caregivers is evidenced in Florence Nightingale's suggestion that every woman was born to become a nurse (Nightingale, 1989). Nightingale's writing was cited in nursing research up until the end of the 1980s (Jorfeldt, 2010), and this notion remains strong today in the profession as well as in broader society. Hence, nursing is constructed based on an idea of what a female body can and should do.

Heteronorms organize how the caring professions are understood, illustrating Hirdman's (2007) observation that the hierarchal gender order positions women as subordinate to men and as not capable of doing men's work. Although not clearly articulated as such, I suggest that this idea provides lines of orientation (Ahmed, 2006a, 2006b) for what is included and excluded in the exhibition. For instance, Karolina Widerström—who in 1888 became the first female physician with a degree in medicine to practice in Sweden—is excluded from the exhibition, as are other women practicing medicine and making discoveries. I will return to Widerström below.

Another example of how sexuality and desire organize nursing is the silenced history of men who entered nursing from the 1950s onwards. The association of nursing as something feminine, for women, and alluding to the relationships of nurses and medical doctors as akin to a heterosexual marriage, often results in male nurses being viewed as feminized and considered gay (Harding, 2007). This misrecognition rests on the idea that men in female-coded and female-dominated professions are unmanly, and that male nurses fail to perform a socially acceptable version of masculinity. These "failed" masculinities have been given negative connotations and used as means of subjugation (Connell, 1996). Heteronormativity has serious consequences in nursing when, for example, homophobic patients refuse being cared for by male nurses because they fear being molested (Harding, 2007).

The role of heteronormativity is further illustrated upstairs in the museum in a display about the Ulleråker psychiatric hospital. This display has a particular resonance as it tells the story of the geographic area and the building in which the museum is situated. The exhibition centers on mental health care and presents a collage of objects such as medical equipment (e.g., straightjackets and belting beds) and tools, kitchen equipment, artwork, and furniture. At the center of the display is the story of life at Ulleråker presented from caretakers' and nurses' perspectives.

Knowledge about mental health care and patients is presented in heteronormative terms. The exhibition explains that Ulleråker was a *home* to staff members, but also to more than 1200 patients. Patients and caretakers are depicted as one big family. Anne-Marie Fortier (2001) and Ahmed (2006b) explain that home is associated with heterosexuality, familiarity, comfort, and belonging. This association is clearly seen in the exhibition: in the stories related, in artefacts like movie projectors and theater costumes used by caretakers and patients, and in pictures showing patients, caretakers, and nurses socializing at different events. Although daily life is described in home-like and family terms, there is a clear division between how able/mentally healthy bodies and disabled/mentally ill bodies are presented. Patients are often pathologized in various ways, which can be seen in a display of keys and stories of how caretakers guarded the patients. It is also exemplified through artefacts used to treat patients before the existence of psychopharmaceuticals, such as with bathtubs for so-called long baths, straightjackets, electrotherapy, and belting beds. The construction of Ulleråker as a home and as one big family stands in stark contrast to these objects.

Yet there is ambiguity in how patients are presented, as seen, for instance, through the experiences of Gustav Fröding and Augusta Strömberg. Fröding is one of Sweden's most famous poets, and his poems are on display in this exhibition. Strömberg was an unknown but talented artist whose work has been acknowledged posthumously and is also on display in the exhibition (Figure 14.4). Illnesses, whether mental or bodily, are exhibited in the museum as something that is wrong and that can be corrected through the use of various treatments (cf. McRuer, 2013; Sandahl, 2003). Working with art, poetry, and so on is portrayed as a way to correct mental illness or to calm agitated patients. Fröding and Strömberg are presented as

Figure 14.4. Augusta Strömberg's artwork in the exhibition on Ulleråker psychiatric hospital (photograph by Patrik Klingborg)

exhibiting artistic excellence, and they are also portrayed as patients with mental illnesses that contributed to their creativity. Patients' abilities are, on the one hand, celebrated and, on the other hand, depicted as abnormal and pathological, revealing that in this museum creativity is seen as a consequence of mental illness. In Fröding's case, sexuality also plays a role in the exhibition as witnessed by medical records that describe his incessant masturbation. His desire and autosexuality are, consequently, made into something pathological.

These objects and narratives articulate power structures that situate the able-bodied caretakers in a dominant position. This relationship, as presented in the exhibition, can be described in familial terms by likening it to a parent-child relation. This, and the fact that institutionalized patients do not fulfill the ideals of an able/mentally healthy body, results in patients being presented as unable to care for themselves, demanding and in need of attention. This discourse is very familiar in colonial narratives where the "other" is described in similar terms by the colonizer (Hall, 1997). It is a powerful narrative where the other—the disabled body, the mentally ill, the black body, the gay body—is portrayed as a less worthy member of society.

The exhibition provides lines of orientation for teaching and learning in the museum, but is this actually the kind of teaching that occurs in the museum? I will answer this question in the following section.

TEACHING NORMS

To study the kinds of teaching made possible in the museum, I followed five guided tours offered to students in the medical and nursing studies programs. The guided tours were given by two retired, male medical doctors working as volunteers in the museum, and two curators with postgraduate degrees in the history of ideas. I used participant observation focused on the teaching, rather than students' responses, to examine how lines of orientation are followed and produced.[5]

The guided tours emphasize the historical development of different medical treatments, from traditional techniques, such as bloodletting, to sanitation, surgery, and anesthesia. Curators and volunteers also take students to the area that displays nursing, and they use the upstairs exhibition to discuss mental health care. The curators and volunteers follow the same direction around the museum and tell a very similar story. In doing so, they create lines of orientation in this otherwise quite disoriented exhibition.

The tours focus on well-known white male medical doctors and their contributions; consequently, the tours create heteronormative, gender-bifurcated lines of orientation that direct students' understanding of medicine. One of the volunteers particularly focuses on the "father" of pediatrics and other male medical doctors who made discoveries in different fields of medicine and won awards such as the Nobel Prize. Volunteers and curators alike use masculinized terms alluding to heterosexual family relations in order to describe the medical profession. Yet, the gender of the medical doctors is not explicitly commented on, other than through names and epithets such as "father." I suggest that this illustrates Eriksson's observation that the medical profession presents itself as nongendered through a discourse with a strong, male-gendered bias. I have already discussed the presentation of medical doctors and the male norm in the exhibition. Since the message of the exhibition is largely echoed in the teaching, I will not dwell further on the subject. Instead, I discuss how nurses and patients are presented and how this is connected to desire and sexuality.

In contrast to surgery, where curators and volunteers dwell on techniques and technical aspects, nursing is presented in terms of its social role in caregiving, which works to exclude methods and technical discoveries within the nursing profession. In fact, one of the volunteers does not talk about nursing at all. The other volunteer and the curators talk about nursing briefly by focusing not on the practices in the profession but on the uniform (Figure 14.5). To them the uniform plays a central role in the narrative, and they discuss and compare it to the general fashion of the time. The length of the dress takes up a large portion of the teaching, and this is connected to nurses' behavior and their role as being, in the words of one volunteer, "a moral example" in society.

Figure 14.5. The history of nursing on display (photograph by Patrik Klingborg)

Sexuality and desire, thus, play a vital role in positioning nurses as those upholding moral values—as those not succumbing to sexual desire despite being physically close to other bodies and expressing emotions in the conduct of their profession. This is also seen in the exhibition that focuses on nurses' daily lives. Here the display text explains that the nurse "left the chaste life of nursing behind" when she entered into marriage. The curators echo this message during the guided tour and tell the group that nursing was an occupation for unmarried women who had to leave the profession when they got married (to a man, it goes without saying). Some few exceptions to this norm exist in the exhibition, where it explains that nurses with "a calling" remained in service for the rest of their lives and did not marry. One of the curators echoes this and says, "to some it was a calling and they never married."

Drawing on Rich (1980) I hold that in this teaching, compulsory heterosexuality works to position female nurses in a role where marriage to a man is considered "normal" and desirable. Remaining a nurse, choosing not to marry, goes against the heteronormative stages of life that Ahmed (2006a) and J. Halberstam (2005) describe as consisting of birth, childhood, adolescence, marriage, reproduction, and death. Remaining a nurse can be seen, drawing on Ahmed (2006a), as not moving forward on this hetero-timeline, but rather sideways, by exhibiting other kinds of desires, such as a passion for the profession and possible lesbian desires (or no desires at all!).

The heteronormative gender positioning of nurses in the teaching at the museum, however, works to marginalize lesbians and others who do not fit the presumed heterosexual norm. This also works to marginalize the history of male nurses. One curator comments on the fact that men were not accepted into the profession until 1953, but makes no comment on why or on how this changed the profession—other than that the uniform had to be transformed and that male nurses could not live in the same house as the other (female) nurses. This sex segregation seems to reflect the assumption that men and women would be sexually attracted to each other and so could not live together, but that nurses of the same sex who lived together would not face such temptations or transgressions. I will return to this later.

In their teaching, the curators and volunteers not only discuss the history of medicine and nursing but also touch upon patients, both mental health patients and others. Patients' experiences and their own stories are, however, not touched upon, a common phenomenon in many museums displaying medical history (Obermark & Walter, 2014). Patients' circumstances play a role in explaining different kinds of treatments and medical discoveries. This is especially evident in the volunteers' teaching, where patients' disabled/mentally ill bodies are the sites in which medicine is deliberated and explained. The body is made sexless, genderless, and without any other identity. Instead the body becomes a piece of flesh that is worked upon and through which discoveries, mistakes, and methods are articulated. Thus it is not the patients' lived experiences that are at the center of the discussion, but the use of the patients' bodies for medical purposes.

One of the students studying nursing picks up on this and interrupts a volunteer to ask if the patient really wanted surgery. The volunteer becomes quiet for a second and says, "Well, they did not have a choice, they were not asked, … only informed that they would die otherwise." He answers fast and shifts back to discuss the technical aspects of the surgery. Although this tendency is present in both curators' and volunteers' teaching, the curators, nevertheless, often focus on patients and the social aspects of medicine. One curator tells the students that "surgeries were considered the last resort" and talks about the physical pain of surgery that the patient must have felt. The volunteers (who are retired medical doctors), on the other hand, seldom touch upon the patients' experiences.

A similar tendency can be seen in how mental health patients are talked about. Mental health patients, in comparison to other kinds of patients, are presented in a more diverse way. The curators and volunteers alike make several attempts to illustrate the diversity of the patients who received mental health care at Ulleråker. The early psychiatric hospital is explained by a curator "as a place where people with illnesses, young and the elderly as well as people that could not care for themselves, lived." A volunteer tells us that "the boundaries between illness, poverty, and mental illnesses did not exist, and it was merely an accumulation of people that did not fit into the society." The volunteers and curators further differentiate among patients based on social position and their tendency toward being "calm" or "worried." Apart from this presentation, the patients are with few exceptions made sexless and genderless. I find this peculiar since the hospital was physically organized in its architectural layout along binary gender lines and, therefore, sex/gender was something that fundamentally organized mental health care. This homosocial environment also opened opportunities for same-sex desires and relations among patients, but there are no such stories told by the curators or volunteers. These stories of same-sex relations do not extend into the museum space, using Ahmed's (2006b) way of putting it.

Some exceptions to the presentation of patients as sexless and genderless exist when volunteers and curators briefly touch upon art and craftwork and other daily activities at the institution. Here the patients are given different identities relating to their creative abilities and life stories. Although the patients are described as skilled in art and craft, the things that they produced are nevertheless often associated with stories of mental illnesses. For example, one of the volunteers refers to Augusta Strömberg's paintings and says, "All of you have surely seen art made by mentally ill patients, but this is so common these days that it is hard to know who is well and who is ill, so we are not going to dwell on the subject." This positioning of the artwork actively discourages students from recognizing the fullness of those with mental illness; it reduces human identity to being a mental health patient.

Yet when the curators and volunteers talk about mental health patients there is an ambiguity in how they are addressed and how the psychiatric hospital is described. The idea of the institution as a home, with the patients and caretakers living together as a family, is a strong image in the teaching. One of the volunteers says that it was "a haven for people with illnesses and mental disorders." Ulleråker is presented as a community within the community were patients and caretakers lived together in harmony. At the same time Ulleråker is also described as a secluded place with dangerous patients. One volunteer explains that the patients "were dangerous to themselves and others" and that the patients "could be seen in corners screaming." The narrative of dangerous patients marks them out as certain kinds of people, and thus marks a difference between caretakers and patients.

The positioning of patients as dangerous also works to produce caretakers as men who embodied hegemonic masculinities. This is seen, for instance, when one of the volunteers tells the medical students that "there were no females working here." He continues to explain that there were only men working there because "you had to be able to wrestle. The patients were violent." The volunteers draw extensively on their own memories of working at other psychiatric hospitals as they tell stories about patients trying to break out of the institution, about radio warnings, murder, surveillance, and the role of keys and locked doors. The positioning of patients as violent and the assumption that women are not as capable and strong as men works to exclude the history of female caretakers from the narrative of mental health care. This teaching, thus, stands in stark contrast to the exhibition depicting mental health, which talks about and shows pictures of female caretakers and offers few examples of violent patients.

This masculine positioning in the volunteers' teaching contrasts with the curators' narratives that present a more ambiguous picture of the life at Ulleråker, where calmer patients were allowed to leave the institution if they promised to return. The need to keep patients calm through different methods remains a central part of the curators' accounts. Whereas the volunteers focus on treatments and medical effects, the curators focus on the social life and consequences of institutionalization. One curator tells us that "many of the treatments were designed to keep the patients calm and easy to handle. There have existed some treatments that would to us seem more like torture. Many people have been treated wrongly by mental health care." Another curator explains that the purpose of the secluded and enclosed psychiatric hospital was "to make sure that the patients did not escape, but also to ensure that curious people from Uppsala did not tease them [the patients]." In their teaching, the curators, contrary to the volunteers, try to question science and issues of power, especially in relation to doctor-patient relationships. There is, however, no recognition of how heteronormativity is fundamental in explaining the caring professions and how it structures and organizes medical science, or any effort to expose compulsory heterosexuality.

Although the curators' and volunteers' narratives vary, both depict patients as victims in unfortunate circumstances. Patients are described as childlike, unable to care for themselves or make logical choices. They are depicted as a burden and emphasis is placed on efforts to keep them calm to prevent them from doing harm to themselves or others. As a consequence this teaching actively works to position the patient as deviant.

QUEERING TEACHING IN THE MUSEUM

I have shown that this museum is a largely heteronormative space in which teaching provided by the volunteers and curators orients students' understanding of nursing and medicine in straight lines. In so doing, the teaching socializes students into the professions by providing examples of how the professions should be understood, and how the students should understand themselves in the professions. Museums, and teaching in museums, can be said to be "straightening devises," to use Ahmed's (2006b, p. 23) concept—a way to align the students in their understanding of nursing and medicine in a heteronormative way. In moving beyond this straight alignment, Susanne Luhmann (1998) proposes that a queer pedagogy can assist in interrupting heteronormative assumptions to produce new ways of thinking about science that lie outside binary models. Ahmed (2006b), in her own way, also argues that we need to move beyond binary models and that the heteronormative and the queer are always intimately connected. Ahmed (2006b) suggests that when we experience disorientation we notice that orientation is something we do not have. So, if orientation is something that is provided in the teaching and in the exhibition layout (although it is admittedly queer), then it is also possible to disorient it by illuminating the lines of orientation for students to see.

The museum exhibitions and teaching orient students by presenting medical doctors as hegemonic and emotionally restrained white men. Moreover, nurses are presented as heterosexual women; surrogate mothers; and chaste, moral examples for society. Disorienting this would suggest including those that have been excluded as a way to un-think heterosexual desires that are used in depictions of the professions. For instance, physician Karolina Widerström, the first women to practice medicine in Sweden, lived in a domestic partnership with a woman for most of her adult life. The masculinized heteronorm of medicine is disrupted when Widerström is a face of the profession. Another example is in Annika Berg's (2009) writing about the homosocial and closed, female community at Vänersborg hospital from 1910 to 1920, where nurses were involved in intimate relationships with each other. Signe Dahl (later Höjer), one of the nurses studying and working there, describes how the nurses relied on each other for "friendship, tenderness and caressing desire" (Höjer, 1978, p. 195). According to Signe Dahl it was also common for "friends" to spend nights in the same bed. Dahl had a long-lasting relationship with an older nurse. She eventually married a man but kept her past relationship no secret. Although their physical relationship had ended, she remained in contact with her past lover who continued working as a nurse (Berg, 2009). This example makes it possible to argue that some nurses remained in service because they fell in love with another nurse or because the identity and work of being a nurse were more desirable than being a wife (to a man).

Alison Oram (2011) points out that by providing possibilities to sexually identify with people from the past, museums can challenge heteronormativity and produce queer possibilities. In following Oram I am, however, not arguing for an inclusion of new identities within otherwise heteronormative frameworks. Although this acknowledges queer people's contributions to science and knowledge and provides opportunities for identification, the focus should be put on how queerness disorients heteronormativity and provides other kinds of knowledge positions and ways of understanding nursing and medicine.

Widerström and the nurses provide more than an inclusion of the excluded and a possibility for identification. Making lesbian women present in science is a "disorienting device" that disturbs the order and offers a way to question heteronormative and patriarchal hierarchies that science is founded upon. Following Austin, Brier, Herczeg-Konecny, and Parsons (2012), I suggest this provides an explicit focus on sex and sexuality and could realize queer possibilities in knowledge productions. Or as Ahmed (2006a) writes: Lesbian desires and the orientation of female bodies towards other female bodies provide contact zones that offer other ways of facing the world. Widerström and the nurses' experiences and desires, thus, provide a way to think about how nursing and medicine present other ways of caring for patients deriving from queer sensibilities. This can provide other ways of practicing the professions, and talking about the professions, beyond heteronormative lines. Moreover, it can provide other ways for students to understand themselves within the professions.

Making lesbian women present in science is, in my view, an answer to what Will Letts (2001) and Mattias Lundin (2014) call for: that the masculinized heteronorm in science needs to be acknowledged and explicitly addressed. Widerström and the nurses offer the insight that medical science, and to a degree also nursing, rest on a masculinized idea of an objective gaze on the world (cf. Haraway, 1988) which seldom, or never, acknowledges that personal lives are explicitly connected to knowledge production. Biologist Donna Haraway (1988) and theoretical physicist Karen Barad (2007) challenge this and show that sexuality, desire, gender, class, and race fundamentally affect the scientific result. Thus, making lesbian women present in science is a disorienting devise that, drawing on Ahmed (2006b), puts science on a different path. Nevertheless, if lecturers and museum staff are to, in the words of Noreen Giffney (2009), start pushing heteroboundaries, they need an awareness of how sexuality and desire influence science and how they orient knowledge about, and within, the caring professions.

We have also seen how the exhibitions and teaching orient students to mental health care through references to home and family. Nurses are called surrogate mothers, the psychiatric hospital is described as a home, and the caretakers and patients are described as a family. Home and family are closely connected to what is considered a livable life (cf. Butler, 1993, 2009), which can be considered a life that is lived within the frame of straight lines of orientations (Ahmed, 2006b). Ahmed (2006a) and Halberstam (2005) write that queer lives do not necessarily comply with the heteronormative life course of birth, childhood, adolescence, marriage, reproduction, and death. Life at psychiatric hospitals in the past may have excluded marriage and reproduction and, in a way, also a conventional death and a recognized loss. I suggest therefore, drawing on Ahmed (2006a) and Halberstam (2005), that psychiatric hospitals in the past can be understood as queer spaces providing alternative temporalities and a life that sometimes lies outside of, or diverges from, the conventional narratives of life. Reading mental illness as a queer temporality can suggest that desire, such as Fröding's autosexuality exhibited in the museum, can be seen as queer desires that move sideways from the assumption of sexuality leading to reproduction. Patients' medical records on display that document this behavior can thus be interpreted as traces of queer lives, spaces of desire and lustful resistance.

If mental illness can be understood in queer terms, and as a way of disorienting heteronormative lifelines and desires, then we need to reconsider the articulation of the psychiatric hospital as a home as portrayed in the teaching at the Museum of Medicine. This articulation is the museum's way of countering popular culture's frightening depiction of mental health care and institutions as seen in films such as *One Flew Over the Cuckoo's Nest* (1975) and *Sucker Punch* (2011). It is a way to move beyond the idea of sadistic medical doctors and nurses, but it is also a way to downplay the suffering of the patients.

The concept of home offers, nevertheless, a possibility to rethink life at the institution and can assist in queering it. Fortier (2001) shows that home is often portrayed as a place of origin, but it is also connected to displacement for many queer people. Ahmed (2006b) discusses home in her concept of "migrant orientations" (p. 10). This orientation is facing at least two directions: a home that has been lost and place that is not yet home. Considering this, it is possible to think about the psychiatric hospital as a queer home—not as a place of origin, but a destination, a place to arrive when being displaced, as such unhinging the idea of an original home. This way of looking at the psychiatric hospital deconstructs the idea that relations must have heteronormative space and time, and validates other kinds of lives, homes, and desires. Moreover, it offers a possibility to think about the patient beyond a charity case or a body that needs correction.

Considering the institution as a site for queer lives assists in recognizing the vulnerable situation of patients within this home and the potential to think about mental health care as domestic violence. A construction of patients as family members, and not merely as bodies that can be worked upon and potentially "corrected," repositions mental health care and can work to disorient the binary relations between patients and medical doctors/nurses/caretakers. This configuration allows medical doctors/nurses/caretakers to assume more multifaceted roles that can recognize hierarchies within and between the professions. The disorientation assists in recognizing that mental health care exists because there are patients, and patients exist because there are masculinized and able-bodied lines of orientation in society. Without such norms—such lines of orientation—there might not have existed medical doctors, nurses, and caretakers in the first place. Medical science and contributions can be seen not through the lens of white male medical doctors but through the patients. It can actively acknowledge their presence in and contributions to medicine.

I have offered ways to disorient the heteronorm and other normalizing behaviors and practices as expressed in the exhibitions and teaching at the Museum of Medicine. I have tried to produce new ways of thinking about science that lie outside binary models and ways that could potentially orient students in new directions. There are many more such opportunities in the museum and, although the museum provides lines of orientation, it remains open for rereadings and disorientations. Such rereading depends, following Ahmed (2006b), on how we orient to, and in, the museum space and how we face directions, how we orient towards that which slips away and towards the oblique. Rereading depends on and cultivates our openness to new directions and what they can bring.

Queer teaching in museums involves revealing discursive silences; it is about interrupting and questioning the very foundation that science is built upon. This work does not start (or end) in the museum but should be seen as an ongoing effort between the scholarly community, activists, and museums. This is also why university lecturers need to reconsider their collaboration with museums. Queering education in museums requires an investigation into how museums are used in education and the purpose for using them. This also involves an investigation of what part of the museum is used and in what ways. Moreover, it requires a rethinking of science, knowledge, and how professions are articulated. Although involving a substantial effort, queering science education in museums is worth the investment and has the potential to provide revolutionary results for teaching and learning.

ACKNOWLEDGMENTS

I am grateful for and want to acknowledge the valuable comments on drafts of this chapter by Steve Fifield, Will Letts, and Hedvig Mårdh.

NOTES

1 The Museum of Medicine, founded in 1980, is one of Uppsala University´s newest museums and came into care of the University in 2012.

2 This project was funded by Uppsala University's "strategiska medel" and Swedish National Heritage Board's "Forskning och utvecklingsmedel" under grant 3.2.2-5161-2016. I would also like to thank the Swedish Institute in Athens, Greece, for providing an excellent research environment.

3 The researcher identifies herself as being feminist, female, and heterosexual with a middle-class background.

4 In this chapter I use "caring professions" to refer to nursing and medicine, and "medical profession" to refer to people practicing medicine, such as physicians and surgeons. I also refer to physicians and surgeons as "medical doctors."

5 Participant observation is a method that involves participating in the daily lives of a group (Joergensen, 1989). I participated in five guided tours with the assistance of Hedvig Mårdh. We recorded the teaching using a mobile camera and took notes during the guided tours focusing on expressions of heteronormativity and queerness. I transcribed the video and the notes. The informants (i.e., curators, volunteers, and students) were told about the study beforehand in writing, and they consented in writing to be part of the study. The identities of informants were coded to ensure anonymity. In this chapter informants are identified only as "curator" or "volunteer." To protect informants, I do not associate particular volunteers and curators with their statements.

REFERENCES

Ahmed, S. (2006a). Orientations: Toward a queer phenomenology. *GLQ: A Journal of Lesbian and Gay Studies, 12*(4), 543–574.

Ahmed, S. (2006b). *Queer phenomenology: Orientations, objects, others*. Durham, NC: Duke University Press.

Ahmed, S. (2007). A phenomenology of whiteness. *Feminist Theory, 8*(2), 149–168.

Ambjörnsson, F. (2006). *Vad är queer?* Stockholm: Natur och kultur.

Austin, J., Brier, J., Herczeg-Konecny, J., & Parsons, A. (2012). When the erotic becomes illicit: Struggles over displaying queer history at a mainstream museum. *Radical History Review, 2012*(113), 187–197.

Barad, K. (2007). *Meeting the universe halfway: Quantum physics and the entanglement of matter and meaning.* Durham, NC: Duke University Press.

Berg, A. (2009). *Den gränslösa hälsan: Signe och Axel Höjer, folkhälsan och expertisen* (Doctoral dissertation). Uppsala universitet, Uppsala.

Bremer, S. (2011). *Kroppslinjer: kön, transsexualism och kropp i berättelser om könskorrigering* (Doctoral dissertation). Göteborgs universitet, Göteborg.

Butler, J. (1993). *Bodies that matter: On the discursive limits of "sex."* New York, NY: Routledge.

Butler, J. (2006). *Gender trouble: Feminism and the subversion of identity.* New York, NY: Routledge.

Butler, J. (2009). *Frames of war: When is life grievable?* London: Verso Books.

Connell, R. W. (1996) *Maskuliniteter* (Åsa Lindén, Trans.). Göteborg: Daidalos.

Davidsson, E., & Jakobsson, A. (2012). *Understanding interactions at science centers and museums: Approaching sociocultural perspectives.* Rotterdam, The Netherlands: Sense Publishers.

Dillon, J. (2012). Science, the environment and education beyond the classroom. In B. J. Fraser, K. Tobin, & C. J. McRobbie (Eds.), *Second international handbook of science education* (pp. 1081–1095). Dordrecht: Springer.

Dufwa, S. (2010). Klass och genus i vården. In H. Strömberg & H. Eriksson (Eds.), *Genusperspektiv på vård och omvårdnad.* Lund: Studentlitteratur.

Eriksson, K. (2003). *Manligt läkarskap, kvinnliga läkare och normala kvinnor: Köns- och läkarskapande symbolik, metaforik och praktik.* Gondolin: Stenhagen.

Falk, J. H., & Dierking, L. D. (1997). School field trips: Assessing their long-term impact. *Curator: The Museum Journal, 40*(3), 211–218.

Falk, J. H., & Dierking, L. D. (2000). *Learning from museums: Visitor experiences and the making of meaning.* Altamira, CA: Altamira Press.

Fifield, S., & Letts, W. (2014). [Re]considering queer theories and science education. *Cultural Studies of Science Education, 9*, 393–407.

Fortier, A. M. (2001). 'Coming home': Queer migrations and multiple evocations of home. *European Journal of Cultural Studies, 4*(4), 405–424.

Geertz, C. (1994). Thick description: Toward an interpretive theory of culture. In M. Martin & L. McIntyre (Eds.), *Readings in the philosophy of social science* (pp. 213–232). Cambridge, MA: MIT Press.

Giffney, N. (2009). Introduction: The "q" word. In N. Giffney & M. O'Rourke (Eds.), *The Ashgate research companion to queer theory* (pp. 1–13). Farnham, England: Ashgate.

Golding, V. (2016). *Learning at the museum frontiers: Identity, race and power.* New York, NY: Routledge.

Halberstam, J. (2005). *In a queer time and place: Transgender bodies, subcultural lives.* New York, NY: New York University Press.

Hall, S. (1997). The spectacle of the 'other.' In S. Hall (Ed.), *Representation: Cultural representations and signifying practices* (pp. 223–290). London: Sage Publications/Open University.

Haraway, D. (1988). Situated knowledges: The science question in feminism and the privilege of partial perspective. *Feminist Studies, 14*(3), 575–599.

Harding, T. (2007). The construction of men who are nurses as gay. *Journal of Advanced Nursing, 60*(6), 636–644.

Hird, M. J. (2000). Gender's nature: Intersexuality, transsexualism and the 'sex'/'gender' binary. *Feminist Theory, 1*(3), 347–364.

Hirdman, Y. 2007. *Genus – om det stabilas föränderliga former.* Malmö: Liber AB.

Höjer, S. (1978). *Läsarungen växer upp.* Stockholm: LT.

Jorfeldt, I. (2010). Vårdvetenskapen ur ett genusperspektiv. In H. Strömberg & H. Eriksson (Eds.), *Genusperspektiv på vård och omvårdnad.* Lund: Studentlitteratur.

Jorgensen, D. (1989). *Participant observation: A methodology for human studies.* Newbury Park, CA: Sage.

Letts, W. (2001). When science is strangely alluring: Interrogating the masculinist and heteronormative nature of primary school science. *Gender and Education, 13*(3), 261–274.

Luhmann, S. (1998). Queering/queer pedagogy? Or, Pedagogy is a pretty queer thing. In W. Pinar (Ed.), *Queer theory in education* (pp. 141–155). Mahwah, NJ: Lawrence Erlbaum.
Lundin, M. (2014). Inviting queer ideas into the science classroom: Studying sexuality education from a queer perspective. *Cultural Studies of Science Education, 9*(2), 377–391.
Macdonald, S. (Ed.). (2011). Expanding museum studies: An introduction. In S. Macdonald (Ed.), *A companion to museum studies* (pp. 1–12). Malden, MA: Blackwell.
McRuer, R. (1997). *The queer renaissance: Contemporary American literature and the reinvention of lesbian and gay identities*. New York, NY: New York University Press.
McRuer, R. (2013). Compulsory able-bodiedness and queer/disabled existence. In L. J. Davis (Ed.), *The disability studies reader* (4th ed., pp. 369–378). New York, NY: Routledge.
Mills, R. (2006). Queer is here? Lesbian, gay, bisexual and transgender histories and public culture. *History Workshop Journal, 62*(1), 253–263.
Mills, R. (2008). Theorizing the queer museum. *Museums & Social Issues, 3*(1), 41–52.
Nightingale, F. (1989). *Anteckningar om sjukvård*. Skellefteå: Artemis.
Obermark, L., & Walter, M. (2014). Mad women on display. *Reflections, 14*(1), 58–80.
Oram, A. (2011). Going on an outing: The historic house and queer public history. *Rethinking History, 15*(2), 189–207.
Rich, A. (1980). Compulsory heterosexuality and lesbian existence. *Signs, 5*(4), 631–660.
Rodéhn, C. (2016). (Re)doing men in museums? Masculinity theory and the democratization of heritage in South Africa. In A. Häyrén & H. Wahlström Henriksson (Eds.), *Critical perspectives on masculinities and relationalities: In relation to what?* (pp. 67–96). New York, NY: Springer.
Romesburg, D. (2014). Presenting the queer past: A case for the GLBT history museum. *Radical History Review, 2014*(120), 131–144.
Rooke, A. (2009). Queer in the field: On emotions, temporality, and performativity in ethnography. *Journal of Lesbian Studies, 13*(2), 149–160.
Rosenberg, T. (2002). *Queerfeministisk agenda*. Stockholm: Atlas.
Sandahl, C. (2003). Queering the crip or cripping the queer? Intersections of queer and crip identities in solo autobiographical performance. *GLQ: A Journal of Lesbian and Gay Studies, 9*(1), 25–56.
Smith, L. (2006). *Uses of heritage*. New York, NY: Routledge.
Steorn, P. (2012). Curating queer heritage: Queer knowledge and museum practice. *Curator: The Museum Journal, 55*(3), 355–365.
Sumara, D., & Davis, B. (1999). Interrupting heteronormativity: Toward a queer curriculum theory. *Curriculum Inquiry, 29*(2), 191–208.
Svensson, S. (2014). *"Det var ju inte illa menat" En longitudinell enkätstudie om genusmedvetenhet, könsdiskriminering och sexuella trakasserier bland Uppsala Universitets läkarstudenter* (Unpublished master's dissertation). Uppsala Universitet, Uppsala.
Tyburczy, J. (2012). All museums are sex museums. *Radical History Review, 2012*(113), 199–211.

ABOUT THE AUTHOR

Cecilia Rodéhn is a researcher in Critical Museum and Heritage Studies at the Centre for Gender Research, Uppsala University, Sweden. In 2009, she obtained her PhD in museum studies at the University of KwaZulu-Natal, South Africa. Rodéhn's research centers on democratization of museum and heritage sites during times of social transformation and the roles of race, ethnicity, gender, and sexuality. From 2014 to 2016, she was funded by Uppsala University to develop cultural heritage as a pedagogical tool in higher education. She is currently (2017–2019) funded by the Swedish National Heritage Board to investigate and critically analyze transformations of the cultural heritage of former psychiatric hospitals.

NICHOLAS SANTAVICCA, JESSE BAZZUL AND
STEPHEN WITZIG

15. CAMPING SCIENCE EDUCATION

A Trip to Camp Wilde and the Queer Nature of Nature

Science education's inextricable entanglement with sex/gender and sexuality has received little attention in recent years, even after repeated encounters with queer theory (see Broadway, 2011; Fifield & Letts, 2014). Research on sex/gender and sexuality often centers on equalizing access to STEM content for "girls and boys" through focused recruitment and/or balancing "sex representation" in curriculum. Few researchers trouble sex/gender and sexuality at the level of subjectivity, that is, how discourses of science and science education actually come to constitute students, teachers, and all who pass through apparatuses of education as subjects of sex/gender and sexuality. According to philosopher Judith Butler's (1993) seminal work, the very stake of feminist—and, like Butler, we mean political—critique is the subject itself. While equity struggles concerning sex/gender and sexuality still follow more liberal orientations of access and equal labor relations, they have also widened to include questions of self and collective transformation, what it means "to be" outside strict categories of race, class, ability, sex/gender and sexuality, humanness, etc.

To face this beast—a monster embodying both the terrific possibilities and potential violence of discourses and practices around sex/gender—science education will need to engage wide-ranging critical theories, such as Karen Barad's (2012) work on ethical-matterings and new materialisms, where diverse abiotic and biotic (including nonhuman and human) entities come to challenge how categories emerge through intra-actions. As will be seen in the sections that follow, our camping trip troubles these subjectivities in science education, as well as what we come to see as entity or category, with narratives that include nonhuman and nonliving entities (even if storied).

Despite its tactical power, the term *queer* has been the victim of its own popularity and is at risk of proliferating to the point of becoming overstretched when used loosely to refer to the transgression of any norm (e.g., queering education, queering science, queering queering). Used in this sense, the term becomes confusing, since it often connotes a homosexuality that may in fact not be at stake when the term is used so broadly. *Queerness* also refers to the multiple ways that sexual practice, sexual fantasy, and sexual identity fail to line up consistently. This definition expresses an

 | DOI:10.1163/9789004331068_017

important insight about the complexity of sexuality, but we suggest it describes a state experienced by everyone, not just "queers." But if everyone is queer, then no one is. And while this is exactly the point some queer theorists want to make, reducing the term's pejorative sting by universalizing the meaning of queer, it also depletes its explanatory power. One way around this, however, is to dislocate *queer* as an identity with a perceived distance from the norm and take up *queering* as a practice and mode of response complete with a queer ethics. Used simply as a departure from the norm, we hide, for example, the political reasons why new laws are made regarding religious freedoms, people are banned from parades, and students face queer bullying. Educators create a queer(ing) response by exposing the rhetorical strategies and irrational assumptions inherent in curriculum and pedagogy (for examples in science education, see Bazzul & Sykes, 2011; Snyder & Broadway, 2004). When teachers, for example, challenge the heteronormative language concerning sexual diversity in animals (Bagemihl, 1999), they provoke students to consider a world where heterosexuality may have never been "natural." However, educators also need to remain critical of invoking arguments related to "nature," like biological concepts such as survival and fitness, when engaging sexual diversity. The unreflective use of scientific discourse—for instance, how "homosexual" animals increase the overall fitness of populations—can work to reify a heteronormative "survival of the fittest discourse" and limit queer becomings to simplistic (and often dangerous) evolutionary logic (McWhorter, 2010). Indeed, science classes, if taught in uncritical ways, can ironically create irrational monsters (much like the beast alluded to above) based on assumptive behaviors, normed observations, and staunch rhetorical structures.

The upcoming camping trip into the "wilde" creates unsettling thoughts. We begin to think about why we turn to "nature" when we have been told time and time again that being "queer" is not natural (Gough et al., 2003). Schools are places that deal with knowledge, yet in this knowledge creation, where mind, body, and spirit are separated, knowledge is connected to the mind only (or worse, reduced to the brain). There is little acknowledgement of the role that bodies play in knowledge creation. There is no recognition of the role that Eros, desire, satisfaction, and pleasure play in learning, especially in fields outside of the humanities. Our camping experience will create new insights into how the natural world/order that we teach in the classroom curriculum fails in comparison to nature's wonderful dis/order.

Our writing process follows aspects of narratology (Patton, 2002; Prince, 1982), which involves the diversification of narratives. In one sense, we share the camping trip as sequenced events in time, while providing literary freedom to express the diverse emotions of the stories told. We have storied our queer world for the purposes of making sense of the many cultures we are imbedded in: emerging queer spaces, our family stories, and cultures of science and science education. We have fictionalized both our personal accounts of our encounters with cultures of science, as well as the very notion of a methodology that is often taken to be synonymous with qualitative research, to blur the lines around what counts as pragmatic narrative research. Effectively, we blur the line between fiction and nonfiction. This is intentional.

Blurring in this work becomes a form that operates in terms of what constitutes truth (games), representations, dialogue and voice, space and time. Another way we blur these lines is around the author. Are the authors real? Are the characters real? Who is speaking? Here we are (re)enacting Foucault's death of the author in that this story is no longer ours. And it is also more than just discourse—it is a monstrous tale.

THE MONSTROUS ENTAILS …

As queer friends, we enable thinking and flight, our questions on the walk spill out onto the path, beside the path. Though we've followed a path through to camp, it is not clear whether it was straight or curved, South or North, familiar or strange, real or fantasy. As we walk across the fallen sign of Camp Wilde (Gough et al., 2003) we feel a surge of love: love of the world, love for our friends (were we alone before?). We are re-membering: The erotic encounters of David Blades (2006) in the name of ethics—with/for the other, with/for animals, with/for abiota. Our work strives to be queer but is never queer enough … never erotic enough … never thoughtful enough … never loving enough. This is why we are here: for new love and new ways of making ontic cuts into reality, in other words, new ways of co-naming and co-making the world around us (Barad, 2007). We smile at the image of disheveled lovers coming out of the forests, with their tools and queer animals—(ex)changing names. We notice our speech has changed—it is slimy, sliding off objects.

Walking into the woods, into a space, to be a part of a space, we imagine the woods of Massachusetts, USA, our shared space, full of questions. As we walk forward toward Camp Wilde, a virtually enacted space created by Gough et al. (2003) as a place for different ways of being (queer), we think about how the world is always ready for queer-becomings (discoveries). As queer campers-in-science we tell marvelous tales, ask indeterminate questions—we enable each to *give attention to*, and gawk carefully at, classification: the categories of/for nature already here and what they do. Do they mask the "nature of things?" Do they become the nature of things? Can we play with "cuts" into nature, and if so, what are ethical ways of making these cuts? In asking these questions, we ask what can be, and what kinds of cuts will be made at Camp Wilde.

We think of the human/inhuman divide, as well as the operation of gawker/gawkee, the conversation about queerness and animality. We are all animals, yet we are not all human. Is our task to include the inhuman into the human world or must we also ask how the very split of human/inhuman, straight/queer, gawker/gawkee sets about a series of distributions, exclusions, and relations that need to be examined, turned over, and rediscovered? How shall we queer and be queer as we think about science and the natural world? Science is a love of questions, a love of evidence-based explanations in response to those questions—a study of the natural world—and are we not in some way in a relation of love as we ask each other? We wonder how to introduce this desire in our thinking and why this desire was absent when we studied the natural world in school. We are heading into the woods as queer(er)s with the aching tension that we are also at home, writing in our beds, meeting in the offices of our institution. And why escape to Camp Wilde? To rediscover nature's queer performativity (that is, how nature is always constructed in terms of ever-changing desire and radically provocative ways), and maybe even to develop a queer-ethical-political-scientific performativity of the researcher/observer—performativities that allow productive assemblages to come forward. These are probably the preoccupations that come with a journey somewhere, boring and blind, unable to see without the help of others. But having asked questions, we feel we are getting close to Camp Wilde.

ARRIVING …

Entering the cabin, Chase sets down his bags, outerwear, cutting tools. He places a stack of papers and his camp book, *Meeting Mr. Universe Halfway*, beside the bed. He lays three picture frames out on the bed and stares at them for a second or two. He feels empty, like there are only faint lines inside, waiting to be filled—organless. Chase walks outside and picks up a thin strip of birch bark and tests its underside with a pencil. The camp is an *assemblage* of *intra-acting* arts (*arts* being a play on *parts*), each bringing another into being. It is ordered by signs and codes, segments of materials, and the presence of other assemblages (e.g., University of Massachusetts, Chase, Dirk, Henri). Chase puts a pencil to the birch bark, thinking about this assembly of "others" (Figure 15.1). Chase sees parts of all kinds, their joining and fading away, the presence of signs (of languages, signs of life in codes—molecule codes), and materials (like our beds and bodies full-on gendered). The camp is already queer, because it has been cast as a site of interrogation in relation to sex/gender and sexuality (in an almost uncanny way). But how do we make new queer cuts—new productive, enabling cuts of signs and relationships?

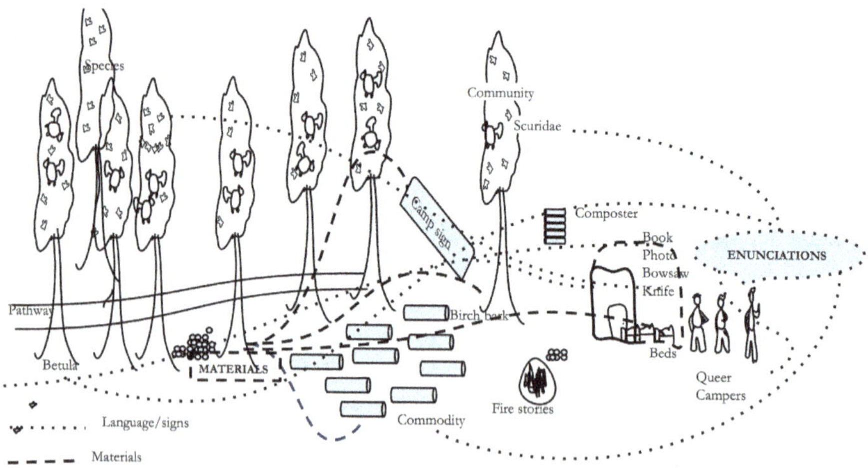

Figure 15.1. Birchbark sketch

Chase: [giving the drawing to Henri] Here, look at this. Does this look like camp?[1]

Henri: This isn't a typical camping map. What the hell is this?

Chase: Ha ha! Labeling is a part of my science training. I guess old habits die hard. So I noticed that there are only a few species of trees, for some reason, and many of the same squirrels.

Henri: But how are you categorizing?

Chase: Just by species—all the trees and squirrels *seem* to be of the same type. Representations like this become messy, an arrangement of words, lines, and signs.

Henri: Okay, just between us squirrel-friends (just kidding), I am curious: Why do you feel the need to label the drawing?

Chase: Actually, I label everything, nothing I draw or make is not overcoded with language, words, pictures. Everything I make seems to be a mixture. I think this is true of people who just draw and people who just write—drawings are representations of concepts to some extent, and writing is a way of getting at, like … like a reciprocal object … a likeness of something.

Dirk: Why did you write "commodity?"

Chase: I don't know, it's just something that came to mind. It seems that as soon as the birch tree falls and is cut, then it becomes something else.

Henri: Yeah, like firewood. I'm getting cold. We'll need them for the fire later.

Chase: Exactly! This whole camp is made with birch, these trees, or logs—whatever you want to call them.

Dirk: It's a double or triple entendre: That thing that's always there, but, like, is not in your drawing.

Chase: What do you mean?

Henri: I think he means something's missing in the drawing … something we've seen around here, on the way in, didn't you …?

Chase: Here then, you add it. I'm tired. If you need me, I'll be alone with myself.

Henri: He's screwing with us. What a dick! Ha ha!

TRUTH GAMES

Later on … that night …

Sitting around a roaring campfire, with a full moon lighting the forest canopy around us, we decide it would be fun to play some campfire games we recall from our childhood. The game truth or dare comes up, and we all instantly become quiet as horrid flashbacks fill our thoughts of when we were awkward young teenagers. What is the lesser of two evils? If you take a DARE, you might have to kiss someone, or strip down naked and run around the camp, or something equally frightening. However, if you choose TRUTH, you can be sure that no one is going to be comfortable with whatever question they have to answer. You surely cannot lie—you would certainly be caught, and the consequences of that will haunt you for the rest of your school-age years. What to do? … oh no! … I'm next … what do I say? After each of us clearly experiences this mild freak-out moment from our childhood, we realize, surely, it's not going to be as bad as that, and we give it a go.

[Chase contemplates] (*As my stomach performs—somersaults and leaps—I consider the moment of truth-telling: Who will require me to speak the truth? Why? How will it be asked? How will it be recognized as the truth? It is the truth that is more fearsome than the dare, as the dare as a command lays out the order and the person who arbitrarily speaks the command! It matters who asks, Dirk and Henri are not the same, so who should I confess to? Who will demand a truth from me? I'd better jump in.*)

Chase: Alright, alright, alright—I'll decide who will go first. (*Dirk and Henri turn and stare at me. Before they can respond, a flock of crows bursts through the trees, obviously startled by something. Taking advantage of the distraction, I begin.*) Eenie meenie miny mo, catch a tiger by its toe, if it hollers, let it go, eenie meenie miny mo. My – mother – said – to – pick – the – very – best – one – aaannnddd – iiittt – iiisss – YOU! [Chase points to Dirk]

Dirk: Oh no—well, let's get this over with!

Chase: Truth or dare? (*I feel myself enjoying the role of the interlocutor—perhaps enjoying it too much. Deliciously I wonder, what exercises more power? Is it to command action or truth?*)

Dirk: Truth. (*There is no way I am doing a dare, especially being the first camper picked. How bad can a truth be?*)

Chase: (*So I am the one who is going to demand a truth. Does this mean that what Dirk says is the truth or is it just a function of what I wish?*) Okay, in all your life, tell us about how and when you questioned your own sexuality. Or … tell us a story when sex freaked you out. (*Why did I choose those questions? Why did I choose to hear this truth? What would it tell me? What do I need to hear?*)

Dirk: Really? Isn't every conversation about sexuality in your childhood awkward?

Henri: I didn't have anyone to talk to about sexuality. [Dirk and Chase laugh]

Dirk: Okay, let me think. I've got one, and I'm sure that all of us can relate to this one. Let me set the scene. I'm a sophomore sitting in a suburban high school health class. Who's teaching the class, you ask? Well, the phys ed teacher of course—Coach! The same person that has us do jumping jacks and sprints around the gym to warm up, and then pits us against each other in a game of dodgeball or full-contact basketball. The same dude that coaches the football team and staffs the weight room, encouraging everyone to get pumped. I think you get the picture. Up until this day, the health class was co-ed. Most of the conversation to date had been about the importance of keeping ourselves clean and our hair washed, with references to why we all had a separate set of gym clothes in our locker, and that we should all shower after gym and before going to class—like that was ever going to happen!—and the importance of diet and exercise.

But this day was different. We could sense it. Why are two phys ed teachers in the room today? Why is Mrs. Smith here too? Well, the topic of discussion today clearly couldn't be handled in a co-ed classroom, so Mrs. Smith took all the girls to a separate room. Today we would learn about making babies! Or so we thought. Then Coach brought out cartoonish pictures of male and female genitalia—like those history-class flip charts that have world maps printed on them—and rested them on easels in the front the classroom. As Coach walked us through what a penis is and what a vagina is, you can imagine the awkwardness that ensued. We wondered what the girls were seeing and why we needed to be separated for this discussion. After all, doesn't it take a male and female to make a baby? Shouldn't we be together for a conversation about sex?

Then what began as a quick overview of the technical terms (like penis, testes, vagina, ovaries, fallopian tubes, uterus, sperm, egg, etc.), quickly, without warning, and with the simple flip of the genital charts, changed into something that is forever etched into my memory: Grotesque images of genitalia bearing the effects of sexually transmitted diseases (STDs, or venereal diseases, VD, as we heard from older kids, parents, and those equally awkward after-school TV specials). This was not a conversation about sex and making babies, this was a conversation about why we shouldn't have sex: Because if we do, our penises will look like bloody cauliflowers and will surely fall off! I'm not sure that any of us really wanted to know what was happening with Mrs. Smith and the girls. We were extremely excited for the bell to ring so we could head to science class—where, as you know, we never talk about sex, at least not human sex. [Henri and Chase are nodding heads and crossing legs as if those grotesque cauliflower penis images are haunting them as well.]

Henri: Why would anyone listen to a coach about sex?

Chase: You know, I was in a similar class, and even though my coach told a lot of dirty stories, he didn't seem to know what he was talking about. He never answered any of the questions I was thinking about. Like why would I get a hard-on in the middle of school? Like where do you find out about that crazy shit? Not from any person or book I've come across.

Dirk: Well, I found a stash of pictures in the bottom of my parents' closet. My friends and I spent days trying to figure out what those photos meant.

Henri: My health textbook said that arousal was spontaneous, sometimes with no immediate explanation.

Chase: Yeah, and that's real helpful! My textbooks just labeled labia and areolas, and all we had to do was figure out which diagram was us!

Henri: None of those diagrams represented me!

Dirk: Okay, kids, let's get our heads out of high school science class and get back to this game of truth or dare. I told my truth, so now I get to pick. Let's see—Henri, you're up. You want a truth or a dare?

Henri: Dare … do I get to decide my own dare? Who decides?

Dirk: I do! I dare you to expose yourself by taking off what you're wearing.

Henri: Everything?! Which parts?
Dirk: The whole kit and caboodle!
Chase: You don't have to do it if you don't want to.
Dirk: Yes he does! Those are the rules.
Henri: Fine!
Chase: What do you want him to do exactly?
Dirk: He's doing it now. Lookin' good!
Chase: I didn't think he'd remove that piece, because I don't want to see that!
Dirk: No, we have to see everything! Like the details!
Chase: How many layers do you have, Henri? It's like you knew this was going to happen!
Henri: I do this a lot. I can do this all day.
Chase: Okay, okay. Let's all do it! … There, I've done it too. I'm totally exposed. So let's call it a night!
Henri: Listen, chub, you're not getting off that easy. That doesn't count. You still have to play the game. Truth or dare?
Chase: Since my exposure doesn't count—truth.
Henri: Have you ever wanted to do something bad? Like, real bad.
Chase: What do you mean? Have I done something bad?
Dirk: No, I think he means have you ever desired to do something, you know, kind of like—
Chase: Like wrong?
Henri: Uh, something you don't tell your mom about.
Chase: Some people may think that anything I desire to do is right, some people may think it's right and wrong. For example, if I go into the woods and come back and say I did something that made you all feel better about yourselves, more secure in yourselves, but actually made me feel worse, would that be bad or good?
Dirk: What do you mean?
Chase: Like tonight—what if I wanted to do something really bad, and I woke up and decided to do it? It would only be bad or good depending on what you guys thought about it.
Dirk: What are you planning? Now I'm afraid to go to sleep.
Henri: Ha ha! I'm excited to see what you're gonna do.
Chase: Ha! But remember we're not talking about doing something but *wanting* to do something. Even if I did it, how could you be sure I wanted to do it, or if I wanted to do it again, or if it was even me?
Henri: Did you hear that?
Dirk: We all have our demons.
Henri: No, seriously.
Chase: Hear what?
Henri: It sounded like the cabin door.
Dirk: The door's been open the whole time.

Chase: Let's be safe and get inside the cabin. We're exposed out here.
Dirk: Why is that safer? Hiding under the covers won't save you. Anything's that out there can get into the cabin easily.
Henri: I think we should all just go to bed. It's time.
Dirk: Time for what? I'll stay and put the campfire out—it's responsible camping.

WORMS AND THE BODY: SEX, IDENTITY, AND "NATURE"

At twilight …

[Henri thinks to himself as he wakes up.] (*Waking up is difficult. I can't remember how the night ended. Who said what? Did we get to the truth or did I dream that part?* [Rolling over, Henri sees Chase standing over something. He appears to be drawing.] *What's he doing? I'm so tired …*)

A few hours later …

Dirk: Hey! Wake up! Why are there worms all over the floor? [see Figure 15.2]
Chase: I'm not sure—and they are not all over the floor … only in the chalk outline for some reason.
Henri: Did you draw that outline, Chase? I saw you drawing something a few hours ago.
Chase: No, I didn't draw it. I've been sleeping the whole time. Don't yank my chain.

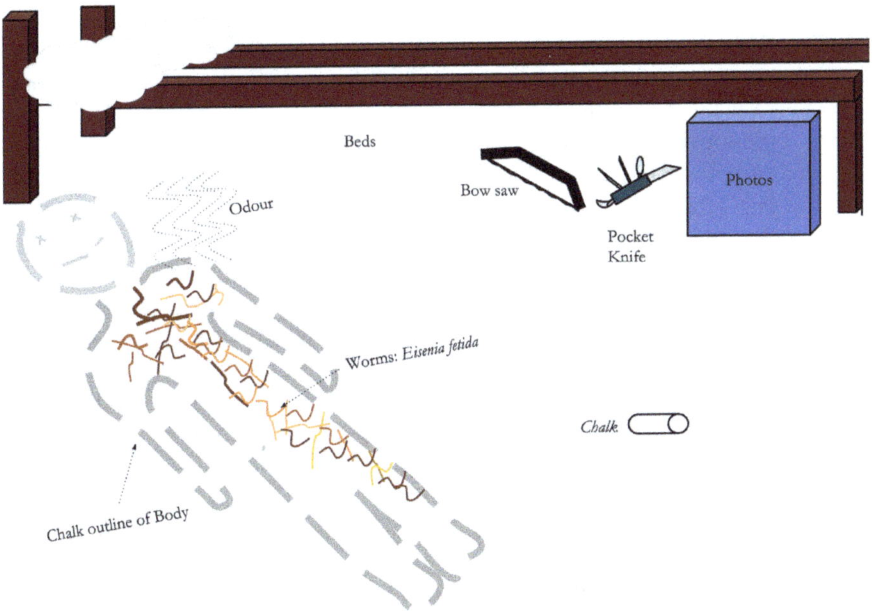

Figure 15.2. Chalk outline on cabin floor

Henri: Are you hiding something, Chase? Just tell me. I'm not really concerned about whether you drew the body or not. I'm more freaked out by these worms. Where did they come from?

Chase: I know you don't believe me about the drawing, I can tell. And those worms? I'm not sure … but you know that pile of sticks shaped like a box over near that maple tree, the one with leaves coming out of it? Well, when I woke up a couple hours ago, I had to pee like a racehorse, and I headed toward that tree. While I was watering the garden, I wondered what was going on in that leaf pile, because the leaves were moving at the top. I investigated and decided it must be an old compost pile left by other campers in the past. I dug around a bit to see and right under the surface of the leaves on top was a crazy amount of worms. And the more I dug down, the more worms I found. They all seemed to be stuck together. [Chase points just beyond the campfire pit] But actually, now that I think about it, the body was always there, you just didn't see it—

Dirk: [interrupting] Well, why the hell are they in our cabin now? Wait a minute, before you answer that, I first want to know why the hell is there a CSI-like chalk outline on the floor? I believe Henri, and that's the majority of us, two against one! What's going on in that queer-ass brain of yours, Chase?

Henri: Easy Dirk, I'm still not sure about anything! I don't remember one way or another whether the body was there. Why would someone draw the body in the first place? Hmm. Wait. [moving closer to the body] What kind of body is it?

Dirk: What do you mean? Do mean female or male? Human or animal? More importantly, is the body some kind of sign or message? I'm freaking out. How did they both get there?

Henri: [now overtop the body] Well, it's not really any *kind* of body, actually, except maybe human-like. But there are real worms inside the drawing, that's for sure!

Chase: By the way, Dirk, what do you mean by "queer-ass"? Is it the opposite of your super-straight brain? Maybe you should really ask why I noticed the body last night and you didn't! It wasn't me who made the body. How could I? How many times do I have to repeat it? It wasn't me who made the body! It wasn't me who made the body!

Henri: Yeah, Dirk, why are you looking for one of us to have drawn the body? Maybe someone else did it. It may even have been drawn by multiple people. Hey! Maybe it was a monster. You know—the one that lurks in every damn camp story. Oh! Maybe the worms did it!

Dirk: [kneeling on the chalk body] Hey, wait, I know these worms. *Eisenia fetida*. They're hermaphroditic.

Chase: Uh … I have to leave for a second.

Henri: Why, Chase, do you feel guilty?

Dirk: [picking the worms up] Get back here, Chase! Take a look at these beauties.

Chase: Are you touch … touching them?

Dirk: Notice how there are some lone worms and some worms clumped together in pairs? See those lightly colored bands on these ones?

Chase & Henri: [in unison] Yeahhhhh.

Dirk: Well, that's their reproductive organs—only visible when mating. The worm pairs, well, are getting it on—the little devils.

Chase: So … then … how do you know if they are male or female?

Henri: Didn't you say they're hermaphroditic? So they're both right?

Dirk: Or neither. Tell me, what is specifically male and what is female for these worms?

Chase: So they perform male and female functions, but what you're saying, Dirk, is that male/female are inadequate categories for these critters. Yeah, I can see that … they're actually completely useless!

Dirk: Right, they both produce sperm and lay eggs. Yet they need a partner to fertilize each other's eggs in a kind of subterranean slime dance.

Henri: So … what concepts work then? How would you talk about the sex of these worms? [nervously shifting] How … er … how would I identify with these worms?

Dirk: [with worms in hand] Well, these two are having sex right now.

Henri: I'm not talking about the act of sex. I'm more interested in what these worms are in terms of sex. What they are … tell me their category!

Chase: Maybe that makes them gay? Lots of animals are gay, I hear.

Henri: But how are they possibly gay? What makes them gay? Is it how they line up? Is it that they have both parts?

Dirk: Why do we have to classify them either way? I just like watching. They're doing what worms do.

Henri: So they are just born that way.

Chase: Maybe it's in their genes—some are born with both sex organs, some are born attracted to one sex or another. Maybe.

Dirk: These worms are naked; they're not wearing jeans.

Chase: Not the jeans that we wear, but the genes that wear us.

Henri: I've heard being gay is in the genes, but what do we really know about this? It has nothing to do with biology anyway. How can biology define us? I'd hate to think that what I liked and what I did and who I did it with are all decided for me ahead of time.

Dirk: Right, but let's get back to why are the worms in this chalk outline! That's pretty queer of these worms. They should be outside in that leaf pile. And why is there no trace of where they came from or where they are going? That leaf pile is easily 20 feet away. [all three go to the cabin window to take a look]

Chase: There may be something on this floor, in the chalk outline, that is attracting these worms here or allowing them to organize in this way.

Dirk: Ha ha ha—like a worm drag bar? You're probably making that up, Chase.

Henri: Let's all take a closer look at the chalk drawing.
Dirk: [putting the worms back down on the floor] Hey, the worms are scattering.
Chase: Are they frightened?
Henri: I'm not sure I can answer that for them, but they're definitely moving away fast.
Dirk: Oh my god! What happened to the chalk outline?
Henri: It's gone!
Chase: Now I'm actually scared. How can it be there one minute and gone the next? Let's just pack up and get out of camp. We've had fun—isn't that enough? The road is just up the way.
Henri: Maybe there's something under the floorboards. I want to check it out before we go.
Dirk: You do that! Keep in touch with yourself. I'm leaving with Chase.

DIARY

Dirk

Dear Diary,

As I reflect on my recent camping trip with Henri and Chase, I want to share with you something that has been troubling me from the beginning. Why is it that sexual identity issues are taboo in science class? And as I reflect back on Chase's birchbark sketch (Figure 15.1), why does everything need a label and what do those labels mean? I mean, yes, in schools we talk about hermaphroditic worms and how some flowers self-pollinate. But when it comes to talking about our own bodies and how we as humans identify as sexual beings, the issue suddenly becomes so black and white. I'm reminded of our school's biology textbook and how sexuality is treated as a body system—the reproductive system. A cartoon of a bisected female figure flanks one page, while the opposite page bears (and bares) a cartoon of the bisected male figure. Carefully drawn arrows point to the vagina, penis, ovaries, testes, etc. However, the diagrams lack what is in between—the grey area that is the continuum of sexual identity. Or is this intentional—a display of matter of fact? Boys have a penis and girls have a vagina. Are penises and vaginas what make "boys" and "girls" as far as biology class is concerned? Is what they use those for, how they feel, and/or what they desire irrelevant given the purpose of those drawings? Where does the trans student fit in this biological story? What about the intersex student who was born with certain expressions of both penis and vagina and raised a boy, but who feels more like a girl in every ounce of her body? Or vice versa? What about students whose genitals were altered at birth, without their knowledge or consent, who might not even have fully functioning sexual organs? Some researchers are exploring these questions that arise from deconstructing textbook images (Bazzul & Sykes, 2011). What knowledge do science teachers need to create a safe place for dialog in the classroom related to sexual identity? How can our science teachers queer science education? If queering is a practice, as we posit, then what does this practice look like, and what does this look like in practice?

The implications for science teacher education are dramatic and demand a transformative approach. We must, as a field, seduce teachers to recalibrate their comfort zones so that they have the tools they need for these conversations to become comfortable, to move past the textbook images, to consider the whole student, the student as a human, the student as a member of society that does not define itself in terms of black and white, right or wrong. But where can we begin?

Let's consider (school) educators writ large: pre-service teachers, in-service teachers, and science teacher educators. Each of these communities has interactions with school students. So let's first consider some actions that can be put in place immediately for the benefit of our students. Research and educational standards documents suggest we should avoid teaching science as a collection of facts and instead teach the processes of science that have led to the development of core scientific ideas (NGSS Lead States, 2013; National Research Council, 2012). The knowledge base that is science is just too vast to expect that every topic can be taught. That outdated approach leads to a curriculum that is "a mile wide and an inch deep" (Michaels, Shouse, & Schweingruber, 2007, p. 1). Instead, we should align instruction around the big ideas in science with the goal of bolstering the scientific literacy of society as a whole (American Association for the Advancement of Science, 1989, 1993). One way to accomplish this is through discussions about the social implications of science—how science influences everyday life and how everyday life influences science. Socioscientific issues-based education is one approach (Presley et al., 2014). While the cartoon images of male and female genitalia serve a particular purpose in the biology textbook, could this be explored in different ways? Science teachers are expected to be knowledgeable about the science content but will likely need support regarding the social and political dimensions of the issue. For starters, what is the issue? It might be: Is there a biological basis for sexuality? Critically addressing this question would include asking what various stakeholders (e.g., scientists, teachers, the public, etc.) mean when they say *biological*, *basis*, and *sexuality*. What meanings and practices around these ideas are in circulation in society and what possibilities do they open and close?

So, for instance, next time we open a biology textbook to cartoon genitalia, perhaps the discussion should seduce us to think more broadly about difference. How then can we put scientific knowledge (in its fullness) to more liberating uses? How has science been used in society to define kinds of people and how might we use science differently? These are the thoughts that I am left with as I reflect on this transformative camping trip. Tomorrow I start the work of transforming how I teach science and how I prepare science teachers to teach science.

Yours Truly,
Dirk

Chase

April 1, 2048

> On Bohr's account, identity is not given, but rather performed. (Barad, 2012, p. 41)

Thinking back to the year of the camping trip, I was keeping a secret from Henri and Dirk—perhaps unavoidably. But, in my line of work, education, many secrets are kept, knowingly and unknowingly. Secrets are even kept by the community at large! It's not a secret that the camping trip was queer, but nothing about the trip after the fact is queer anymore. What is queer is always laid-out on the horizon. The way the camp was drawn on my mind—the species, the game of truth—was queer as-it-happened: cut queer. My educator friends would want me to talk more about this, why and how—I do not want to, but I will for them.

Jesse Bazzul and Heather Sykes, camping friends pre-dating my camping trip with Henri and Dirk, wrote a text early on in their careers called "Secret Identity of a Biology Textbook: Straight and Naturally Sexed" (Bazzul & Sykes, 2011), which was queer in one moment, but not in the next—people told them this. But they also knew that an argument for inclusive textbooks does not problematize the very normative conditions that frame debates of inclusion. Jesse, for example, failed to see the transformative struggle in humanist forms of queer struggles. He also knew queering wasn't about the integrity of a "universalizing" subject, but the (bio)politics of communities. That is, that all social struggles are necessarily queer struggles. They all have as their stake, at least in part, the question of being/being different and/or the material or discursive conditions that enable this "being different" (Bazzul, 2014). And remembering the camping trip, the maps, exposés, the events of the cabin, I realize Bazzul and Sykes missed something else: Things or beings or entities in the world are mutually constitutive. Could we have been queer if it wasn't for these entities, the fictions that bled into the camp, virtual and actual? As feminist scholar and quantum physicist Karen Barad (2012) argues, "individual entities are not separate but bounded and propertied objects" (p. 32). That is, they are mutually constitutive through *intra*-actions (*inter*- signifies separate entities). Recognizing this means things are constantly in flux, their composition is a matter of questioning and becoming. Barad insists that identity is historic, multidirectional, indeterminate, and temporal—it is the result of boundary-making processes, while boundaries are the effects of intra-actions. Consequently, an ethic of responsibility that allows queer entities and processes to emerge in the world "entails providing opportunities for the organism to respond" (p. 38). The ethic here is one of mutual coemergence that allows emergent entities in the world a more diverse and distributed agency, rather than one imposed by knowledge systems that strictly categorize, define, and optimize. The camp set the conditions of exteriority—for things to emerge; it was a configuration that allowed for a becoming. To put it more lucidly, "Identity is a phenomenal matter; it is not an individual affair" (p. 32).

If I could do anything at camp again, I would replay the truth or dare game and allow it to unfold in another way. Maybe it would have changed things. Maybe it wouldn't have. Maybe the conditions of possibility were set. I have this uncanny feeling that the events of camp could be different at each moment of iteration, when things can be done differently. But maybe our trip was set from the get-go. I guess it depends on the individual moments—the moment we set down the path to Camp Wilde, or the moment the body disappeared … when the drawings were made.

Henri

December 2050

What I learned at Camp—

My camp, my monstrous experience, my feeling of betrayal, is formed/mutilated through the prestige given to certain forms of sexuality—in their assemblages they claim and retain authority—and how sociocultural practices associated with sexuality enable authority to be claimed and retained from my camp to the classroom and to all educational practices. My camping trip demonstrated a critical stance on these categories, reflecting on their vulnerable potential and questioning their ontological status by showing, for example, that actual linguistic choice and sociocultural practices in science cannot be invariably or neatly related. In this way, I echo a fundamental argument from Eve Sedgwick's (1990) *Epistemology of the Closet*: "Categories presented in a culture as symmetrical binary oppositions [like] heterosexual/homosexual … actually subsist in a more unsettled and dynamic tacit relationship" (p. 10). We have learned to uncover the conventional terms of viewing: One half of the binary is always subordinated to the other, rather than being equal to it. And this camping trip shows, in turn, that "the question of priority between the supposed central and the supposed marginal category of each dyad … is irresolvably unstable" (p. 10). We possibly found the monster is omnipresently unstable.

What I did learn from camp is that understanding the monster emerges "outside" of the heterosexual/homosexual dyad; academic and vernacular references often acknowledge the irregular, ambiguous location through phrasings like "neither man nor woman," "third sex," or "trans-identity." And as research continues to explore linguistic and social practices outside of the normative binary of educational practices, a richer profile of this terrain emerges, and so do the limitations of such labels at camp, in the science classroom, and beyond.

XOXO, Henri

NOTE

[1] Text in square brackets is stage directions of a sort, indicating what characters are doing. Italicized text in parentheses is inner thoughts of characters, their self-talk.

REFERENCES

American Association for the Advancement of Science. (1989). *Science for all Americans*. New York, NY: Oxford University Press.

American Association for the Advancement of Science. (1993). *Benchmarks for science literacy*. New York, NY: Oxford University Press.

Bagemihl, B. (1999). *Biological exuberance: Animal homosexuality and natural diversity*. New York, NY: St. Martin's Press.

Barad, K. (2007). *Meeting the universe halfway: Quantum physics and the entanglement of matter and meaning*. Durham, NC: Duke University Press.

Barad, K. (2012). Nature's queer performativity. *Kvinder, Koen og Forskning, 21*(1–2), 25–54.

Bazzul, J. (2014). Science education as a site for biopolitical engagement and the reworking of subjectivities: Theoretical considerations and possibilities for research. In J. L. Bencze & S. Alsop (Eds.), *Activist science and technology education* (pp. 37–53). Dordrecht: Springer.

Bazzul, J., & Sykes, H. (2011). The secret identity of a biology textbook: Straight and naturally sexed. *Cultural Studies of Science Education, 6*(2), 265–286.

Blades, D. W. (2006). Levinas and an ethics for science education. *Educational Philosophy and Theory, 38*(5), 647–664.

Broadway, F. S. (2011). Queer (v.) queer (v.): Biology as curriculum, pedagogy, and being albeit queer (v.). *Cultural Studies of Science Education, 6*(2), 293–304.

Butler, J. (1993). *Bodies that matter: On the discursive limits of "sex."* New York, NY: Routledge.

Fifield, S., & Letts, W. (2014). [Re]considering queer theories and science education. *Cultural Studies of Science Education, 9*(2), 393–407.

Gough, N., Gough, A., Appelbaum, P., Appelbaum, S., Doll, M. A., & Sellers, W. (2003). Tales from Camp Wilde: Queer(y)ing environmental education research. *Canadian Journal of Environmental Education, 8*(1), 44–66.

McWhorter, L. (2010). Enemy of the Species. In C. Mortimer-Sandilands & B. Erickson (Eds.), *Queer ecologies: Sex, nature, politics, desire* (pp. 73–102). Indianapolis, IN: Indiana University Press.

Michaels, S., Shouse, A. W., & Schweingruber, H. A. (2007). *Ready, set, science! Putting research to work in K–8 science classroom*. Washington, DC: National Academies Press.

National Research Council. (2012). *A framework for K-12 science education: Practices, crosscutting concepts, and core ideas*. Washington, DC: National Academies Press.

NGSS Lead States. (2013). *Next generation science standards: For state, by states*. Retrieved from http://www.nextgenscience.org/next-generation-science-standards

Patton, M. Q. (2002). *Qualitative research and evaluation methods* (3rd ed.). Thousand Oaks, CA: Sage Publications.

Presley, M. L., Sickel, A. J., Muslu, N., Merle-Johnson, D., Witzig, S. B., Izci, K, & Sadler, T. D. (2013). A framework for socio-scientific issues based education. *Science Educator, 22*(1), 26–32.

Prince, G. (1982). *Narratology: The form and functioning of narrative*. New York, NY: Mouton Press.

Sedgwick, E. K. (1990). *Epistemology of the closet*. Berkeley, CA: University of California Press.

Snyder, V. L., & Broadway, F. S. (2004). Queering high school biology textbooks. *Journal of Research in Science Teaching, 41*(6), 617–636.

ABOUT THE AUTHORS

Nicholas Santavicca is an assistant professor of English at the University of Massachusetts Dartmouth (USA), and founding director of the American Language &

Cultures Institute. Dr. Santavicca has held faculty and administrative positions in K–12 and higher education in the United States and abroad. He can be reached at nsantavicca@umassd.edu

Jesse Bazzul is an associate professor of science and environmental education at the University of Regina, Saskatchewan, Canada. He feels that comprehensive attention needs to be given to the way typically depoliticized fields of study, such as science education, work to constitute political and ethical forms of life. He can be reached at jesse.bazzul@uregina.ca

Stephen Witzig is an associate professor of science education at the University of Massachusetts Dartmouth where he teaches in the doctoral and masters of arts in teaching (MAT) programs in the Department of STEM Education and Teacher Development in the School of Education. He studies the development of teachers' specialized knowledge for teaching science. His work focuses on sources of teachers' content and pedagogical knowledge; how experience shapes knowledge; scientific practices; socioscientific issues-based education; and areas of student learning, including the roles of students and teachers in learning science and in bridging research relationships among scientists, classroom teachers, and science teacher educators. Contact email: switzig@umassd.edu

JAMES SHELDON

16. TOWARDS A QUEER CURRICULUM OF INFINITY

WHAT IS THE BIGGEST NUMBER YOU CAN THINK OF?

A recent television commercial starts in a school library. An adult in a suit and tie asks a group of children, "What's the biggest number you can think of?" One child answers, "a trillion billion zillion!" The next child answers quite matter-of-factly, "ten." The third child answers, "infinity." The adult asks another child, "Can you top that?" That child immediately replies, "infinity plus one." When the adult corrects them and states that the correct answer is "infinity plus infinity," another child quickly interjects with "infinity times infinity." The adult responds with hand signals and sound effects to indicate that his mind is exploding; suddenly his preconceptions are blown away and he realizes that there is more to the subject of infinity than he had originally imagined.

This kind of "one-up-personship" is to Howes and Rosenthal (2001) one of the most delightful ways in which people engage with infinity. Howes and Rosenthal offer a feminist case for the inclusion of infinity into the curriculum. They suggested that "infinity's innate, inescapable contradictions make it not merely the ultimate postmodern subject-object … but a superb article of study for kids from 1 to 92" (p. 178). Looking through the lens of gender, they sought to "scan the airwaves for instances of when conceptualizations of the infinite have been vividly related to beliefs and practices about sex and gender" (p. 178). Infinity's exclusion, in Howes and Rosenthal's conception, is a gendered process, and much like how feminism suggests that we as a society need to welcome women into professional jobs, into politics, and into civil society, so must we as teachers "welcome infinity into the classroom" (p. 188).

That last paragraph sums up fairly succinctly a more traditional feminist take on infinity. Infinity in its nature, in this schema, has something that is too transgressive to be included in a K–12 curriculum. Thus, a feminist response would be to include back in what has been excluded. But I want here to turn things in a queer direction, to move beyond a politics of inclusion and representation to uncovering and then challenging the mechanisms that produce the mathematical canon. With mathematical inqu[ee]ry (Rands, 2009) and queer curriculum theory (Sumara & Davis, 1998, 1999) as my interlocutors, I argue in this chapter that mere inclusion of nonnormative topics in the mathematics curriculum falls short and that, instead, mathematics education researchers need to take up an inqu[ee]ry into why these

 | DOI:10.1163/9789004331068_018

topics are left out and how they get "straightened" when presented in a formulaic, formalized, and/or rote algorithmic fashion.

TOWARDS A QUEER CURRICULUM THEORY FOR MATHEMATICS

In this chapter, I use the concept of *queer curriculum theory* as a theoretical framework for my inquiry. Queer curriculum theory, a phrase first used in print by Dennis Sumara and Brent Davis (1998, 1999), is not a theory about queers per se, but rather about the way sexuality is reflected in the production of knowledge.[1] Sumara and Davis contended that they "are interested in showing how all educators ought to become interested in the complex relationships among the various ways in which sexualities are organized and identified and in the many ways in which knowledge is produced and represented" (p. 203).

How do these relationships appear in mathematics? Kai Rands (2009) argued that in queer studies, mathematics is "the subject that dare not speak its name" (p. 182). One might ask what sexuality and queer studies, and even queer curriculum theory, have to do with mathematics. There is a difference between studying sexuality and looking at how sexuality underpins the foundations of curriculum. Mathematics (at least of the pure sort) may not be about sexuality per se, but as I will demonstrate in this chapter, libidinal dynamics underpin mathematics not just at the level of mathematical applications but at the basic level of mathematical foundations. Sumara and Davis (1998) argued that queer curriculum sees sexuality "not as an object of study but as a necessary valance of all knowing" (p. 215). Sexuality forms a foundation for knowledge, and in proper oedipal fashion, we learn to rely on authority figures in mathematics classrooms; the teacher or the (anthropomorphized) textbook decides if an answer is correct, rather than students relying on their own ability to make arguments, evaluate evidence, and produce knowledge.

Queer curriculum studies scholars see desire as forming the foundation for curriculum. They make an analogy between a physical body and a body of knowledge in order to demonstrate how our own relationship to desire forms the basis for curricular decisions. William Pinar (1998) argued in this vein that

> systems of knowledge production and distribution, such as school curricula, are likewise systems, or in the present context, codifications of desire. The knowledge we choose for presentation to the young is in one sense like the parts of our bodies we allow them to see. Both the physical body and the body of knowledge are cathected objects, and decisions and policies regarding them follow from our own organization and repression of desire. (p. 231)

Curriculum encodes our desires, hopes, and fears and turns them into legitimated truths that are then taught unquestioningly, without acknowledgment of this reification. Certain topics are covered while others are excluded, and the process of constructing curriculum acts to reify certain knowledge as legitimate and other knowledge as illegitimate.

Some math topics are too queer for the K–12 math classroom and are thus excluded from the curriculum. In addition to infinity, topology, fractals, and non-Euclidean geometries are topics that challenge mathematical normativity and so are not generally included in K–12 curricula. Challenging mathematical normativity is not always synonymous with challenging heteronormativity, but both offer a similar critique of categorical boundaries (Rands, in press), challenging the traditional notion of mathematics as being composed of discrete, bounded objects (Sheldon, 2017).

In the rare cases that such queer topics show up in the K–12 curriculum, they tend to be de-queered by being taught in an algorithmic or formulaic way rather than by allowing students to engage with them conceptually and to articulate their own ideas. An example from the high school curriculum is infinite series, where students are often taught formulas to find the sum of an infinite series of numbers (e.g., ½ + ¼ + ⅛ + …) instead of having opportunities to inquire into the nature of such a strange, counterintuitive mathematical object. Students, rather, are taught that there is one right way to approach such topics, ignoring the long histories of contestation and desire that form what today are considered accepted truths.

Mathematical topics are not necessarily inherently queer, but I do believe it is possible to develop a generally agreed upon list of topics that can be considered queer. In doing so, I am inspired by Susan Sontag's (1964) list of what constitutes the canon of campy cultural expressions, such as King Kong, Flash Gordon comics, or women's clothes of the 1920s. Rands and I (Sheldon & Rands, 2013) put together a list of mathematical topics that we thought had a certain queer potential, aesthetic, or history to them and that we felt were particularly good starting points for mathematical inqu[ee]ry. On our list, we included

> time, infinity, space, topology, knots, numbers, measurement, statistics, place value, alternate number bases, modular arithmetic, composition and decomposition, cardinality, counting, differentiation, integration, zero, irrational numbers, rational numbers, dimension, binary, polygons, polyhedra, spheres, tori, double tori, gluing, imaginary numbers, complex numbers, the complex plane, complex analysis, [and] linear algebra. (p. 1371)

These topics are not inherently queer, much like women's clothing of the 1920s is not inherently campy, but there is often value to asserting queerness as a property rather than as a process. When a mathematician asks me why I think queer theory is relevant to mathematics, it is often far more poignant to challenge their normative ideas of mathematics by saying, "Well, complex numbers sure are queer," rather than to explain that I am choosing to ascribe queerness to them in a form of strategic essentialism. This essentialism is not intended to reify queerness; rather, I intend it as a way of opening up uncomfortable questions about mathematical normativity.

For instance, Rands and I once arranged to meet a graduate student at a math education conference who had expressed curiosity about the idea of queer mathematics after having read Rands (2009). For a half hour, he asked question after question, trying to place what we were saying into some other framework. He asked if we were talking about constructivism, if we were talking about sociocultural theory. Finally, what we had said about queer epistemology started to click. His next words after finally seeing what we were getting at were: "That makes me really uncomfortable." Engaging in mathematical inqu[ee]ry ought to be uncomfortable; it is uncharted territory for many. Think of alternate number bases. Imagine spending your whole life in base ten, only to find out that arithmetic can be worked in the base of any positive integer. You would probably feel curious and uncomfortable at the same time, much like that graduate student did when he learned about Dr. Rands's work on queer mathematics.

Many of the topics on our list are deferred until more advanced courses, ostensibly because students first need a foundation in algebra before they can consider them. In reality, though, it is possible to explore them without formal foundations; for example, *Heart of Mathematics* (Burger & Starbird, 2012), a mathematics textbook for nonmajors, has an entire chapter that introduces topology without relying on formal mathematical proof. Perhaps these nonnormative topics are considered too transgressive to study before one has first been indoctrinated into normative mathematical ways.

The nonnormativity of these topics is in some ways relative to one's mathematical background; alternate number bases to a sixth grader represent uncharted territory while to a number theorist seem banal and almost not worthy of mention. The challenge then is to convince mathematicians of the nonnormativity of mathematics that to them seems routine, *and* to expose mathematics educators and mathematics students to these queer bodies (of knowledge) that provoke uncomfortable questions and open up new (queer?) possibilities.

As I consider the status quo, in which mathematical topics are de-queered by the K–12 curriculum's insistence on formulas, algorithms, and procedures, I am reminded of Gunckel's (2009) suggestion that a queer pedagogy of science necessitates moving beyond knowledge transmission to allow students to enter into inquiry, drawing upon "curiosity and passion" (p. 71) in the exploration of scientific possibilities. This idea of exploring nonnormative possibilities through inquiry leads nicely into the next section, where I elaborate further on mathematical inqu[ee]ry.

DEFINING MATHEMATICAL INQU[EE]RY

Gunckel (2009) proposed that students and teachers inquire not just into the natural world but into the nature of science itself, exploring the ways in which normativity becomes embedded into the scientific tools that we teach our students. Rands (2009) offered a similar inquiry, or as ze calls it, "inqu[ee]ry," into the normative constructions of mathematics. Styling the term as *inqu[ee]ry* highlights the ways in which queerness and inquiry coconstruct each other, while at the same time playing with the normative conventions of grammar. To illustrate inqu[ee]ry Rands offered an initial thought of including families with two moms or two dads in word problems ze teaches in hir classroom. Ze then further suggested that we need to go beyond this add-queers-and-stir approach in order to help students queer how "time is measured and regulated within classrooms" (p. 189), to explore how mathematics is used to make rhetorical arguments, to look at alternative kinship structures, and to question normative temporalities.[2] Rands makes a case, informed by hir work on mathematics for social justice as well as careful reading of queer theory, for approaching mathematics curriculum and teaching praxis through a queer lens, engaging not merely in tactics of inclusion, but instead taking the entire curriculum in a different direction of "scribbling graffiti over its texts" (to borrow a concept from Bryson & de Castell, 1993, p. 299).

Sheldon and Rands (2013) continued this discussion of how to queer mathematics curriculum by suggesting that a queer analytical lens be used to "critique what is normative … and how inquiry/inqueery might move beyond what is normative" (p. 1369). Mathematical inqu[ee]ry suggests the need for a mathematics curriculum to explore nonnormative topics and how these queer subjects affect our own subjectness and sense of self. Applying mathematical inqu[ee]ry to our mathematics curriculum means seeing how sexuality and desire underpin our curricular decisions and teaching praxis.

INQU[EE]RY: INFINITY IN AN UNDERGRADUATE MATHEMATICS TEXTBOOK

Given that considerations of the infinite are rare in K–12 curricula, my inqu[ee]ry into questions of the infinite will start with where these discussions are most prolific, in undergraduate mathematics texts. Discussions of the infinite occur at a variety of points in a typical undergraduate curriculum. The most common place is in courses on mathematical foundations, which teach techniques of mathematical proof as well as the basics of mathematical logic. Material about infinity is often found, too, in the prefatory sections of upper-division texts, as well as in textbooks specifically about set theory or mathematical logic. It also (although rarely) is found in textbooks for nonmajors that are designed to introduce them to topics of mathematical interest. I focus here on the study of mathematical foundations, given that set theory and the associated study of cardinality are generally taught in modern courses only as a subset (pun intended) of mathematical foundations courses. Cardinality is the study of the sizes of sets, which is a key foundational topic for studying infinity. Foundations courses, in addition to being taken by mathematics majors, are often required for those seeking to become secondary school mathematics teachers. So this inqu[ee]ry is relevant not only to the study of higher education but also to those interested in further developing secondary-level teacher knowledge and classroom curricula.

The text under consideration is *The Art of Proof: Basic Training for Deeper Mathematics* (Beck & Geoghegan, 2010), part of the seminal Springer series, Undergraduate Texts in Mathematics. This book is written in an unconventionally conversational style, offering quotations about mathematics and a variety of comic strips, as well as advice on mathematical study skills and frequent marginal notes and pointers for students. I conducted this analysis to examine how this text explores nonnormative topics such as infinity and how the authors construct and produce knowledge in both queer and "straight"/linear modalities.

Cardinality is covered in the final chapter of the book. The word "cardinal" comes from the Latin *cardinalis*, meaning "principal, chief, or essential," yet the material on these topics is relegated to the last chapter. The map of section dependencies (p. xix) shows that chapter 13 depends on the material in chapter 12 (decimal expansions), chapter 9 (embedding $\mathbb{Z}$ in $\mathbb{R}$, in other words, rigorously proving that the integers are a subset of the real numbers), and chapter 11 (rational and irrational numbers). Being last, though, none of the material depends on what is in chapter 13, and so, even though in a sense everything in the book might be building up to this topic, given that chapter 13 depends on everything before it, it would not be uncommon for a time-strapped teacher to omit the topic entirely; it is typical for teachers not to reach the end of the textbook by the end of the semester in mathematics courses.

I began my analysis with the chapter epigraph. Chapter 13 starts with a quotation from T. S. Eliot's (1917/2017) poem "La Figlia che Piange":

> Sometimes these cogitations still amaze
> The troubled midnight and the noon's repose. (stanza 3)

Placing the quotation in the context of the original poem, I find that the poem is about a narrator who witnesses a breakup between a man and a woman. It starts in the present tense with the narrator commanding her to "weave the sunlight in your hair" as she turns "with a fugitive resentment in your eyes" (stanza 1). Then the poem moves to the conditional as the narrator tries to find a way to mark what just happened, looking for some way to share in the experience in a way that is both "light and deft" (stanza 2). The narrator then moves into the past tense, describing his fascination with the haunting past, exploring what might have happened over and over again in his head. And even now, in the present, he sometimes is still amazed by what comes up when he stays up all night and is not able to find rest until morning.

This poem raises interesting questions about the construction of identity, something that is critical to doing a queer reading of a curricular text. In this poem, we know nothing about the narrator other than his obsession with this woman, and we know nothing of this woman except a single slice of time. Much as how Sumara and Davis (1999) suggest that we must "wonder what circumstances lead to different identification experiences" (p. 204), I suggest we must ask what it means to constitute your identity based on a single moment in time. As Anne Carson (1999, cited in Halberstam, 2005) says about memory, "Remembering brings the absent into the present, connects what is lost to what is here. Remembering draws attention to lostness and is made possible by emotions of space that open backward into a void" (p. 47). The narrator brings this scene from the past into the present, constantly shifting temporality through the changing use of tense, drawing the reader into the queer temporality not only of the poem but of infinity itself.

The narrator's obsession with the woman in the poem is to me reminiscent of Georg Cantor's (1845–1918) obsession with a mathematical theory he invented known as the continuum hypothesis. The continuum hypothesis is a theory that there are no sizes of infinity between the size of the set of the integers (called *countable infinity)* and the size of the set of the real numbers (*uncountable infinity*). Thus, the continuum hypothesis proposes that despite there being an infinite number of sizes of infinity, there are no infinities in between the countable and uncountable infinities. Cantor struggled with both depression and psychosis, and spent much of his later years in mental institutions. Amir Aczel (2000) conjectured (perhaps oversimplistically) that trying to prove the continuum hypothesis literally drove Cantor insane.

At the turn of the 20th century, the continuum hypothesis topped mathematician David Hilbert's (1862–1943) list of the most significant unsolved mathematical problems. In 1963, it was proven that the continuum hypothesis can never be proved or disproved under the Zermelo-Fraenkel set theory axioms, which are the standard set of axioms used by most mathematicians when working with set theory. It is still not agreed upon whether or not this proof constitutes a solution to the problem, and the continuum hypothesis, as well as questions of cardinality in general, continue to leave mathematicians with both amazement and troubled nights, much like the narrator's memories of the woman with sunlight in her hair.

The textbook authors, though, rather than using the epigraph from T. S. Eliot as the foundation for what they are about to present about cardinality, immediately jump into a comic strip discussing cardinality through the eyes of a child whose teacher has been reading him a little too much of a fictional book called *Adventures of Ed the Actuary*. Although the precocious child and (possibly) campy sense of humor in the comic could be seen as queer, the juxtaposition with poetic lines about the amazing and haunting nature of memories (?) and infinity (?) seems disjointed. Some might see this disjointedness as queer; queer could be, perhaps, playful and yet haunting at the same time. Sumara and Davis (1999) discuss how juxtaposition is key to queering curriculum, explaining that when "things not usually associated with one another are juxtaposed, [it allows] language to become more elastic, more able to collect new interpretations and announce new possibilities" (p. 205). I hoped to see some of this campy playfulness, haunting deepness, and new interpretations and possibilities in the text that follows.

Disappointingly, however, the textbook proceeds into a rather traditional presentation by stating: "The goal of this chapter is to compare the sizes of infinite sets. More generally, how should we measure the size of infinite sets?" (Beck & Geoghegan, 2010, p. 121). In doing so, it reduces the contemplation of the infinite to merely a question of measurement, setting aside the "cogitations [that] still amaze" in favor of showing only the simplest of mathematical results. Section 13.1 begins on the following page, where the text retreats from haunting amazement and campy humor into a swamp of mathematical formalism offering a slew of propositions, theorems, and proofs with little sense of where the authors are taking the students.

In section 13.2 the authors present on Cantor's most famous proof, the diagonalization proof, which shows that the real numbers constitute a larger infinity than the natural numbers. The book presents this topic, though, as a rather obvious result rather than showing how much struggle and effort went into its development. Showing the messy, complex, and socially constructed nature of mathematics, as well as the human side of history (cf. Aczel, 2000), are some ways the topic of this section could be further queered. By showing the entire proof, the authors also miss out on the opportunity of having students try to prove the same result for themselves. The authors then, in rapid succession, prove that there is an infinite hierarchy of infinities, with only a brief mention of how revolutionary these ideas were in the history of mathematics. Their exposition in this section is certainly intriguing but still fails to live up to the promise of "cogitations [that] still amaze."

The chapter's opening poem set a high bar, and I expected to be kept up all night pondering the infinite. Instead, the chapter straightened out infinity into a clear set of theorems and proofs, preventing the reader from becoming entangled with infinity like the poem's narrator became inexorably obsessed, confused, and fascinated with the woman that he witnessed in the first stanza. In the undergraduate curriculum, we present infinity within a carefully constructed walled garden; students are not to ask questions or make their own conjectures; there will be no inqu[ee]ry on Beck and Geoghegan's watch. In the next section, I will address how to construct a queer curriculum of infinity with space for passion and history and which compels one's imagination in the way the woman in Eliot's poem captivates its narrator.

CONSTRUCTING A QUEER CURRICULUM OF INFINITY

A queer curriculum of infinity would, above all, take seriously the libidinal dynamics that underpin all mathematics. More than just starting a traditional presentation with a quotation or a comic, or augmenting it with sidebars about famous mathematicians, a queer curriculum of infinity would consider Pinar's (1998) arguments about a "body of knowledge" (p. 231) and would be open to the possibilities that queerness raises for this "body." Even in the more modern and unconventional text just discussed, the topic is presented rather formulaically, organizing (queer?) desire into a straight, linear presentation. A queer curriculum of infinity would guide students to have key moments of "cogitations [that] still amaze" that would be so exciting they would want to stay up all night working on the questions raised. Cantor spent his entire life struggling with the continuum hypothesis, and yet we expect to spend 10 minutes presenting this hypothesis in lecture and somehow believe that does justice to the continuum hypothesis? Queer bodies of knowledge captivate, engross, and puzzle in a similar fashion to the way that literal bodies do.

This queer curriculum would also engage with questions generally thought to be relegated to literature and philosophy. It would draw upon literary references much like Beck and Geoghegan's (2010) textbook attempted to do, and it would also focus on the philosophical study of infinity, particularly with reference to religion. Cantor himself was devoutly Jewish. Some of the terminology he uses comes from Jewish mysticism, such as the word *aleph* (א), used to denote different sizes of infinity (Aczel, 2000). If we are going to seriously explore the queer potential and possibility of infinity to amaze our students, we need to tell the fascinating tales of mysticism and demonstrate the ways in which infinity (despite the historical attempts to squelch it) has fascinated philosophers and mystics throughout the centuries. We need to bring the mind, the body, and the spirit all into the classroom in order to fully grasp the complexities, subtleties, and libidinality of these curriculum topics.

Returning once again to Sumara and Davis (1998, 1999), we need to look at how knowledge is produced and represented. I suggest that a queer curriculum of infinity would bring the historical events and characters to life (much more than a mere sidebar showing a picture of a key mathematician) and show the dramas and passions behind key players in historical encounters with infinity: for instance, the wars between Cantor and Leopold Kroenecker (1823–1891) concerning the possible existence of infinite sets,[3] Cantor's obsession with the continuum hypothesis, and even the legendary tales of the Kabbalists' encounters with infinity. Aczel (2000) recounts the story of a group of Kabbalists, led by Rabbi Joseph ben Akiva, who around 100 CE attempted to visualize the intense light of God's infinitely bright robe as he greeted Moses on Mount Sanai. This visualization exercise led to death, insanity, and loss of faith, respectively, in three out of the four who attempted the exercise. As Aczel concludes, "Only Rabbi Akiva survived this experience" (p. 27), at least if we count physically, mentally, and with his faith in God intact, the things which the other three lost. Infinity in this story is thought to be powerful and dangerous, something that only the most skilled can afford to harness. This story has an interesting parallel to the way that mathematicians deal with infinity; it is presented only in the most carefully chosen, formal ways, which I suggest reflects our fear not just of infinity but of all topics that challenge mathematical normativity.

As an example of what a queer curriculum of infinity might look like, I offer a five-week class that I taught about counting, modular arithmetic, set theory, and infinity to adults interested in learning about approaches to mathematical problem-solving. I used Dr. Tova Brown's (2015) series of videos about the Hotel Infinity thought experiment. (I suggest you take a few minutes now to watch one of the videos before you proceed. See https://www.hotel-infinity.com) Each video ends with a thought-provoking puzzle for students to solve. I asked students to consider each of the problems as a group with only light facilitation by me during the discussion. I did not offer any answers and instead showed the next segment of the video, which gave the solution, during the next class. I asked them not to watch the videos at home before the next class so that they would have the entire week to process what we discussed without rushing ahead to the next concept or idea. The ideas in the videos were so captivating that my students begged and pleaded with me to show the next section; the idea of waiting a week to see the next segment was almost inconceivable to them.

I supplemented Dr. Brown's videos by telling historical stories from Aczel's (2000) book, such as the aforementioned Rabbi Akiva story, in order to provide extra context and color to the ideas I was asking students to consider. Heeding Aczel's warning, I did not ask students to attempt the visualization themselves, but asked them to consider the power that such an exercise might entail. I also told the story of the conflict between Cantor and Kroenecker, drawing on the narratives from Aczel (2000) and Theoni Pappas (1997). Infinity became the one topic that formed the common thread throughout the course rather than being "dispensed with" at the beginning or "relegated" to the end. This gave infinity its "cardinal" importance in the curriculum that I designed and implemented, and invited students to feel the sense of the mystery, fascination, power, intrigue, drama, and contradictions that infinity entails. By rejecting the role of the teacher as the Oedipal father that presents straight mathematical narratives and then enforces conformity to them through grades, I instead built affective connections with my students, premised on entangling them in infinity's web rather than straightening out their experience of queer mathematics.

CONCLUDING REMARKS

Infinity is by no means the only mathematical topic filled with these mysteries, fascinations, and contradictions. If one was to pick any of the topics with queer potential that were proposed by Sheldon and Rands (2013), one would find ways in which the topic is both elided within the standard curriculum and "straightened" by rote instruction. Mere inclusion of queer topics in the curriculum is not enough to challenge mathematical normativity, much like how including LGBTQ folks in mainstream societal institutions fails to adequately challenge heteronormativity.

In this chapter, I demonstrated how to construct an alternative to the status quo of (hetero)normative mathematics, first by queerly reading a curricular text, and then by discussing how to construct a curriculum that puts inqu[ee]ry into practice. I offer this demonstration as a case study for those who might choose to tackle other topics on the list of queerable topics. Mathematics is no longer the subject that dare not speak its name in queer studies, and the possibilities for further inqu[ee]ry here are, well, rather infinite.

NOTES

1 I asked Brent Davis about the origins of the phrase "queer curriculum theory." Davis said that similar phrases were utilized by other attendees of the Bergamo Conference on Curriculum Theory and Classroom Practice around the time they first used it in print, and that their choice of term was prompted by discussions with Patti Lather, Deborah Britzman, Liz Ellsworth, and Janet Miller (personal communication, November 15, 2017).

2 For a more thorough introduction to theories of queer temporalities, see Sheldon and Rands (2017).

3 Kroenecker spent years attempting to suppress Cantor's research into infinite sets, working behind the scenes to keep him from presenting and publishing, and blocking him from more prestigious positions.

REFERENCES

Aczel, A. (2000). *The mystery of the aleph*. New York, NY: Pocket Books.

Beck, M., & Geoghegan, R. (2010). *The art of proof: Basic training for deeper mathematics*. New York, NY: Springer.

Brown, T. (2015). *Hotel infinity* [Video]. Retrieved from http://www.hotel-infinity.com

Bryson, M., & de Castell, S. (1993). Queer pedagogy: Praxis makes im/perfect. *Canadian Journal of Education, 18*(3), 285–305.

Burger, E., & Starbird, M. (2012). *The heart of mathematics: An invitation to effective thinking*, (4th ed.) Hoboken, NJ: Wiley

Carson, A. (1999). *Economy of the unlost*. Princeton, NJ: Princeton University Press.

Eliot, T. S. (1917/2017). *La figlia che piange*. Retrieved from http://www.poetryfoundation.org/poem/177122

Gunckel, K. L. (2009). Queering science for all: Probing queer theory in science education. *Journal of Curriculum Theorizing, 25*(2), 62–75.

Halberstam, J. (2005). *In a queer time and place: Transgender bodies, subcultural lives*. New York, NY: New York Press.

Howes, E. V., & Rosenthal, B. (2001). Chapter seven: A feminist revisioning of infinity: Small speculations on a large subject. *Counterpoints, 137*, 177–192.

Pappas, T. (1997). *Mathematical scandals*. San Carlos, CA: Wide World Publishing.

Pinar, W. F. (1998). Understanding curriculum as gender text: Notes on reproduction, resistance, and male-male relations. In W. Pinar (Ed.), *Queer theory in education* (pp. 221–244). Mahwah, NJ: Lawrence Erlbaum.

Rands, K. (2009). Mathematical inqu[ee]ry: Beyond 'add-queers-and-stir' elementary mathematics education. *Sex Education, 9*(2), 181–191. doi.org/10.1080/14681810902829646

Rands, K. (in press). Queering mathematics pedagogy: Mathematical inqueery. In C. Mayo & N. Rodriguez (Eds.), *Queer pedagogies: Theory, praxis, politics*. New York, NY: Palgrave Macmillian.

Sheldon, J. (2017). *"How do you even do this?": Queering mathematics and mathematics education*. Manuscript in preparation.

Sheldon, J., & Rands, K. (2013). Queering, en-gendering, and trans-forming mathematics and mathematics education. In M. Martinez & A. Castro Superfine (Eds.), *Proceedings of the 35th annual meeting of the North American chapter of the International Group for the Psychology of Mathematics Education*. Chicago, IL: University of Illinois at Chicago. Retrieved from http://files.eric.ed.gov/fulltext/ED567771.pdf

Sheldon, J., & Rands, K. (2017). Whatever will be will be: Queering disabled subjects' temporality. *Philosophical Inquiry in Education, 24*(4), 368–378. Retrieved from https://journals.sfu.ca/pie/index.php/pie/article/view/947

Sontag, S. (1964). Notes on camp. *Camp: Queer aesthetics and the performing subject: A reader* (pp. 53–65). Edinburgh: Edinburgh University Press.

Sumara, D., & Davis, B. (1998). Telling tales of surprise. In W. F. Pinar (Ed.), *Queer theory in education* (pp. 197–220). Mahwah, NJ: Lawrence Erlbaum Associates.

Sumara, D., & Davis, B. (1999). Interrupting heteronormativity: Toward a queer curriculum theory. *Curriculum Inquiry, 29*(2), 191–208.

ABOUT THE AUTHOR

James Sheldon is a doctoral student in the math and science focus of the Teaching and Teacher Education PhD program in the Department of Teaching, Learning, and

Sociocultural Studies at the University of Arizona, in Tucson, Arizona (USA). Their research involves bringing preservice and in-service mathematics teachers into dialogue with frameworks such as queer theory and critical disability studies in order to help them critically engage with reform mathematics from a curriculum theory perspective. James can be reached at jsheldon@jamessheldon.com

AMY E. SLATON, ERIN A. CECH AND DONNA M. RILEY

17. YEARNING, LEARNING, AND EARNING

The Gritty Ontologies of American Engineering Education

INTRODUCTION

This essay considers the self-reproducing character of U.S. engineering knowledge as a significant source of oppression and exclusion in STEM higher education. We see the situation this way: It is only persons of certain bodily experiences and desires who are eligible for authentic participation in American engineering occupations, but, crucially, that which is legible to these fields and their audiences *as authentic participation* delineates the eligible aspirant. This is a loop in which doing what is understood/felt/intuited to be good engineering not only enfolds unmarked assumptions about eligible bodies but re-produces eligible and ineligible bodies and desires as the very practices of engineering. In this way, engineering specifies certain bodies/desires as necessary complements to its more familiar material and conceptual products. That which is thought or performed by, for example, the nonheterosexual[1] or the nonsexual person, by the gender-nonconforming person, by the person of physical or intellectual "disability," the old or the poor person, and often the person of color or the female-gendered person, is frequently illegible as rigorous engineering—its impossibility assured before and aside from whether any engineering-directed thought or action has been undertaken by that person.

A possibility of inclusion currently exists, whereby persons identified as minority or disadvantaged enter, and in different senses thrive, within American STEM settings; this has been the case since de jure segregation and other explicit discriminatory practices faded in the 1960s. But the construction of inclusion *as structural change* is precisely what we want to critique here. On one level, inequities in U.S. engineering proceed as the field systematically deems some groups ineligible for participation (as happens, for example, through reliance on standardized tests that determine many persons of lower socioeconomic status to be suboptimal engineering school enrollees). The discipline does not see those faculties and excitements it detects in these students as appropriate qualifications for engineering. On a second level, when members of marginalized groups are admitted to engineering programs or workplaces, their coursework or labor may not be seen as promising or legitimate (as when instructors specify that so-called adaptive lab instruments used by visually or aurally differentiated students cannot generate accurate data; Slaton, 2013). On both levels, that which is defined as engineering performs the work of discrimination.

 | DOI:10.1163/9789004331068_019

Engineering is not unique in this regard, as all models of education-as-credentialing reproduce prevailing social structures (Bourdieu, 1977; Bowles & Gintis, 1976). But engineering engages and sustains political and economic structures very directly, not only materially but also epistemically and ontologically. We are trying to imagine an engineering education freed from the taxonomic social projects that have shaped it to date.

We arrive at this outlook through a process-ontological view of engineering higher education, including of that discipline's admissions standards, instructional content, classroom and lab interactions, and prevailing conceptions of student self-efficacy or tenacity. Such a view suggests that empirical findings about the world do not derive from an ordered, if socially constructed, revelation of a singular cosmos, but, rather, that "practices generate orders" (Law & Lien, 2013, p. 364; Woolgar & Lezaun, 2013). In the case of engineering, we come to understand that no material object or engineering intervention into such an object, no technical problem or solution, no evidence of successful engineering learning or practice exists apart from its enactment by interested, positioned persons.[2] This stress on ontology radically disrupts customary depictions of equity in STEM fields as being a matter of social inclusion. To the contrary, we do not seek an extension of eligibility for any existing set of STEM activities or opportunities, productions which already represent a distribution of privileges we do not wish to reproduce. Perhaps more challenging to conventional STEM diversity work, there is also no *preexisting* set of excluded others *to* be included—only enactments of selves, minority or majority, welcome and unwelcome, in relation to enactments of engineering knowledge. Bodies, identities, and knowledge always and invariably coproduce. Categories of difference, of course, have salience and power at both the structural and institutional level; it is in the interest of understanding such salience and power that we suggest they may derive from denials of coproduction. Such denial is the essence of "diversity" when posed as a solution to oppression (Slaton, 2015). We want to interrogate such solutions and also illuminate the nature of disruptive resistance to oppression. We thus propose that there can only ever be bodies whose engineering knowledge (or lack thereof) springs from identity (and vice versa) (Britzman, 1995).

This is a perspective that builds on the analytical dispositions of intersectionality that Grzanka and Brian (2015) helpfully describe as foregrounding the power relations that constitute categories themselves (see also Grzanka, 2015; Shields, 2008). If considered ontologically, this pedagogical world further reveals its basis in a driving conceptual split between "selves" and "knowledge," between knowers and the known-about (Fifield & Swain, 2002). An individual is seen to acquire and/or use knowledge if successful as a learner and to fail in those endeavors if not.[3] Disrupting that depiction, Tierney (1997), Britzman (1995), Haraway (1988), and others who draw on a Deweyan legacy are concerned to show, instead, that identity is knowledge; the very effort to separate being and knowing is what constitutes cultural norms in places of learning and working. We add to that a further contingency surrounding identity by referring, instead, to *personhood*: any ascriptions (such as race, age, sexuality, IQ, etc.) or combination of ascriptions that are thought to matter about an individual in a given time and place. This concept helps dispel the determinacy of ascribed identities and may help us bear in mind that beings are not self-contained but distributed in relations of knowing, feeling, and doing.

Fifield and Swain (2002) and Fifield and Letts (2014) clarify further that only the proper person (or proper personhood, if we want to stress the taxonomic nature of these distinctions) admits of the possibility of proper knowledge. In the case of STEM disciplines, that is presumed to be proper knowledge about a naturally occurring world and the means by which humans might intervene in the world. Engineering promises always to secure this connection between good people and good things. But seen relationally, the mere construction of such knowledge as "out there" waiting to be apprehended primarily serves to create prospective possessors of the knowledge (Law & Lien, 2013). Disinterest in that knowledge, or desire for something other than the acquisition of that knowledge, takes on the character of a culturally transgressive attitude. At the same time, the possibility of such purportedly useful knowledge manifesting itself in any particular person might well approach zero because, to make sense, this ontology requires the existence of ineligible aspirants. In fact, a low level of educational promise and attainment detected in any specific individual can only corroborate the overall project of associating legible personhoods with reasonable knowledge.[4] That overall project is what we want to expose here by trying to see knowledge-and-bodies as one, a reconnection that is still an unusual adventure in education theory and nearly unheard of in STEM education (Fifield & Letts, 2014).

Resistance in the academy to seeing in this way is an important part of our subject. It strikes us that the critique of Western educational commitments to knower/known-about binaries is discouraged in manifold ways by the very definition of rigorous scholarly inquiry itself as an empirical pursuit (Riley, 2014; Slaton, 2015). We risk vexing our disciplines and troubling our own reputations when we abandon the normal bifurcation between expert selves and subject matters. Thus we are excited to work with queer theory that simultaneously "offers methods to mark the reception of normalcy as a structure and as a pedagogy" (Britzman, 1995, p. 153). The positivism undergirding conventional demarcations between knower and known-about is nothing if not consoling in its normalizing potential, promising not *knowledges* but a singular, stable *knowledge*. It is that singularity that enables a comparative judgment regarding each student or worker: How much knowledge has each individual attained? The projection of future knowers' potential attainments is also dependent on the to-be-known-about: How much knowledge might this individual attain over time? What's more, the agendas of STEM inclusion and attainment based on those comforting demarcations seem remote in our view from authentic inquiries about power or democratic alternatives to current social structures. The commitment of this volume to following productions and denials of desire in STEM practices (rather than outlining existing and potential paths of STEM knowledge acquisition, as is customary in STEM education literature), seems very promising as a way of deposing static, dualistic ideas of students-encountering-opportunities. We believe that shift makes room for the relational understanding that "technical mastery" is not just a matter of the actions of worker upon world but also those of world upon worker (Law & Lien, 2013; Slaton, 2015).

GRIT

Our specific subject for this chapter is a recent celebration by STEM educators, theorists, and policy makers of personal desire as it is expressed in the form of *grit* (Duckworth, Peterson, Matthews, & Kelley, 2007; U.S. Department of Education, 2013). In multiple academic and government studies of the last seven or eight years, this characteristic is taken to signal students' personal sturdiness in the face of discomfort or challenge, their willingness to move onward towards pre-set goals, and it is linked by analysts to school and workplace achievement. It is in some ways a very old idea; self-discipline was seen to be central to the development of callow youths into reliable adult practitioners from the earliest days of American higher education (Miller, 2013; Oldenziel, 1999; Slaton, 2001). But grit has lately gained new adherents. In the 21st century, it is gritty students who succeed, while ungritty (apathetic? cowardly? *smooth*?) students do not, and contemporary educational researchers at prestigious universities and government agencies have taken up the task of figuring out just what grit looks like and how much grit one needs.

For roughly the last 20 years, advocates for greater diversity in STEM fields overall, seeking to increase the presence of underrepresented groups in schools and workplaces, have urged attention to self-efficacy as a necessary trait for student success (e.g., Cosgrove & Blaisdell, 1996; Hackett & Betz, 1981). A detectable sense of self-efficacy is seen to reflect a sufficient level of belief in one's potential to take effective action and is recognized by researchers as a desirable governing impulse in the STEM student (Slaton, 2011b). Good conduct and proper attitude have always been seen by American educators as necessary ingredients for classroom success, but invocations of tenacity and self-actualization refine the notion of compliance to include a particular sort of mettle directed towards the completion of given bodies of effort. This is not a metric that displaces, but rather one that complements, broader social structures. For example, students with sufficient educational preparation or highly developed social networks may attain educational success by marshaling those other resources. Educators' stress on self-efficacy disguises those structural conditions by explaining achievement through merit. The perceived absence of such capacity renders the student ineligible for attainment in the classroom and world beyond. Even when schooling takes on the task of improving students' self-efficacy, in apparent sympathy with inclusive social aims, this formulation is one of many means by which educational thinkers ascribe individuals' experiences of higher education to personal attributes, including cognitive capacity, physical ability, or energy, rather than drawing our attention to socioeconomic structures (Bowles & Gintis, 1976; Slaton, 2011b). But giving weight to self-efficacy also foregrounds personal agency in ways that other descriptions of individual potential (say, through comparisons of students' standardized test scores) cannot.

When racial and gender diversity became widespread concerns for U.S. STEM educators and policy makers in the 1970s and 1980s, the "support" of "disadvantaged" students commonly included the provision of social programming and psychological services. Notions of self-efficacy that gained traction in the neoliberal era towards the end of the 20th century frequently shifted responsibility onto students for such emotional (self-)sustenance. We should of course not draw clear lines between these older societal and newer individualistic approaches to cultivating student self-confidence: Both assert an arbitrary demarcation between social and intellectual experience, and both reproduce a deficit model of minority educational experience. What does emerge around the year 2000, however, is a nearly unalloyed claim for the value of overcoming one's self-doubt and enduring suffering as such, a more complete banishment of any ethos of care that education might have come to entail over the course of the 20th century (Pawley & Phillips, 2014; Pawley & Slaton, 2015; Riley, 2013). Today's student with high self-efficacy believes in herself *in the face of a hostile or discouraging setting*, in a sense learning to tolerate and transcend intolerance. Any experience of the self as inadequate must itself be doubted.

Favorable invocations of grit by educators in the last few years articulate this individualistic ideology by delineating education and the "world" at large as social settings demanding such traits, while still pegging students' success to aspects of their characters:

> Grit, tenacity and perseverance … are essential to an individual's capacity to strive for and succeed at long-term and higher-order goals, and to persist in the face of the array of challenges and obstacles encountered throughout schooling and life. ... As the world becomes increasingly complex, technical, multicultural, and competitive, children and adolescents also face a weakening of the family and informal community support that was once available. ... These issues contribute to dramatic gaps in achievement that are detrimental to individuals and corrosive to society as a whole. (U.S. Department of Education, 2013, pp. v, 2)

The authors' identification of "weakening" family and community networks carries significant resonance with racialized understandings of poverty and obscures economic and environmental inequities (Greenbaum, 2015).

But even more than in formulations regarding student self-efficacy, in the elevation of grit and tenacity, satisfaction is to be attained *by way of* the difficulties produced by a degenerate and daunting world. Education today thus optimally equips students with "perseverance to accomplish long-term or higher-order goals in the face of challenges and setbacks, engaging the student's psychological resources, such as their academic mindsets, effortful control, and strategies and tactics" (U.S. Department of Education, 2013, p. vii). The passage of time or accretion of difficulties only confirms the presence or absence of grit: "The gritty individual approaches achievement as a marathon; his or her advantage is stamina. Whereas disappointment or boredom signals to others that it is time to change trajectory and cut losses, the gritty individual stays the course" (Duckworth et al., 2007, p. 1088). If self-efficacy implies confidence in one's own agency as that is deployed in the service of self- and world-improvement, the idea of grit indicates that such capacity is actively enhanced through adversity. In this perfect expression of compassionate conservatism, bent on both acknowledging and preserving adversity, pain is to be welcomed as an expedient to personal fulfillment in the STEM life; desire is driven by suffering, suffering by desire.[5]

In one Foucauldian sense, grit is clearly one more form of knowledge that modern institutions generate about the individual in their efforts to produce knowable, productive participants, much like IQ or metrics of physical and mental health (McRuer, 2006; Townley, 1994). Firms, government, schools, hospitals, and prisons maximize their functionality through such assessments. As Schull (2013) explains, however, modern institutional participation may involve not merely control as Foucault saw it, let alone a Marxian sort of coercion, but also the cultivation of a kind of collusive impulse among the voters, consumers, students, prisoners, or workers required for institutional perpetuation (or, as in Schull's case studies, on the part of Las Vegas machine gamblers). As sexuality is purported to inhere in the individual and require acknowledgement (if not correction) (Waidzunas, 2015), so too must grit be self-discovered and self-monitored. For those who invoke its societal value, grit represents a proactive impulse, gaining meaning for subjects from the educational and labor contexts whose interests such proactivity serves. We take up grit in particular—among the many personal traits seen to be salubrious for American learners and workers—because it helps us understand very directly the social structures associated with modern industrial capitalism as those enlist citizens in productive regimes.

In turn, STEM, among modern systems of knowledge, has a special instrumentality for postindustrial capitalism, and we want to focus here on the utility of the knower/known-about split for capitalist labor relations. That hierarchical system, predicated on some citizens attaining more, and more useful, knowledge about the world than others, and thereby providing labor of greater value to the collective, makes exquisitely precise use of citizens' STEM capacities to distribute occupational opportunities. As Townley (1994) captures the relationship crafted between capacity and individuality in industrial management: "Traditional approaches in personnel have taken the individual as a self-evident phenomenon—an observable reality, a unit continuous in time, possessing an essential personal identity—to be recruited, appraised, remunerated, etc." (p. 11).

Because the possession of grit implies a welcome harnessing of energy and yearning to the advantage of existing social structures and power relations, we may see it as an instrumental approach to desire on the part of its proponents. Grittiness, as an expression of students' willingness to move forward on a prescribed path of occupational preparation, can be seen as striving-within-a-structure, willing oneself forward via unflagging desire for STEM success. Without grit, one is not achieving and one cannot achieve STEM knowledge. One should *want* to be gritty—to desire a driving desire. Crucially, the presence of willful unwillingness in students (expressed, say, as a consistent and sustained disinterest in STEM-directed effort) does not have value as another sort of grit or as an effortful STEM-critical project, but rather stands as a lack of productive yearning.

To be clear, grit is not a free-floating quality that can manifest in any individual, however insistently recent scholarship may have articulated it as a personal trait to be sought among all students. It is rather like every other constitutive element of our supposedly meritocratic educational system in that the engineering student with potential to perform grittiness must *also* be aesthetically bearable to the field in both mind and body (Carlson, 2010; Cech & Waidzunas, 2011; Fifield & Letts, 2014; McRuer, 2006; Riley, 2013; Waidzunas, 2015). Constructions of grittiness enfold these aesthetic norms and travel unmarked in its "objective" measures. Long-standing coconstructions of masculinity, heteronormativity, self-discipline, and productivity that have configured engineering under industrial capitalism continue in today's deployments of grittiness. For instance, Sally Hacker (1989) describes the physical fortitude and heterosexual vigor that male engineering students have always been expected to display in the United States (see also Frehill, 2004; Herzig, 2006; Oldenziel, 1999; Slaton, 2001).

We now want to ask how it is that male heterosexual longing remains foundational to engineering education settings today, as both a fulfilled and an unfulfilled desire, while other expressed social desires remain stubbornly unspeakable, let alone inessential. And more broadly, we want to follow how the desir*able* engineering student's ascribed capacity and that student's experience of desire are conflated in one fabulous, laudatory package; educators' detection of meritorious intellect in engineering students and their recognition of students' grit-driven striving towards prescribed behaviors (demonstrating both mental and physical, factual and sexual prowess) are mutually confirming. Meanwhile those of unruly conduct, unpalatable pleasures, and unbearable bodies are rendered as incapable of meaningful engineering action, let alone virtue. These characteristics are not only the causes of exclusion but also the effects. The same conducts and bodies might be read differently in a different discipline, such as comparative literature or performance art, though we should presume each discipline has its own social ontology. The reluctant or doubting, the ungritty, can never fully participate in the simultaneous ordering of wills, talents, and futures we call engineering, a sector universally (and, we emphasize, not coincidentally) celebrated for its material and intellectual contributions to human welfare (Slaton, 2011a).

There are many ways in which all STEM education sectors share a worldview that weds the science, technology, engineering, or math student's attainment of empirical understanding and practical skills to the conjoined aims of self- and world-improvement. But engineering forwards bodily hopes and consequences of special weight among those sectors. The readiness to move both the material of industrial capitalism *and* the computer cursor distinguishes the gritty engineering student from more theory- or calculation-focused, weaker-limbed, and thus weaker-willed, fellow learners.

THE GRITTY ENGINEER

We can articulate further how recent invocations of grit continue old commitments in engineering education to engage so-called noncognitive capacities and how they craft new power relations in the university (see applications of grit in engineering education, including Bennett, Schelter, Olsen, Guffey, & Li, 2013; Bottomley, 2015; Jaeger, Whalen, Payne, & Freeman, 2010; Lerner, 2013). Continuities running from late 19th-century through contemporary engineering settings are numerous: Traditionally, and in current practice, to be read as having talent, the engineering student must be willingly present in the classroom, behaviorally and politically compliant, and intellectually and physically productive. The aspiring engineer, like the biology or statistics student, must be ready to remain awake yet appropriately fatigued from steady scholastic effort, alert though silent and still in the classroom except when bidden by an instructor to speak or act. This is much as Bourdieu (1977) has described historical experiences of modern Western education overall. Needless to say, any matters of personal health, family security, and emotional well-being that might preoccupy the student, for either positive or negative reasons, are intended to be dealt with in strict, self-abasing silence, as if they were not there at all (Cech, 2013; Cech & Waidzunas, 2011; Fifield & Letts, 2014; Hacker, 1989; Pawley & Phillips, 2014). From the very early days of "equal opportunity" rhetoric in U.S. higher education, starting after the 1954 U.S. Supreme Court decision in *Brown v. Board of Education*, the concept of "soft skills" also took hold as a central element of inclusive, uplifting education in STEM fields and beyond. Not entirely distinct from assimilationist Americanization programs aimed at immigrant students and workers starting in the late 19th century, or from the even older disciplining techniques used to refine rural workers entering the nation's first factories after 1800, these skills encompassed the development of such personal traits as punctuality, grammatical speech, and appropriate dress and hair styles in the workplace (Slaton, 2010).

What was new in the civil rights era was a modernization agenda that promised infinite upward mobility through paired acquisition of new behaviors and new knowledge: a democratic, if only rhetorical, erasure of the identifications that might once have held minority Americans back from economic inclusion. Crucially, the proposition of an imminent postracial equality across identities meant that the possession of the right social skills could be seen to automate inclusion, in keeping with meritocratic claims about the nation and its educational systems. This remains today a means of denying the existence of different levels of opportunity, silencing discussion of structural racism, sexism, ableism, and other power relations (Ahmed, 2012). In schooling intended to bring economically disadvantaged communities into technical fields, the idea still holds that "interview techniques," "communication," and even "manners" are to be taught in parallel with academic topics; conduct is held to trump identity, whether identity is conceived as an expression of cultural heritage, an outward appearance, or some other personal features left strategically unnamed to facilitate their ostensible trivialization.

There is an identifiably oppressive social instrumentality in the rhetorical separation of conduct and identity; as with ex-gay reparative therapy, certain categories of identity remain innate but are rendered moot by behavioral changes (Waidzunas, 2015). In STEM programming, behavioral capacities are treated as distinct from cognitive endowments while integral to the successful deployment of intellect in the labor marketplace. This construction helps further the idea that there is something to be called "STEM knowledge" that is devoid of social intentionality, further conserving existing distributions of privilege in the places where such knowledge is deployed: in virtually all places of modern life (see endnote 1).

In a related and very ironic sense, the promotion of personal grit also elides any distinction between individual and collective commitments in students' experience of American education. Individuals' self-formation along gritty lines actually serves a wage-based industrial economy ideally, as each person internalizes the alienating features of work and comes to feel proud or even merely comfortable with that modern emotional accommodation (Townley, 1994). As Ahmed (2014a) beautifully puts it, working off Raymond Williams, we would do well "to explore 'structures of feeling,' but also consider 'feelings of structure'; feelings might be how structures get under our skin" (para. 17). We would say that prescriptive literature focused on students' sense of tenacity fits quite neatly today's specific social-structural instrumentalities of STEM knowledge and capacity. For one thing, anxious workforce policy discussions regarding the "skills gap" supposedly afflicting American industry today routinely invoke a reluctance among young people to embrace the rigor and complexity of technical occupations or, at the very least, a reluctance to approach those as viable personal undertakings (National Academy of Sciences, 2007). Greater grit would presumably be a solution to this problem. On a deeper level, however, both the notion of a "STEM crisis" and the celebration of student perseverance add further directionality to individualizing projects of industrial capitalism, demarcating the willing from the unwilling citizen, the legible from the illegible economic participant (Schull, 2013).

The construction of a sovereign self, capable of making the right choice (as opposed not just to making the wrong choice, but to making no choice at all!), is thus utterly central to contemporary STEM education theory of this kind (Britzman, 1995; Fifield & Letts, 2014). Litfin (1999) writes, "Knowledge and sovereignty are conceptual kin; both sorts of claims are fundamentally about whose voice is to be regarded as authoritative" (p. 95). Grit is a concept that seems destined to forward precisely this kinship. That is, if a sufficient allotment of self-confidence allows the student to feel good about themselves in the course of learning, grit seems to be further specified as determined conduct in the face of obstacles or "bad" feelings: an experience of "working strenuously towards challenge" (Duckworth et al., 2007, p. 1087). Alongside the greater proactivity of grit relative to mere self-efficacy, there is a pronounced teleology here as grittiness is treated as a means of achieving the "fruition of human talent" (p. 1088). The idea of some human talent existing as a singular destination apart from any political utility would be essentialist enough; to add a possibility of fruition (as if that talent awaits in embryonic form beyond obstacles and doubt) redoubles the commitment here to a social singularity.

It is worth emphasizing the particular understandings of personal and social responsibility at work here. Implicit in the formulation of recent educational programming around grit is the idea that environment matters; students' life conditions may discourage or encourage increased levels of grit. In this way, the proposition of grit as a real and meaningful factor in student achievement may seem to justify the accrual of social resources around disadvantage (Duckworth et al., 2007; U.S. Department of Education, 2013). But in this formulation, self-discipline and perseverance, however we may claim they originate, exist only in the form of individual agency, ultimately mooting any real significance for environmental factors. Thus, those who "succeed" may be commended *as individuals*, and those who "fail" cannot ever evade blame and responsibility for their own life circumstances.

THE GRITTIEST BODIES

As long as identity and skill are rhetorically cast as discrete objects, the political instrumentality of merit is secured: The individual is seen to apply their capacities to an object of attention, to greater or lesser effect depending on those exertions. For example, the historical trope of the curious scientist or assiduous engineer is one of brain being successfully brought to bear on world, never the other way around, never a problematic of social or cultural settings bringing value to certain ideas or intellectual exertions. Bodies, as assemblages of flesh, skin, hair, bones, muscles, metabolisms, diseases, strengths, sensory organs, nerve endings, pleasures, pains, and desires, etc., are so selectively admitted into contemporary discourses of technical merit, while at the same time so completely inescapable as elements of all human experience, that we make that selective logic central to our exploration here (Carlson, 2010; Hacker, 1989; Herzig, 2006; Riley, 2013).

As a student, one puts oneself perceptually and bodily in the path of challenges and rewards. But there is good yearning and bad, desirable desire and regrettable desire. This distinction is possible for analysts because these human experiences are first positioned by researchers as things both recognizable and measurable in students. Tellingly, Duckworth et al. (2007) line up grit as a factor similar to IQ—that is, as a trait that is impactful upon human learning, subject to measurement, and deserving of the authors' and readers' attention:

> In particular, we know a great deal about intelligence, or general mental ability, a construct for which formal study was initiated by a British contemporary of

> [William] James, Sir Francis Galton. Notwithstanding vigorous debates over the dimensionality and origins of intelligence, we know more about IQ—how to measure it reliably and precisely and what outcomes it predicts—than any other stable individual difference. In contrast, we know comparatively little about why, as James put it, most individuals make use of only a small part of their resources, whereas a few exceptional individuals push themselves to their limits. (p. 1087)

Notice how many uncertainties and value-laden choices must be dismissed in this passage in order for the systematic study of student tenacity to be justified. Why we would ever want to ignore "vigorous debates" is not clear. Even more worryingly, reliable and precise measurement, and proposed causal relationships between data and outcomes, are understood by historians and sociologists of science invariably to be subject to experimenter's regress (Collins, 1981): A reliable scientific or social scientific measurement is that which confirms the existence and variable expression of the phenomenon being measured. Such denials of subjectivity were the heart and soul of Galton's anthropometry, and invoking his work as an inquiry devoid of political implications is itself a politically laden assertion cloaked as an objective description. But even without reference to Galton as an antecedent, the authors' confident assertions that "formal study" transcends "debates" about IQ, and that both IQ and grit are human qualities subject to reliable and precise measurement, argue for the reality of "stable individual difference" (Duckworth et al., 2007, p. 1087). We begin to see how the knower/known-about split directly buttresses the idea of differently qualified personhoods, producing those equipped for success and those of errant, regrettable traits and aims.

The systematic study of personal traits itself falls under the purview of positivist STEM techniques: the demarcation, quantification, and ranking of human experiences in service to predetermined outcomes (Riley, 2014). In acknowledging this epistemic home base for educational research, we find a socially instrumental use of empiricism underway in the development of a Grit Scale. In this metrological project, as in much educational research, Duckworth et al. (2007) proceed in circular fashion (Biesta, 2007; Riley, 2014). The problem is not that the researchers define meaningful life achievement with extraordinary narrowness, although that is certainly the case and a recognizably sociopolitical choice: "Our overarching goal for scale development was to capture the attitudes and behaviors characteristic of the high-achieving individuals described to us in early, exploratory interviews with lawyers, businesspeople, academics, and other professionals" (Duckworth et al., 2007, p. 1090).

The more inequitable impacts of this project derive from the inquiry's research design, which appears to be strongly evidential when in fact it is self-justifying. The authors design a "cross-sectional study for which the major purpose was to develop and validate a self-report measure of grit in a large sample of adults aged 25 years or older," in which "the predictive validity of grit was assessed by its association with higher levels of lifetime schooling among individuals of identical age" (Duckworth et al., 2007, p. 1090). But if grit is what enables educational achievement, and educational achievement is what confirms the presence of something to be called grit, are we not trapped in a circular definition? Elsewhere in the study, in a similar self-confirming process, subjects' higher or lower SAT scores are used as "a measure of general mental ability" (p. 1093) against which to test findings of relative grittiness. Put simply—the grit measurers seem destined to detect a significant quantity of the stuff where they seek it and find none where they do not.[6]

Seen ontologically, the project of measuring grit necessarily produces both more-gritty and less-gritty STEM students, just as SAT testing produces high scorers and low. If all human levels of grittiness were of equal value, and all SAT scores of equal merit for college aspirants, one would not need spectra along which to place each individual. But there is also a unified function for readings taken along any of these spectra, in confirming that *this is a measure by which persons should in fact be judged.* Thus the Grit Scale constructs a failure to yearn for and embrace personal challenge as a volitional omission among some students, precluding the attainment of STEM knowledge on these students' part. But we could also see that the Grit Scale projects the possibility of improper yearning—a desire for something that is not grit, perhaps to be measured on a different scale or not measured at all. Unlike the identification of insufficient desire for the right goals (in other words, the affliction of those suffering from low grit), students' yearning for the wrong things, or an absence of yearning altogether, can never be *directly* inscribed on the grit researchers' instruments. And yet, that alternative, not-on-the-scale state is not left *unconsidered* in the world in which grit is measured; rather, it comes to represent an unfortunate or unpalatable desire. This surprises no one who has read Foucault's (1990) *History of Sexuality*: "the law [or the scale] is what constitutes both desire and the lack on which it is predicated" (p. 81). Thus such non-STEM yearning might well constitute a "companion knowledge" or "shadowy reality" in environments of learning and work, as an ontological stance might recognize (Fifield & Swain, 2002; Law & Lien, 2013). Here we want to get *ourselves* off the Grit Scale (or the Kinsey Scale) to think the vitally unthought. We want to imagine a commendable practice among students that is neither grit nor its absence. Queer theory and ontologically-informed science and technology studies are of value in helping us consider the student who is thinking about the unthought.

MAKING UNCOMMON SENSE

In many respects, the American project of distinguishing between likely and unlikely intellects by firewalling knowledge and its mastery is common not just in various global settings, but also across science, technology, engineering, and mathematics fields. Possibly that distinction characterizes all places where positivism explicitly undergirds mental labor. For many generations, engineering has popularly carried a reputation as "applied" scientific knowledge, positing and buttressing a duality which may tidily enact institutional notions of prestige (with science above engineering), but which, in fact, draws many false distinctions among occupational niches. In practice it is as difficult to find an engineer who does not undertake discovery in the course of her work as it is to find a scientist who proceeds without the use of existing tools, technologies, and metrological routines. All STEM activity forwards some established sense of what is worth doing and how it should be done in our engagements with nature, bodies, and other materialities; as the editors of this volume have reminded us, data-table templates and toolboxes full of sanctioned investigative techniques guide STEM practitioners' sense of what needs to be known next. But we think that engineering disciplines traffic in certainty in ways that scientific fields generally do not: The expert scientist pursuing discovery is praised for not knowing what she may be doing three steps hence in the lab; the unpredictable engineer would be seen as deficient in knowledge—and dangerous, too. To the extent that uncertainty is acknowledged, it must be formally and quantitatively characterized with probabilistic and numerical methods that go by telling names such as *bootstrapping*, *Monte Carlo simulation*, and *finite element method*. The common sense that makes up engineering in this regard is inseparable from the question of who can be/do engineering (Slaton, 2010).

This construction of cognitive rigor disallows not merely technical uncertainty and any deep questioning of what the engineer should do next, but also strange selves, selves remote from the directed sovereignty of gritty, forward-facing STEM students. This definition of rigor is integral to the pervasive depoliticization of engineering, which Cech (2013) indicates works partly by rendering so-called social and political concerns like equality as tangential to "real" engineering work, while treating LGBTQI lives as more problematic than heterosexual experience for the furtherance of good technical conduct. That production of conventionally unwelcome personhoods is precisely the sexual "strangeness" that Fifield (2012) captures powerfully as bringing not merely alienation for its subjects, but potentially a uniquely constructive disruption of *all* our familiar selves. Building on Broadway (2011), he suggests:

> When we inquire from intentional standpoints as sexual strangers, we discover ways to deny heteronormativity its claims to common sense, which suggests that other unmarked assumptions in our thoughts and practices can be thrown into relief when we take up different interpretive perspectives, ask new kinds of questions, and evaluate answers using new criteria. (Fifield, 2012, p. 15)

Fifield also articulates that the more familiar progressive normativities today supporting gay persons' "coming out," accompanied by the goal of increased acceptance of the "out" educator by heterosexual students and teachers, satisfy our liberal impulses much more reassuringly than do such denials of common sense, "which is why the potential pedagogical and epistemological costs of exchanging strangeness for normalcy are so rarely explored" (p. 14). The "transcendental unity" implied by that easy liberalness, we would add, is deeply imbricated throughout engineering, which disdains strangeness and shadows at least as efficiently as does experimental science, banishing any doubt that the prospect of human-generated explanations and interventions upon the cosmos is a reasonable one and that this is what all reasonable people will believe (Fifield & Letts, 2014).

That banishment of doubt is, of course, predicated on the idea that there is an optimally knowable cosmos to be optimally acted upon by optimized actors. But again, who may be optimized? In the STEM classroom or workplace, the nonheterosexual often remains odd and vexing in a way necessary for heterosexuality to be made normal and reassuring; other categories enact the reassurances to be offered by youthful, male, conventionally-abled, white or Western bodies (Cech & Waidzunas, 2011; Fifield & Letts, 2014; Waidzunas, 2015). As long as taxonomies of personhood are important to our culture, any of these subjects, as agents, will require objects to render them as such; knowers cannot exist without the to-be-known-about.

WHAT IS TO BE DONE, UNDONE, AND NOT DONE?

The Grit Scale and broader understandings of grit as a metric of potential are tools to produce, measure, sort, transform, and/or discard learners, not tools for transforming educational practice. These offer strong evidence for us that STEM education is not geared toward producing unpredictable experiences, as yet unidentified participants, or as yet unknown outcomes. So … what is to be done? Can we queer the structure of yearning, learning, and earning? Can we queer yearning so that the objects of our desire are deviant? Can we queer STEM learning so that it is freed from its socially constraining focus on depicting a singular world? Can we queer earning away from capitalist creature-comfort goals to values and economies of difference?

One kind of queering across these sites occurs simply in paying attention to grit as an object of others' attention and abiding desires. In a stratified labor system, differentiated learning opportunities are always required, but effort must nonetheless be made to seem reasonable to those unlikely to reach the highest levels of STEM attainment, security, or salary. This is the landscape in which personhoods are functioning. It is a landscape of technical knowledge and practice that involves a depoliticization of all constituent elements: engineering, business, and achievement. Grit is an exquisite instrument for this social reproduction as it pretends that success inheres in the individual. Even when educators speak of enhancing the conditions of education to encourage students to get in touch with their grit potential, it falls upon the individual to make something of those enhancements. Exposing grit as a depoliticization of learning means breaking the conspiracy of silence by refusing to comply with rhetoric of meritocracy.

Most critiques of meritocratic systems concerned with equity worry that a focus on merit hides the social conditions in which people learn and work, obscuring privilege that leads to achievement (Slaton, 2010). But a study of grit indicates that we have an epistemic commitment to inequity: The measuring tool produces the thing up against which it is held. Focusing only on the differences among rulers, not on the ways in which measurement itself enacts power, ensures the durability of privilege.

Where the Myers-Briggs Type Indicator® was discredited as measuring false binaries (Pittenger, 1993), the StrengthsFinder® has taken its place. If taxonomies of personhood like grit and other psychometrics render each person in the room (or not in the room) known or knowable, what are the alternatives? Some people are already there as strangers: the disabled student, the queer student, the trans or genderqueer student who deflects tabulation, the double major who is simultaneously an engineer *and* not an engineer. While queer existence is an act of resistance to power, it need not dismantle that power in order to be disruptive.

To queer STEM education would require queering epistemics: to out the politics of knowing; to question who can know and what can be known; to take up othered ways of knowing, and ways of not knowing. Queering resides in drawing attention to the manifold intentionalities that normally circulate in STEM settings. As we have indicated, the grit that is cultivated in the college or community college classroom is in obvious ways a preamble to the focus required for a successful working life: a willingness to show up, remain alert, and subordinate other life goals to that of earning money (Weeks, 2011). In this sense, the moment-by-moment doing that is represented as grit (residing in both the learning/working subject and the object of that person's attention) fills time and holds the subject at a remove from the big picture: values, ethics, and the question of what ought and ought not be done. We offer the idea that in some moments engineering might be the thing that is not done. For instance, if a standard civil engineering class asks students to consider the type of dam that should be built in a particular location, a more open inquiry might include the possibility that no dam should be built and, instead, a more sustainable environmental option—or no option—be pursued. Students learning engineering ethics often worry that their designs might end up killing people, not realizing that, in many engineering fields, that is exactly the point (H. Bernstein, personal communication with D. Riley, ca. 2003). To not engineer is to queer engineering.

This active choice to not engineer is not to be confused with passivity, and it is emphatically not a position that need be outside one's schooling or professional life in engineering. For one thing, inactivity approached mindfully can be generative: Boredom, disappointment, and frustration can cast a raking light on striving and the systems that demand that we strive. Dulled aversion is not an improvement on mindless acceptance of challenge, and mindfulness may put the engineering student in a position of deeply noticing the conditions of practice, as this book's editors have suggested. Buddhism would say that desire and suffering are locked together, and there is no doubt that in a world that makes virtues of accumulation and exhaustion, releasing oneself from that cycle may be liberating. But it is not so easy to live in the moment if you seek a redistribution of power; equanimity can forestall equity. Yet at the same time, we must be careful at all moments not to treat effort as itself virtuous, in the vein of Calvinism. This would not only depoliticize grit but mistake suffering for virtue. As Ahmed (2014b) describes the effortful social life of the marginalized, "a desire for a more normal life does not necessarily mean identification with norms, but can be simply this: a desire to escape the exhaustion of having to insist just to exist" (p. 149).

CONCLUSION

Underlying our discussion has been the idea that engineering historically persists in demarcating the knower (or aspiring knower) and that which is to be known about: The learner and the world that awaits the learner's attention. Is this ambitiously maintained cordon sanitaire between personhood and knowledge itself perhaps that which we should be calling engineering? To answer "yes" would mean that engineering is constituted of the logic that renders it most crucial and most reasonable among human actions to demarcate identity (defined as bodies and their desires, or as distributions of power, position, or privilege) from knowledge-making or material aims. Certainly the social instrumentalities of manufacturing, infrastructure creation, biomedicine, war-making, and other enterprises in which engineering is widely deployed gain much from that demarcation. Strictly committed to preventing reflection on the operative limits to reasonable knowledge and yearning under industrial capitalism, all those sectors do better when strangeness is kept at bay.

So, if strangeness is a powerful enemy of privilege, how best to engage with it? How best to queer our own scholarly inquiry? On some level, the very distance we perceive between our concerns as critical observers of STEM education and those of our subjects in this paper makes us want to ask about still another unity: How can the same institutions produce both our perspective and that of the uncritical grit measurers, diversity proponents, and personhood definers? How can it be that, as far apart as the political commitments of these two perspectives would appear to be, both are utterly compatible with the aims of enlisting scholarly audiences and receiving paychecks from American universities? Given that seeming conundrum, and thinking ontologically, we might go on to ask: What if we are not at all incompatible with those whom we critique in this essay, but rather apparent opposites who simply render one another meaningful? That is, what if the grit-critical (us) and grit-believing (those about whom we have written here) function in tandem in exactly the same way that "gritty" and "not gritty" STEM students work to jointly produce a reasonable universe of possibilities? To *together* banish the strangeness of uncertain and unmeasurable pedagogical conducts? Staying with this suspicion would mean neither more nor less than rendering problematic our own common sense, of asking: What possibilities for a more equitable and just engineering pedagogy lurk as yet unthought beyond the reason of our own critical practice?

Our point in this chapter is that the academic programs, organizations, and businesses that have historically operated under the rubric "Engineering" in the United States are not uniquely oppressive or exclusionary among American disciplines, but they do express in particular ways the dominant U.S. culture's reluctance to embrace liberatory change. Ideas about *who counts as engineering* (to us, the only possible phrase to use here) enact neoliberal distributions of power with exquisite precision, even as or precisely because the field evacuates any discussion of power from that which is seen as legitimate teaching, learning, and practice (Cech, 2013; Pawley, 2012; Riley, 2008; Riley & Claris, 2006; Riley et al., 2013; Slaton, 2010).

We recognize that many individuals and organizations have devoted time and effort to ending discrimination along lines of identity in American engineering (following categories of race, ethnicity, sexuality, gender, [dis]ability, and other historically mutable demarcations). However, those efforts tend to reinscribe meritocratic claims for identity or personhood as things distinct from the bodies of knowledge and labor relations called engineering. That is, most STEM equity efforts imply an encapsulated world of technical practice into which unfairly excluded "missing talent" should now be allowed access. This expanded access, it is commonly said, will invariably benefit both our industrial society and its individual citizens. Yet the very idea of some persons being regrettably absent from a sphere of practice forecloses examination of those practices as themselves expressions of exclusion. We believe that only by recognizing the ontological character of technical knowledge and practice—seeing those commitments as indeterminate and relational—is a reparative move possible.

There are already models for us in this move (Riley, 2014). For example, Biesta (2007) endorses Dewey's (1929) resistance against mind/materiality dualisms and his critique of the sovereign self. If ways of knowing are understood in Dewey's theoretical action framework as ways of *doing,* then knowledge can no longer be inert, prescriptive, and sovereign, and "no conclusion of scientific research can be converted into an immediate rule of educational art" (p. 29). Engineering educators' current interest in "experiential learning" tends to valorize projects with known outcomes and lean towards affirmations of the productive personhoods that are of most direct value for military, industrial, and other globalizing projects. But in attending to experience, STEM instruction may also provide the pedagogical site where strangeness is most readily allowed to enter. It strikes us that in admitting into the engineering classroom not just the possibility of multiple knowledges, so that nonengineering questions and solutions gain airtime in those settings, but also a legitimate option for strategic *in*action in the face of STEM-derived demands, the edges of engineering may become clear. The personhoods traditionally eligible to participate in engineering production, a group historically constituted of desiring and desirable bodies, may also finally be marked as such and new eligibilities imagined. Grit may cease to become either a necessary or sufficient indicator of technical talent, and aspirants' yearning itself may be newly aimed at infinitely many objects and persons. Both who and what counts as engineering may thus be made strange, in every wonder-inducing sense of that word.

If grit is the "known known" in Rumsfeldian parlance, then same-sex marriage, Lockheed's LGBT drone builders, and other assimilations that produce "knowable others" are the known unknown.[7] Queers and strangers, the unknowable others, are the unknown unknowns that threaten to disrupt the status quo. But in our efforts to queer engineering, we would be wise to remember Slavoj Žižek's (2004) warning that the real danger is the unknown known, the Abu Ghraib to Rumsfeld's weapons of mass destruction: "the disavowed beliefs, suppositions and obscene practices we pretend not to know about, even though they form the background of our public values" (para. 10). In STEM, the unknown known, we suggest, is the belief in the superiority of white, able-bodied, cis-hetero masculinity; of logical positivism; of capitalism; and of the fictitious meritocracy that masks all these epistemic and ontological hegemonies.

NOTES

1 Nonheterosexual (and nonsexual) for us do not binarily oppose heterosexuality but rather move in a plurality of unorthodox directions: asexual, queer, nonlabeling, polyamorous. In STEM, fortitude and attainment (or what we will recognize in this chapter as comprising ascriptions of *grit*) are about pulling all in one direction towards capitalist productivity and the requisite denial of other selves. The plural desires of queerness, bisexuality, polyamory, unmarried sexuality, asexuality, etc., all are centrifugal forces showing that the center cannot hold.

2 For this critique to work, *positioned* must be understood relationally, ephemerally, and indeterminately—that is, positioned in a shifting network.

3 Continuities and resonances between STEM education and STEM-related activities occurring in government, industry, commerce, home, combat, and other noneducational settings are innumerable and incessant. Critical studies of educational history, philosophy, and theory predicated on demonstrating the social function of education generally can help us break down our analytical boundaries between school and not-school, a task beyond the scope of this chapter (Bourdieu, 1977; Bowles & Gintis, 1976; Weeks, 2011). But we would also stress that the study of STEM pedagogy supports critical studies of science, engineering, and related fields, wherever they may undertake their practices, by queering the insistent lack of reflexivity inhering in those sectors (Cech, 2013; Riley, 2013; Riley, Slaton, & Pawley, 2013).

4 In this study focused on engineering, we could here say simply, "with knowledge," as there is, intriguingly, no such thing as "unreasonable engineering." An activity lacking earmarks of technical empiricism or uncritical commitments to numeracy simply ceases to be legible as engineering, as does any project integrating explicit political claims or self-reflection.

5 As we write, we are following strong reactions against new provisions being made for college students grappling with personal trauma, such as "trigger warnings" or "safe spaces" for students who may be upset by exposure to confrontational speech or otherwise disturbing classroom content. This caretaking is seen by at least one observer as infantilizing students, while earlier eras that made no such provisions are said to have produced "hardier souls" (Shulevitz, 2015, para. 16). Students' experiences of personal security or contentment, if achieved through institutional caretaking, are nearly contemptible in this depiction, an extension of emergent conservative ideas regarding the destructive nature of conflict around identity (Ahmed, 2012).

6 For further critiques of such circular deployments of evidence in engineering education research see Pawley and Phillips (2014), Pawley and Slaton (2015), and Riley (2014).

7 The idea of Lockheed's LGBT drone builders arose from Donna Riley's participation in a Capital Pride Parade in Washington, DC, where she watched an army of Lockheed employees marching with their corporate logo on light green polo shirts paired with khakis. Lockheed is a leader in "unmanned" systems such as drones (http://www.lockheedmartin.com/us/what-we-do/aerospace-defense/unmanned-systems.html). Drone manufacturers, including Lockheed Martin and Northrup Grumman, are corporate sponsors of mainstream LGBT advocacy groups such as the Human Rights Campaign (http://gawker.com/haaay-to-the-chief-the-military-industrial-complex-con-486133694).

REFERENCES

Ahmed, S. (2012). *On being included: Racism and diversity in institutional life*. Durham, NC: Duke University Press.

Ahmed, S. (2014a, July 21). *Feminist hurt/feminism hurts* [Web log post]. Retrieved from http://feministkilljoys.com/2014/07/21/feminist-hurtfeminism-hurts/

Ahmed, S. (2014b). *Willful subjects*. Durham, NC: Duke University Press.

Bennett, R. M., Schleter, W., Olsen, T., Guffey, S., & Li, W. (2013, June). *Characteristics of students who do not do homework*. Paper presented at the meeting of the American Society for Engineering Education, Atlanta, GA. Retrieved from https://peer.asee.org/19300

Biesta, G. (2007). Why "what works" won't work: Evidence-based practice and the democratic deficit in educational research. *Educational Theory, 57*(1), 1–22.

Bottomley, L. (2015, June). *Assessing the GRIT of incoming engineering students*. Paper presented at the meeting of the American Society for Engineering Education, Seattle, WA. Retrieved from https://peer.asee.org/assessing-the-grit-of-incoming-engineering-students

Bourdieu, P. (1977). Cultural reproduction and social reproduction. In J. Karabel & A. H. Halsey (Eds.), *Power and ideology in education* (pp. 487–510). Oxford: Oxford University Press.

Bowles, S., & Gintis, H. (1976). *Schooling in capitalist America: Educational reform and the contradictions of economic life*. Chicago, IL: Haymarket Books.

Britzman, D. (1995). Is there a queer pedagogy? Or, stop reading straight. *Educational Theory, 45*(2), 151–165.

Broadway, F. S. (2011). Queer (v.) queer (v.): Biology as curriculum, pedagogy, and being albeit queer (v.). *Cultural Studies of Science Education, 6*, 293–304.

Carlson, L. (2010). *The faces of disability*. Bloomington, IN: Indiana University Press.

Cech, E. A. (2013, June). *The veiling of queerness: Depoliticization and the experiences of LGBT engineers*. Paper presented at the meeting of the American Society for Engineering Education, Atlanta, GA. Retrieved from https://www.asee.org/public/conferences/20/papers/6540/view

Cech, E. A., & Waidzunas, T. (2011). Navigating the heteronormativity of engineering: The experiences of lesbian, gay, and bisexual students. *Engineering Studies, 3*(1), 1–24.

Collins, H. M. (1981). Son of seven sexes: The social destruction of a physical phenomenon. *Social Studies of Science, 11*(1), 33–62.

Cosgrove, C. R., & Blaisdell, S. L. (1996, June). *A theoretical basis for recruitment and retention interventions for women in engineering*. Paper presented at the meeting of the American Society for Engineering Education, Washington, DC. Retrieved from https://peer.asee.org/6348

Dewey, J. (1929). *The sources of a science of education*. New York, NY: Horace Liverlight.

Duckworth, A. L., Peterson, C., Matthews, M. D., & Kelley, D. R. (2007). Grit: Perseverance and passion for long-term goals. *Journal of Personality and Social Psychology, 92*(6), 1087–1101.

Fifield, S. (2012). *Teachers as sexual strangers* (Unpublished manuscript). Retrieved from https://www.researchgate.net/publication/324276952

Fifield, S., & Letts, W. (2014). [Re]considering queer theory and science education. *Cultural Studies of Science Education, 9*(2), 393–407. doi:10.1007/s11422-013-9509-4

Fifield, S., & Swain, H. (2002). Heteronormativity and common sense in science (teacher) education. In R. Kissen (Ed.), *Getting ready for Benjamin: Preparing teachers for sexual diversity in the classroom* (pp. 177–189). Lanham, MD: Rowman & Littlefield.

Foucault, M. (1990). *The history of sexuality, Volume I: An introduction* (R. Hurley, Trans.). New York, NY: Random House.

Frehill, L. M. (2004). The gendered construction of the engineering profession in the United States, 1893–1920. *Men and Masculinities, 6*(4), 383–403.

Greenbaum, S. D. (2015). *Blaming the poor*. New Brunswick, NJ: Rutgers University Press.

Grzanka, P. R. (2015). *Intersectionality: A foundations and frontiers reader*. Boulder, CO: Westview Press.

Grzanka, P. R., & Brian, J. D. (2015, April). *"Analytical dispositions" and the shared affinities of science & technology studies and intersectionality*. Paper presented by P. Grzanka at Drexel University, Program in STS, Philadelphia, PA.

Hacker, S. (1989). *Pleasure, power and technology: Some tales of gender, engineering and the cooperative workplace*. Boston, MA: Unwin Hyman.

Hackett, G., & Betz, N. E. (1981). A self-efficacy approach to the career development of women. *Journal of Vocational Behavior, 18*, 326–339.

Haraway, D. (1988). Situated knowledges: The science question in feminism and the privilege of partial perspective. *Feminist Studies, 14*(3), 575–599.

Herzig, R. (2006). *Suffering for science: Reason and sacrifice in modern America*. New Brunswick, NJ: Rutgers University Press.

Jaeger, B., Whalen, R., Payne, R., & Freeman, S. (2010, June). *Successful students: Smart or tough?* Paper presented at the meeting of the American Society for Engineering Education, Louisville, KY. Retrieved from https://peer.asee.org/16281

Law, J., & Lien, M. E. (2013). Slippery: Field notes in empirical ontology. *Social Studies of Science, 43*(3), 363–378.

Lerner, A. A. (2013, June). *Gritty students: The effect of perseverance on retention for traditional and nontraditional students.* Paper presented at the meeting of the American Society for Engineering Education, Atlanta, GA. Retrieved from https://peer.asee.org/19664

Litfin, K. T. (1999). The status of the statistical state: Satellites and the diffusion of epistemic sovereignty. *Global Society, 13*(1), 95–116.

McRuer, R. (2006). *Crip theory: Cultural signs of queerness and disability*. New York, NY: New York University Press.

Miller, J. (2013). Pathways and purposes of the "French tradition" in American engineering education: The case of the Virginia Military Institute. *Engineering Studies, 5*(2), 117–136.

National Academy of Sciences. (2007). *Rising above the gathering storm: Energizing and employing America for a brighter economic future*. Washington, DC: National Academies Press.

Oldenziel, R. (1999). *Making technology masculine*. Amsterdam: Amsterdam University Press.

Pawley, A. L. (2012). What counts as "engineering": Towards a redefinition. In C. Baillie, A. L. Pawley, & D. Riley (Eds.), *Engineering and social justice: In the university and beyond* (pp. 59–85). West Lafayette, IN: Purdue University Press.

Pawley, A. L., & Phillips, C. M. L. (2014, June). *From the mouths of students: Two illustrations of narrative analysis to understand engineering's ruling relations as gendered and raced.* Paper presented at the meeting of the American Society for Engineering Education, Indianapolis, IN. Retrieved from https://www.asee.org/public/conferences/32/papers/9343/view

Pawley, A. L., & Slaton, A. E. (2015, June). *The power and politics of STEM research design: Saving the "small n."* Paper presented at the meeting of the American Society for Engineering, Seattle, WA. Retrieved from https://peer.asee.org/the-power-and-politics-of-stem-research-design-saving-the-small-n

Pittenger, D. J. (1993). Measuring the MBTI … and coming up short. *Journal of Career Planning and Employment, 54*(1), 48–52.

Riley, D. M. (2008). *Engineering and social justice*. San Rafael, CA: Morgan and Claypool.

Riley, D. M. (2013, June). *The island of other: Making space for the embodiment of difference in engineering*. Paper presented at the meeting of the American Society for Engineering Education, Atlanta, GA. Retrieved from https://peer.asee.org/the-island-of-other-making-space-for-embodiment-of-difference-in-engineering

Riley, D. M. (2014, June). *What's wrong with evidence? Epistemological roots and pedagogical implications of "evidence-based practice" in STEM education.* Paper presented at the meeting of the American Society for Engineering Education, Indianapolis, IN. Retrieved from https://peer.asee.org/what-s-wrong-with-evidence-epistemological-roots-and-pedagogical-implications-of-evidence-based-practice-in-stem-education

Riley, D. M., & Claris, L. (2006, June). *Power/knowledge: Using Foucault to promote critical understandings of content and pedagogy in engineering thermodynamics.* Paper presented at the meeting of the American Society for Engineering Education, Chicago, IL. Retrieved from https://peer.asee.org/power-knowledge-using-foucault-to-promote-critical-understandings-of-content-and-pedagogy-in-engineering-thermodynamics

Riley, D. M., Slaton, A. E., & Pawley, A. L. (2013). Social justice and inclusion: Women and minorities in engineering. In A. Johri & B. Olds (Eds.), *Cambridge handbook of engineering education research* (pp. 323–356). Cambridge: Cambridge University Press.

Schull, N. D. (2013). *Addiction by design: Machine gambling in Las Vegas*. Princeton, NJ: Princeton University Press.

Shields, S. A. (2008). Gender: An intersectionality perspective. *Sex Roles, 59*(5), 301–311.

Shulevitz, J. (2015, March 11). In college and hiding from scary ideas. *New York Times*. Retrieved from http://www.nytimes.com/2015/03/22/opinion/sunday/judith-shulevitz-hiding-from-scary-ideas.html

Slaton, A. E. (2001). *Reinforced concrete and the modernization of American building, 1900–1930*. Baltimore, MD: Johns Hopkins University Press.

Slaton, A. E. (2010). *Race, rigor, and selectivity in U.S. engineering: The history of occupational color line*. Cambridge, MA: Harvard University Press.

Slaton, A. E. (2011a, June). *Note to self: Save humanity (a social and cultural history of the "grand challenges")*. Paper presented at the meeting of the American Society for Engineering Education, Vancouver, BC, Canada. Retrieved from https://peer.asee.org/note-to-self-save-humanity-a-social-and-cultural-history-of-the-grand-challenges

Slaton, A. E. (2011b). *Metrics of marginality: How studies of student self-efficacy hide structural inequities*. Paper presented at the meeting of the American Society for Engineering Education Vancouver, BC, Canada. Retrieved from https://peer.asee.org/metrics-of-marginality-how-studies-of-minority-self-efficacy-hide-structural-inequities

Slaton, A. E. (2013). *Body? What body? Considering ability and disability in STEM disciplines*. Paper presented at the meeting of the American Society for Engineering Education, Atlanta, GA. Retrieved from https://peer.asee.org/body-what-body-considering-ability-and-disability-in-stem-disciplines

Slaton, A. E. (2015, October). *Human/human-made*. Paper presented at the Society for the History of Technology, Albuquerque, NM.

Tierney, W. G. (1997). *Academic outlaws: Queer theory and cultural studies in the academy*. Thousand Oaks, CA: Sage Publications.

Townley, B. (1994). *Reframing human resource management*. London: Sage Publications.

U.S. Department of Education. (2013). *Promoting grit, tenacity and perseverance: Critical factors for success in the 21st century*. Retrieved from http://pgbovine.net/OET-Draft-Grit-Report-2-17-13.pdf

Waidzunas, T. (2015). *The straight line: How the fringe science of ex-gay therapy reoriented sexuality*. Minneapolis, MN: University of Minnesota Press.

Weeks, K. (2011). *The problem with work: Feminism, Marxism, anti-work politics and post-work imaginaries*. Durham, NC: Duke University Press.

Woolgar, S., & Lezaun, J. (2013). The wrong bin bag: A turn to ontology in science and technology studies? *Social Studies of Science, 43*(3), 321–340.

Žižek, S. (2004, June 21). What Rumsfeld doesn't know that he knows about Abu Ghraib. *These Times*. Retrieved from http://inthesetimes.com/article/747/what_rumsfeld_doesn_know_that_he_knows_about_abu_ghraib

ABOUT THE AUTHORS

Amy E. Slaton is a professor of history at Drexel University. She has written on the history of racial and other identities in American technical education from the early 20th century to the present. Her current projects focus on delineations of physical and intellectual capacity in STEM learning and labor, and the role of diversity ideologies in industrial capitalism. She is also co-editor of the journal *History+Technology*. slatonae@drexel.edu

Erin A. Cech joined the Department of Sociology at the University of Michigan as an assistant professor in 2016. Prior to that she was a postdoctoral fellow at the Clayman Institute at Stanford University and was a faculty member at Rice University. She earned BS degrees in electrical engineering and sociology from

Montana State University. Cech's research seeks to uncover cultural mechanisms of inequality reproduction—particularly around gender, sexual identity, and racial/ethnic inequality in STEM. Her research has been cited in the *New York Times*, *Wall Street Journal*, *Chronicle of Higher Education*, and the news sections of *Science* and *Nature*. ecech@umich.edu

Donna M. Riley is Kamyar Haghighi Head of the School of Engineering Education and professor of engineering education at Purdue University. Her research interests include engineering and social justice, engineering ethics, social inequality in engineering education, the liberal education of engineers, and engineering studies. riley@purdue.edu

CHARLOTTE BOULAY

THE BARGAIN

Then an old harrower of the dark ...

—Beowulf

Above the clouded glass ceiling at work
shadows flit over my plastic desk.
I travel back and forth in indigo,

lumps of frozen buildings squatting
on the streets. I stride block after block,
past the life-size animatronic T. rex,

dusted with snow; three days in a row
I jump as he yells at me, waving his tiny arms
and gnashing his teeth. Later, I'm lying

on the table for the thousandth time,
ultrasound screen blurry with eggs
as if the nurse is tuning into a lost TV station

or searching for that missing plane under the sea.
How can this be my story? I'm not the thief
who crept into the dragon's cave,

stole the golden cup and started the war
all over again, just when everyone was getting old
and fat and used to drinking their mead in peace.

And yet I'm made of want: paperwhites listing
in a too-small bowl, pollarded lindens'
scarred fists of stumps, the dragon

taunting me in thirty-second clips—
motion camera triggers and the lizard roars.
Robot dinosaur, I know your relentless rage,

 | DOI:10.1163/9789004331068_020

as if you've never loved or lay down to sleep.
You're the perfect monster, and we've given you
a vengeful, hungry, human heart like ours.

I trip the switch, plow the darkness.
I challenge you who make me jump out of my skin,
put me back, put me back in it again.

ABOUT THE AUTHOR

Charlotte Boulay's poetry has been published in the *New Yorker*, the *Massachusetts Review*, the *Boston Review*, the *Michigan Quarterly Review*, *Beloit Poetry Journal*, *The Pinch*, *Cyphers*, and other journals, and has appeared on the *PBS NewsHour* poetry blog and on NPR's *Morning Edition*. Her book, *Foxes on the Trampoline*, was published by Ecco Press/HarperCollins in 2014. She worked at the Franklin Institute science museum in Philadelphia, Pennsylvania (USA), for five years. She is currently director of development at the Miquon School.

INDEX

A

Ability/able-bodied, 89, 90, 214, 270, 271, 276, 277, 279
Ableism, 215, 326, 335
Abnormality, 4, 17, 18, 25, 61, 90, 95, 96, 193, 230, 277. *See also* strangeness
Aczel, Amir, 313–316
Adewunmi, B., 96
Adsit-Morris, Chessa, 24
Adults, 223, 235, 249
Adventure education, 98
Adversity, 324
Aesthetic knowledge, 258
Aesthetics, 26, 111–114, 119, 136, 202, 309, 325
Agassiz, Louis, 91
Agency, 250, 251, 323, 328
Agential realism, 130
Agyeman, J., 103
Ah-King, Malin, 22
Ahmed, Sara, 5, 6, 14, 24, 31, 60, 118–120, 270–273, 275, 276, 279, 280, 282–285, 326, 327, 333, 336n5
Akiva, Joseph ben, 315, 316
Alaimo, S., 130
Aldea, E., 224
Alexander, B. K., 6
Allen, D. E., 77
Allen, G., 14
Allum, N., 161
Alsop, S., 201
Althusser, Louis, 18
Ambiguity, 6, 12, 27, 31, 32, 53, 72, 75, 113, 137, 155, 276, 280, 281, 303
Ambjörnsson, F., 271
American Association for the Advancement of Science, 77, 264, 301
American Association of Colleges & Universities (AACU), 238, 239
American Educational Research Association (AERA), 65–67, 73
American Physical Society (APS), 127, 131, 133–135
Anatomy, 9, 261
Anderson, A., 47
Anderson, D., 166
Andersson, L., 54
Androgyny, 215, 219n
Angelou, Maya, 63
Angular Acceleration (film), 140
Animals, 22, 100, 230, 290, 291, 297–299
Anxiety, 43, 53, 71, 110, 154, 193
"Applied" scientific knowledge, 330, 331
Arbitrariness, 121
Archilochus, 255
Ardoin, N., 162
Aristotle, 257
Art of Proof, The (Beck & Geoghegan), 312
Assemblages, 11, 24, 232, 235, 292, 303
Assessment, 182, 183, 200. *See also* standards, educational; tests
Assimilationist approach
  engineering and, 326, 335
  environmental education and, 90, 91, 96, 98, 99
  queer theory and, 149, 150
Atherton, T. J., 134, 135
Atkinson, E., 262

Atoms, 137, 138. *See also* quantum physics
Attitudes, 199–201, 322. *See also* interest in science
Atwater, M. M., 218
Au, K. H., 162
Austin, J., 283
Authorial intention, 41, 44
Autobiography, 42, 54, 55, 59–61
Azzarello, R., 7

B

Bagemihl, B., 290
Baker, D., 212, 213
Baldwin, James, 44, 47, 49, 50
Ball, A. F., 162
Balzter, N., 127
Bang, L., 193
Banilower, E. R., 154, 155
Banks, J. A., 162
Barad, Karen, 60, 127, 129, 130, 137, 138, 140, 142n8, 250, 283, 289, 302
Baraka, I. A., 54
Barchi, Paola, 230, 231
Barnhardt, R., 47
Basu, S. J., 218
Bauer, J. E., 260
Bauer, R., 128
Bauman, Z., 53, 54
Bayer, R., 150
Bazzul, Jesse, 21, 24, 32, 82n4, 110, 259, 290, 300, 302
Beauty, 112, 113, 119
Beck, M., 312, 313, 315
Behavior, 8, 9, 17, 21, 22, 96, 102, 103, 116, 140, 165, 173, 210, 217, 284, 325–327
Belgrade International Workshop on Environmental Education, 91
Bell, D., 47
Bell, P., 162
Bennett, J., 10
Bennett, R. M., 326
Bentley, M. L., 92
Berg, Annika, 282
Berlant, L., 27
Berlin, Isaiah, 255
Berlyne, D. E., 165
Bernstein, H., 333
Betz, N. E., 322
Beville, M., 53
Biddle, B. J., 165
Biesta, G., 329, 335
Big Bang Theory (television show), 116
Bill Nye the Science Guy (television show), 156
Binaries, 18, 34n14, 35n22, 63, 303, 336n1. *See also* dichotomies; dualism
  able/disabled, 270, 271. *See also* ability/able-bodied
  adult/child, 235, 249
  contesting, 68, 102, 109, 118, 148
  heteronormative, 29, 137, 213–215, 260, 261, 270, 271, 273, 274. *See also* heteronormativity
  human/inhuman, 291
  knower/known-about, 321, 329
  macro/micro, 90
  normal/abnormal, 90, 95, 96, 188
  normative/nonnormative, 12, 100, 172, 188. *See also* normativity
  physics and, 128
Biology, 20–22, 83n21, 259–261, 299–302
Black identity, 49–55, 55n. *See also* race
Blades, David, 291
Blaisdell, S. L., 322
Bodies, 34n8, 297–300
  engineering and, 319–322, 331–335
  grit and, 328–330
  identities and, 165. *See also* identities
  knowledge production and, 151–153, 290

museum exhibits on, 251, 252
orientation and, 6, 271, 272
physics and, 129, 130
queer theory and, 68, 152, 153
sexualized violence against, 269, 270
Body of knowledge, 151, 169, 171, 246, 308, 314
Bohr, Niels, 130
Boles-Friscia, Mical, 98, 99
Bolte, C., 263
Bond, D., 224
Bose, Satyendra Nath, 131
Bosons, 131
Bottomley, L., 326
Boulay, Charlotte, 3, 25, 59
Bourdieu, P., 320, 326
Bowden, G. A., 197, 198
Bowers, C. A., 95
Bowles, S., 320, 322
Boys. *See also* gender; masculinity
gender roles, 187, 195–197, 209–219
interest in science, 29, 192
Bredekamp, S., 234
Bremer, S., 269
Brian, J. D., 320
Brickhouse, N. W., 219
Brier, J., 283
Britzman, Deborah, 3, 4, 9, 10, 63, 77, 79, 80, 116, 148, 149, 151–154, 168, 169, 172, 224–225, 262, 316n1, 320, 321, 327
Broadway, Francis S., 20–22, 25, 42, 55n, 82n5, 83n19, 110, 116, 149, 154, 225, 230, 236, 259, 289–290, 331
Brody, M., 91
Brookey, R. A., 261
Brookfield, S., 170
Brother to Brother (Evans), 50
Brown, J., 115
Brown, M., 169
Brown, Tony, 77, 79
Brown, Tova, 315, 316
Brown, Wendy, 110, 203
Bryson, Mary, 96, 97, 311
Buffy the Vampire Slayer (television show), 116
Bullard, R. D., 93
Burger, E., 310
Butler, Judith, 18, 33n1, 49, 67, 68, 75, 81, 109, 113, 116, 129, 140, 152, 193, 203, 205, 211, 271, 273, 283, 289
But Still, Like Dust, I Rise: Reflections on Sexual Minority Issues in Science Education (Letts), 63
Butterfield, S., 49
Buxton, C. A., 210
Bybee, R., 110
Byers, N., 131

C

Calabrese Barton, Angela, 215, 216, 218, 264
Caldwell, O. W., 198
Callahan, G. N., 260
Camp (aesthetic style), 309, 313
Campbell, G., 218
Camp Wilde, 23, 31, 32, 289–303
Cantor, Georg, 313–316, 317n3
Capitalism, 90, 94, 100, 102, 138, 324–327, 336n1
Cardinality, 311–313
Cardy, Terri, 230, 231
Carlone, H., 115
Carlson, Dennis, 14, 236
Carlson, L., 325, 328
Carpenter, H. A., 197
Carson, Anne, 312
Carson, J., 195
Carson, Rachel, 91, 113–115
Carter, L., 264
Carter, R. L., 91
Casper, V., 218

Categories, 303. *See also* binaries; identities
of identity, 191–206, 215, 237
Catullus, G. V., 51
Cech, Erin A., 24, 30, 32, 132, 152, 203, 325, 326, 331, 332, 334
Certainty, 110, 239. *See also* uncertainty
Chamberlin, J. E., 197
Chaos, 186, 247
Chapman, T. K., 101
Châtelet, Èmilie du, 132
Chen, M. Y., 205
Children. *See also* elementary school science
adult/child binary, 235, 249
agency of, 250, 251
curiosity of, 154, 185, 186, 227–229, 249, 252
desire and, 11, 30, 154, 194, 195, 223–240, 250, 251
developmental stages, 233–238
essentialism and, 236
play and, 244–247
rhizomatic learning and, 248, 249
sexuality and, 119, 153–155, 157
ways of moving, 223–239
wildness of, 119
wonder and, 118, 119, 186, 229, 238, 251
Children's museums, 30, 243–252
exhibits, 250–252
transgression and, 244–247, 251, 252
Christianity, 73, 74, 265
Cisgender identity, 82n8, 99, 104n1, 210, 213–215
Citizenship, 264
Clair, J. M, 51
Claris, L., 334
Clark, B. M., 195, 198
Class, socioeconomic, 214, 215, 323. *See also* middle-class identity; poverty
educational access and, 178–180
environmental degradation and, 93
environmental education and, 89, 90
physics and, 131
science and, 42
Classical mechanics, 133, 138
Cleaver, Eldridge, 50
Climate change, 93, 94
Cline Cohen, P., 201
Closet, politics of, 64, 74, 134. *See also* coming out
Codere, S., 150
Coffield, F., 161
Cofield, C., 133
Cognition, 109, 151, 162
Cohen, Jeffrey J., 7, 10, 26, 43, 53
Cole, M., 162
College curriculum
mathematics, 311–314
science, 261, 262
Colley, H., 162
Collins, H. M., 329
Coming out, 71–76, 83n11, 83n17, 131, 132, 134, 188, 189, 331
Communication, 166, 229
Complexity, 99, 102, 314
Compulsory heterosexuality, 279, 281
Conlon, T. J., 164, 170
Connection, 119, 247, 248
Connell, R. W., 274
Constructivism, 6, 194, 310
Consumerism, 94, 263
Continuum hypothesis, 313–315
Cook-Gumperz, J., 128
Copenhagen Interpretation, 130
Copes, H., 166
Copple, C., 234
Cosgrove, C. R., 322
Costa, V. B., 64
Cox, Brian, 113–115, 120, 121
Crenshaw, Kimberlé, 96
Critical orientation, 6, 7
Critical pedagogy, 94, 95, 99–101

Critical queer environmental pedagogy, 26, 89, 90, 94, 95, 99–101
Critical science literacy, 262, 263
Critical theory, 6, 289
Critical thinking, 103, 104, 170–173
Crowley, M. F., 200
Crozier, K. D., 50
Cruze, J., 41
Cuevas, A., 115
Cultural Studies in Science Education (CSSE), 69, 70
Culture, 4. *See also* multiculturalism
  nature and, 130
  sexualization of, 8
Curie, Marie, 131
Curiosity, 26, 111, 112
  children and, 154, 185, 186, 227–229, 249, 252
  desire and, 151–155. *See also* desire
  mathematics and, 310
Curriculum. *See also* standards, educational
  desire and, 229, 234–238
  human flourishing and, 265
  knowledge base, 301
  powerful knowledge and, 258–262, 265
  promoting human flourishing, 256–258
  queer curriculum theory, 307–310, 314–316
  science kits, 154, 155
  on sex, 255–265
  sexuality and, 245–247, 255–265
  social justice and, 262–265
  state-mandated, 181–183
Cvetkovich, A., 110, 111, 118
Czollek, L. C., 129

D

Dahl, Signe, 282
Dahsah, C., 150
Damiani, E., 164
Dangerous Knowing, A (Epstein & Sears), 262
Danielsson, A., 140
Danziger, K., 193, 200
Daring, C. B., 89, 95, 96
Darmstadt Technical University, 126, 127
DasGupta, S., 24
Daston, L. J., 139, 194
Data, 114, 150–152, 161, 162, 165, 168, 199–201, 255
Davidsson, Eva, 271
Davis, Brent, 18, 70, 73, 237, 272, 307, 308, 312, 313, 315, 316n1
Davis, C. G., 77
De Castell, Suzanne, 96, 97, 311
Deconstructivist perspectives, 6, 68, 125–130, 133
Defosse, Dana, 82n8
Degele, N., 129
Delany, S. R., 42
Delauney, G., 195
Deleuze, Gilles, 10, 11, 24, 30, 34n8, 224, 226, 229–233, 237, 239, 247, 248
Denes, R., 218
dePalma, R., 262
Depoliticization, 203
Desire
  children and, 11, 30, 154, 194, 195, 223–240, 250, 251
  curiosity and, 151–155. *See also* curiosity
  Deleuzo-Guattarian, 10, 11, 30
  engineering and, 324, 325
  "feelings for the organism," 150–155
  flourishing and, 257
  learning and, 290
  learningscapes and, 161, 165, 170
  normative, 170
  nursing and, 275
  pedagogies of, 11, 27, 230, 231
  physics and, 136, 137, 139, 140

queer theory and, 68, 80, 81, 129, 149, 283, 284
STEM education and, 8, 9, 12
technologies of, 33
wonder and, 117, 119
Developmentally Appropriate Practice in Early Childhood Programs (Copple & Bredekamp, eds.), 234–239
Dewey, J., 166, 231, 335
Diagonalization proof, 314
Dichotomies, 68, 217. *See also* binaries; dualism
Dierking, Lynn D., 166, 167, 170, 271
Dilley, P., 149
Dillon, Justin, 91, 271
Dinshaw, C., 165
Disabled bodies, 271, 276, 277, 279. *See also* ability/able-bodied
Discomfort, 117
Discovery, 19, 117, 152, 183–186, 197, 244, 249, 331
Discrimination, 71, 93, 96, 150, 172, 204, 216, 262, 264, 274, 319, 335. *See also* equity projects; inclusion agendas; oppression
Disorientation, 272, 273, 282, 284
Dissection, 22
Diversity
categories of, 192, 193, 215
in classrooms, 187, 188
coming out and, 75
engineering and, 320, 322
in environmental education, 97, 98
interest in science and, 191–193, 203–205
of knowing, 66, 67
mental health patients, 280
nature and, 290
in physics, 125, 127, 130–133
Doctors, 24, 273, 274, 277, 282
Doering, Z. D., 165
Dominant student groups, 212–215, 218
Donald, C. A., 24
Donovan, C., 166
Double consciousness, 51
Doubt, 110, 332
Downing, E. R., 195, 198
Dreger, A., 24
Dualism, 12, 14, 94. *See also* binaries; dichotomies
in/out (of the closet), 74
mind/materiality, 335
teacher/student, 246, 247, 249
Dublin, R., 47
Du Bois, W. E. B., 43, 50, 51
Duch, B. J., 77
Duckworth, A. L., 322, 324, 328, 329
Dufwa, S., 275
Duggan, L., 83n17

E

Earth Day, 91
Earth sciences, 92
Eckstrand, K. L., 24
Ecology, 91, 92
Ecopedagogy, 26, 95, 100
Ecotourism, 92, 93, 102, 103
Educators Queering Academia (Miller & Rodriguez, eds.), 13, 14
Einstein, Albert, 113, 117, 121, 132
Eisenberger, R., 166
Eisenhart, Margaret A., 219
Electrical engineering, 125
Elementary school science, 147–157, 224–229, 262. *See also* children
kindergarten students, 182, 183, 236–238
learning in, 184–186
sexuality and, 27, 148, 153–157
Elementary school teachers, 27, 77–79, 147–157, 184–186
Elhuff, L., 195
Eliot, T. S., 312, 313
Elite, science as, 43, 48
Ellenbogen, K., 166

Ellsworth, Liz, 316n1
Emerson, Ralph Waldo, 91, 98
Emotions, 26, 111–114, 117, 119, 161. *See also* feelings
Empiricism, 31
Engaged pedagogy, 100
Engineering, 32, 33, 125–141, 319–336
  bodies, eligible and ineligible, 319–322, 331–335
  culture of, 133
  grit and, 322–326
Enke, A. F., 82n8, 210
Enthusiasm, 78, 114, 184, 185, 204, 229
Environmental degradation, 93–95
Environmental education, 23, 24, 26, 32, 89–104
  assimilationist approach, 90, 91, 96, 98, 99
  curriculums, 101–103
  defined, 91–93
  queering, 89–91, 97–103
Environmental Education Act (1970), 91
Epistemology, 66, 67, 333
Epistemology of the Closet, The (Sedgwick), 7, 303
"Epistemontology," 142n8
Epstein, Debbie, 262
Equity projects, 66–68, 81, 205. *See also* gender equity; inclusion agendas
  engineering, 320, 335
  gender inclusiveness, 214–216
Erickson, Bruce, 3, 4, 102
Eriksson, Kristina, 274, 278
Erlemann, M., 128
Eros, 9, 10
Essentialism, 62, 67, 149
  children and, 236
  countering, 109
  gender and, 210, 211, 218
  mathematics and, 309
  sexuality and, 260
  talent and, 328
Estrada, F., 24
Estrogen, 260
Ethics, 246, 257, 258, 289–291, 302
Ethnic minorities, 93, 214. *See also* race
  in engineering, 319–323, 326, 335
  interest in science, 191–206
  in physics, 130–133
Evans, M. M., 75, 98
Evans, Rodney, 50
Evolutionary theories, 195–199, 290
Excitement, 53, 114, 118, 151, 156, 229, 319. *See also* desire
Experiential learning, 335
Exploration, 225–227, 244, 250
"Explorimenting," 225–227

F

Facts, 156, 161
Failure, 111
Falk, John H., 165–167, 170, 271
Fallace, T. D., 195
Fanon, F., 50, 51
Fara, P., 133
Faraday, Michael, 131
Fausto-Sterling, A., 211
Feather, N. T., 165
Feder, M. A., 162, 164, 167
Feder, T., 134
Feelings, 118, 327. *See also* emotions
"Feelings for the organism," 150–155
Feinstein, N., 161
Felsenstein, D. R., 24
Femaleness. *See also* gender; women
  in biology, 79, 80, 83n21, 259–261
Femininity, 136, 209–219, 262
Feminist critique, 289, 307
Feminist standpoint epistemology, 61, 62, 66, 67
Feynman, R. P., 47
Fifield, Steve, 20, 21, 23, 25, 52, 53, 60, 65–67, 69, 71–73, 75–78, 82n5, 90, 101, 120, 128, 149, 154, 155,

203, 214, 217, 218, 225, 264, 271, 289, 320, 321, 325–327, 330–332
Finkel, Elizabeth, 219
flessas, blake m. r., 24, 26
Flint, Michigan, water crisis, 93
Ford, D., 77, 79, 155
Fordham, S., 43, 46, 52
Formal learning, 161, 162, 167, 170, 173, 243
Fortier, Anne-Marie, 276, 284
Foucault, Michel, 7, 16, 18, 29, 192, 193, 198, 202, 205, 243, 245–247, 251, 252, 255, 291, 324, 330
Foutz, S., 170
Framework for K-12 Science Education (National Research Council, 2012), 27–29, 150, 153, 193, 201
Frank, J. R., 42
Freedom of thought, 110
Freeman, S., 326
Frehill, L. M., 325
Freire, Paulo, 47, 94, 100
French Committee for Weights and Measures, 138
Freud, Anna, 154
Freud, Sigmund, 9
Frizzle, Ms. (fictional character), 156
Fröding, Gustav, 276, 277, 284
Funk, L., 77
Fuss, D., 4, 17, 73

G

Galison, P., 139, 194
Galton, Francis, 329
Gametes, 83n21
Gao, J., 230
Garber, L., 264
Garmon, Savannah, 134
Gay and lesbian studies, 129
  exclusion and, 262
  queer theory and, 23, 65, 68–70
Gay men, 48, 180, 181. *See also* LGBTQQI+ individuals
  biology and, 261
  race and, 49–55
Geertz, C., 272
Gender. *See also* boys; cisgender identity; femaleness; femininity; girls; maleness; masculinity; transgender/trans* identity; women
  binary. *See* gender binary
  defined, 33n1
  ecotourism and, 93
  environmental education and, 89, 90
  genetics and, 79, 80
  interest in science and, 191–206, 212, 213
  moving beyond, 216–218
  names and, 177, 178, 181
  performativity of, 109, 129, 140
  physics and, 125, 135–137, 139, 140
  queer theory and, 68, 129, 149
  scientific identities and, 115
  sexualized violence and, 269, 270
  STEM education and, 3, 4, 8, 12, 29
  teacher identities and, 156, 157
  in textbooks, 135, 136
Gender binary, 129, 137, 209–219, 257–261, 270–281
  collapsed with sex binary, 210, 211. *See also* sex
  students and, 209–219
Gender conformity, 210, 216
Gender discrimination. *See* discrimination
Gender equity, 70, 71, 82n9, 83n10, 115, 125, 139, 140, 232. *See also* equity projects; inclusion agendas
Gender inclusiveness, 214–216
Gender minorities. *See* genderqueer identity; intersex identity; transgender/trans* identity

Gender nonconformity, 209, 210, 212–219
Genderqueer identity, 29, 135, 209, 210, 215, 216, 219n1, 333
  defined, 219n
Gender roles, 252
  single-sex classrooms and, 186–188
  theories on, 195–197
Gender studies, physics and, 127–130, 133, 140, 141
General Science Quarterly (GSQ), 28, 193, 197, 203
Genetics, 9, 79, 80, 152
  sexual orientation and, 73, 74, 260, 261, 299
Geoghegan, R., 312, 313, 315
Geopolitics, environmental degradation and, 93
Germany, 125–127, 131, 134
Gerry, H. L., 197
Gieryn, T. F., 65
Giffney, Noreen, 3, 10, 12, 147, 163, 171, 237, 256, 283
Gilbert, Andrew, 26, 81, 110–112, 151, 235, 251
Gilbert, Jen, 118, 119
Gilbert, John K., 164
Gill-Peterson, J., 233
Gilman, S. L., 197
Gintis, H., 320, 322
Girls. *See also* gender; women
  cisgender, 210, 213, 214
  gender equity and, 71
  gender roles, 187, 195–197, 209–219
  interest in science, 29, 191–206
  learning and, 181
  names and, 177, 178
Girod, M., 113
Glasser, H. M., 82n9
Goeppert-Mayer, Maria, 131
Goldberg, L., 24
Golding, V., 271
Goldstein, L. S., 78, 155
Gonsalves, A. J., 140
Gordon, E. W., 162
Gordon, L. R., 50
Götschel, Helene, 26, 35n31, 127–129, 137, 140
Gottfried, M., 24
Gough, Annette, 31, 32, 96, 98
Gough, Noel, 23, 24, 31, 32, 96, 98, 290
Gould, Stephen J., 195, 204, 264
Gräber, W., 263
Graham, Mark, 7, 9
Grant, M., 197
Gray, Emily M., 26, 151, 235, 251
Greenbaum, S. D., 323
Greene, Kathleen, 82n5
Grier, N. M., 198
Grimsley, J., 47
Grit, 32, 33, 322–336
Grit Scale, 33, 329, 330, 332
Groh, S. E., 77
Grossmann, Lisa, 131
Gruenewald, D. A., 92, 95
Grumet, M. R., 42, 43
Grusenmeyer, L., 77, 155
Grzanka, P. R., 320
Guattari, Felix, 10, 11, 24, 30, 34n8, 224, 226, 231–233, 237, 247, 248
Guba, Egon, 6
Guffey, S., 326
Gunckel, Kristin L., 10, 12, 22, 23, 27, 149, 310
Gutiérrez, K., 212

H

Hacker, Sally, 325, 326, 328
Hackett, G., 322
Hacking, I., 192, 200
Hadzigeorgiou, Y., 111–113
Hager, P. J., 161
Halberstam, J., 41, 111, 115, 116, 121, 166, 167, 279, 283, 312
Hall, Donald, 17, 19, 34n1, 34n7

Hall, S., 277
Halliday, J., 161
Halperin, David, 13, 15, 16, 62, 95, 152, 244
Hamilton, R., 48
Hands-on learning, 78, 244
Hannover University of Applied Sciences and Arts, 126
Hanson, T., 197
Happiness, 119, 120
Haraway, Donna, 5, 26, 31, 34n4, 34n11, 82n6, 283, 320
Harding, Sandra, 31, 42, 61, 62, 64, 67, 115
Harding, T., 275
Hark, S., 129
Harris, K., 256, 257
Hart, T., 233
Hartsock, Nancy, 67
Hasse, C., 128, 133
Haussmann, G., 135
Hayes, C., 50
He, M. F., 48
Heaphy, B., 166
Heart of Mathematics (Burger & Starbird), 310
Hegemonic masculinity, 125, 128, 274, 281
Heimlich, Joe E., 28, 162, 164, 168, 170
Hekman, S., 130
Herczeg-Konecny, J., 283
Herdt, G., 35n29
Hermaphroditism, 102, 298–300
Herschberger, Ruth, 259
Herzig, R., 325, 328
Heteronormativity. *See also* normativity
  binaries and, 29, 137, 213–215, 260, 261, 270, 271, 273, 274
  coming out and, 72, 73
  critiques of, 9, 68, 81, 331, 332
  disorientation and, 284
  environmental education and, 89, 98
  evolutionary logic and, 290
  grit and, 325
  learning and, 167, 168
  literature on, 19–24
  marriage and, 279
  museums and, 272, 282
  Other and, 17
  physics and, 125, 127–131, 135, 139–141
  queer theory and, 256
  science and, 22, 48, 64, 154, 157, 198, 259–261, 281, 283
Heterosexism, 93
Heterosexuality
  ambient, 73
  compulsory, 279, 281
Heterosexual matrix, 129
Hewitt, N. M., 131
Heyes, C. J., 165
High school science, 42, 155, 186–188
  general science courses, 192
  heteronormativity in, 64, 259–261
  sexuality and, 70, 71
  stratification in, 198
Hilbert, David, 313
Hillman, B. L., 49
Hird, M. J., 10, 269
Hirdman, Yvonne, 273, 275
Hirschauer, S., 54
Hirschfeld, Magnus, 260
History of Sexuality, The (Foucault), 7, 330
History of the present, 193
HIV/AIDS, 245, 246
Hodkinson, P., 162
Hodson, D., 264
Höjer, S., 282
Holland, J., 256
Hollenbach, A. D., 24
Holtz, P., 54
Homogenizing projects, 62, 82n8
Homo-/heterosexual binary, 7
Homonormativity, 74, 81, 83n17

Homophobia, 45, 72, 147, 215
Homosexual, as identity, 15. *See also* LGBTQQI+ individuals; queer identity
Honeychurch, K., 14
hooks, bell, 100, 217
Hormones, 260
Hotel Infinity, 315
Howes, E. V., 307
Hubbard, Ruth, 140
Huegel, K., 219n
Hughes, B. E., 24
Hulsebosch, P., 218
Hume, David, 258
Hungerford, H. R., 92
Hunter, G. W., 195, 198
Hursh, D., 91
Hutson, Tommye, 28

I

Identifications, 17
Identities, 11–15, 17, 42
  black, 49–55, 55n. *See also* race
  categories of, 191–206, 215, 237
  conduct and, 327
  conflict around, 336n5
  of elementary preservice teachers, 155–157
  environmental education and, 90
  gender and, 115. *See also* gender
  intersectionality of, 215
  knowledge and, 320, 321
  learning and, 165, 169–173
  names and, 177–178, 181
  power and, 192, 193, 198
  science, 47, 48
  skills and, 328
  time and, 312, 313
Identity politics, 12–15, 18
  coming out and, 74, 83n17
  queer theory and, 62–69
Ignorance, 171
"Illyria" (Boulay), 3
Imagination, 114
Imitation Game, The (film), 116, 167
Imperialism, 90, 95, 100
Impetus, theory of, 138
Implication, 7–11
Inclusion agendas, 5, 24, 63, 147, 148, 302. *See also* equity projects; gender equity
  elementary schools and, 147, 148
  engineering and, 319–322, 326, 327
  environmental education and, 90, 97, 98
  heteronormativity and, 316
  for LGBT learners, 167, 168
  physics and, 128, 134
Individual learning, 164–166
Industrial management, 324
Inequity, 332. *See also* equity projects; inclusion agendas
Infinity, 32, 307, 309, 311–316
Informal learning, 28, 162–166, 170–173, 173n1, 243, 244
Inquiry, 11, 149, 150, 154, 227–229, 321. *See also* questions, learning to ask
  inquiry-based learning, 244
  pedagogies, 77–79
Intentional learning, 161, 163
Intentions, 161, 163
Interest in science, 28, 29, 110, 112, 113, 117. *See also* desire; enthusiasm; passion
  categories of identity and, 191–206
  data on, 199–203
  depoliticization and, 203, 204
  gender and, 29, 194–199, 212, 213
  race and, 194–199
Interior exclusions, 4, 17, 73, 76
Intersectionality, 5, 59, 60, 68, 90, 96, 101–103, 215, 260, 320
Intersex identity, 211, 260, 300. *See also* LGBTQQI+ individuals

Intra-action, 24, 34n2, 82n2, 130, 250, 289, 302
Invisibility, 214. *See also* visibility
IQ tests, 199, 200, 328, 329
Irreverence, 251, 252
Irvine, J. M., 265
It's Elementary (Chasnoff), 150

J
Jackson, A., 239
Jaeger, B., 326
Jafri, J., 218
Jagose, Annamarie, 7, 8, 34n7, 68
Jakobsson, Anders, 271
James, I., 131
Jarvis, P., 171
Jefferson, Thomas, 55n
Jenkins, E. W., 264
Jenkins, H., 45
Jennings, Kevin, 74
Jewish mysticism, 315
Johannes, A., 233, 239
Johnson, A., 115
Johnson, B., 72
Johnson, Barry, 82n5
Johnson, M., 151
Jones, L., 54
Jones, T. J., 197
Jordan, L., 53
Jorfeldt, I., 275
Journal of Environmental Education, The, 24
Joy, 111, 113, 115, 117, 119, 120, 225, 244, 247
Juck, M., 77
Justice. *See* social justice
Juxtaposition, 313

K
Kabbalists, 315
Kahle, J. B., 201
Kahn, Richard V., 26, 95, 100, 104n4
Karakayali, N., 53
Keith, Jourdan Imani, 93
Keller, Evelyn Fox, 31, 115, 150, 152
Kelley, D. R., 322
Kellner, D., 95
Kennelly, Jackie, 23, 89
Kenny, J. D., 110
Kessler, S., 128
Kesson, K., 230
Killermann, S., 211
Kilpatrick, W. H., 193, 199
Kincheloe, J. L., 94
Kindergarten, 182, 183, 236–238
Kinsey Scale, 330
Kirby, V., 10, 12, 34n14, 35n22
Kirchgasler, Kathryn L., 28, 29, 205
Klenk, K., 127
Knaier, Michelle L., 29, 201
Knopp, L., 169
Knowledge. *See also* learning
  aesthetic, 258
  "applied" scientific, 330, 331
  body of, 151, 169, 171, 246, 308, 314
  as enlightenment, 117
  environmental education and, 89, 90
  grit and, 324
  learningscapes and, 171
  mathematical, 258
  physics and, 136–139
  queer theory and, 68
  scientific, 258–262
  sedimented, 127
Knowledge production, 150–153, 315
  bodies and, 151–153, 290
  personhood and, 334
  sexuality and, 151–155, 308
Knowles, J., *A Separate Peace,* 43–50
Knowles, M. S., 168
Koerner, M. E., 218
Kohlberg, L., 233
Kollmus, A., 103
Kovács, A., 137

Kovacs, P. E., 110
Kozulin, A., 166
Krajcik, J., 150
Krass, A., 129, 140
Kroenecker, Leopold, 315, 316, 317n3
Krywanczyk, L., 96
Kuhn, T. S., 45
Kumashiro, K. K., 53

L

Labels, 42, 50, 55, 63, 79, 138, 188, 210, 215–218, 293, 300, 303
Lake, V. E., 78, 155
Lakoff, G., 151
Lamb, Y. S., 51
Lassén-Seger, M., 54
Lather, P., 237
Lather, Patti, 316n1
Law, J., 320–322, 330
Laws of nature, 198
Leafgren, Sheri, 11, 25, 30, 52, 112, 114, 226, 227, 229, 230, 239
Leap, William L., 64, 65, 67
Learning. *See also* curriculum; knowledge; pedagogy; standards, educational
- ecosystems of, 164
- experiential, 335
- formal, 161, 162, 167, 170, 173, 243
- individual, 164–166
- informal, 28, 162–166, 170–173, 173n1, 243, 244
- intentional, 161, 163
- LGBT learners and, 167, 168
- life-deep, 162, 163, 173
- lifelong, 162, 173
- life-wide, 162, 163, 173
- nonformal environments, 243–245
- as pleasurable, 154, 155
- social, 166, 167
- STEM literacy and, 161
- transformative, 171, 172

Learningscapes, 28, 161–174
- defined, 161
- mapping and, 169, 170
- queering, 161, 168–173

Lee. *See* Swain, Howard (Lee)
Lee, O., 210
Lee, Y.-J., 97
Leicht-Scholten, C., 127
Lemke, Jay, 8, 21, 90, 101, 120, 168, 258
Leopold, A., 91
Lerner, A. A., 326
Lesbians. *See also* LGBTQQI+ individuals
- biology and, 261
- nursing and, 279, 282, 283

Lesko, N., 195
Letts, Will, 19, 23, 25, 60–69, 71, 73, 82n5, 90, 101, 120, 128, 147, 149, 154, 203, 211, 213, 214, 217, 218, 225, 262, 264, 271, 283, 289, 321, 325–327, 332
Levine, D. N., 53
Lewenstein, B., 162
Lewis, B. F., 204
Lezaun, J., 320
LGBTQQI+ (lesbian, gay, bisexual, transgender, queer, questioning, intersex, etc.) individuals
- coming out. *See* coming out
- identity, 95, 96, 188, 189. *See also* identity politics; queer identity; sexual orientation
- inclusion in STEM education. *See* equity projects; inclusion agendas
- as learners, 167, 168
- physics workplace culture and, 127, 131–135
- political movements, 14, 16, 17, 261
- social and political status, 68, 261
- use of term, 34n3, 142n2

Li, W., 326
Lien, M. E., 320–322, 330

Life-deep learning, 162, 163, 173
Lifelong learning, 162, 173
Life-wide learning, 162, 163, 173
Likert scale, 200, 201
Lincoln, Yvonna, 6
Literature mud map, 19–24
Litfin, K. T., 327
Liu, X., 161
Livingstone, D. W., 161, 163
Lockheed Martin, 335, 336n7
Logic, 148
Long, Elena, 133
Longino, Helen, 31
Lorde, Audre, 17, 90
Lowery, P., 219
Loy, D., 15
Lucht, P., 137
Luehmann, A. L., 155
Luhmann, Susanne, 15, 17, 65, 73, 76, 77, 79, 96, 147–149, 167, 170–172, 258, 282
Lundin, Mattias, 23, 69, 70, 283
Lynham, Susan, 6
Lyon, G. H., 218
Lyon, H., 199
Lysis, 250, 251

M

MacDermut, Anna, 30
Macdonald, S., 271
Macro systems, 90, 92, 93, 95, 97, 103
Madsen, J., 77
Magic School Bus (television show), 156
Maleness. *See also* boys; gender; masculinity
  in biology, 79, 80, 83n21, 259–261
  normalization of, 99
  physics and, 130, 131
Mann, H., 194
Mapping, 169, 170
Marett, C. B., 200
Marginalities, 13, 63, 66, 67, 109
Marginalized groups, 319
Marotta, V., 53
Marsick, V. J., 164, 170
Martin, Emily, 259, 260
Martin, L. M. W., 162, 165
Martin, P., 92
Martina, C. A., 91
Martino, W., 260
Mascazine, John, 82n5
Masculinity, 209–219. *See also* boys; gender; maleness
  "failed," 275
  grit and, 325
  hegemonic, 125, 128, 274, 281
  heteronormative, 214, 262. *See also* heteronormativity
  heterosexual, 45
  physics and, 125, 128, 135, 136
  science and, 43, 48
  STEM classrooms and, 213
Material objects, 129, 130. *See also* new materialism theory
Material practice, 140
Mathematics, 22, 307–317
  knowledge, 258
  mathematical inqu[ee]ry, 22, 32, 307, 310–314
  normativity in, 309, 310, 315, 316
Matthews, M. D., 322
Mavalvala, Nergis, 132
Mayer, Robert, 132
Mayfield, Kerrita, 23
Mayo, C., 15, 83nn11, 12
Mazzei, L., 239
McClintock, Barbara, 152, 153
McClintock, M., 98
McKay, J., 50
McKenna, W., 128
McRuer, Robert, 271, 276, 324, 325
McWhorter, L., 290
McWilliam, Erica, 32, 156
Mechanical engineering, 125

Medical museums, 31, 269–285
critical ethnography, 272–277
desire and, 269–271
gender binaries, 272–281
patients and, 279–281
queering teaching in, 282–285
sexuality and, 269–271
teaching norms, 277–281
Medical students, 269–271, 277–281
Men. *See* boys; gender; maleness; masculinity
Mental health care, 275–277, 279–281, 283
Mental illness, 276, 277, 279, 280, 283, 284
Merit, 328, 332, 335
Messy, learning as, 244, 246, 255, 314
Metzl, J. M., 24
Michaels, S., 301
Microaggressions, 103, 104n7
Micro systems, 90, 92, 93, 103
Middle-class identity, 214
normalization of, 99
physics and, 130, 131
Middle school science, 42, 178–180, 186–188
"Migrant orientations," 284
Milgram, D., 212, 213, 216
Millar, R., 263, 264
Miller, A., 238
Miller, Janet L., 25, 41, 52, 61, 75, 80, 316n1
Miller, Jon D., 263
Miller, Jonson, 322
Miller, L. C., 165
Miller, S., 161
Miller, sj, 13, 14
Mills, Robert, 271, 273
Milne, C., 140
Milne, I., 110, 112, 118
Minorities. *See* ethnic minorities; LGBTQQI+ (lesbian, gay, bisexual, transgender, queer, questioning, intersex, etc.) individuals; nondominant student groups
Misogyny, 212, 215
Mononormativity, 128, 129, 137, 142n4
Monsters/monstrousness, 43, 53, 54, 291, 303. *See also* strangeness
Mony, R. S. P., 166
Morgan, A., 44
Morris, M., 13, 15, 34n10, 35n16, 63, 165
Morrison, C., 218
Mortimer-Sandilands, Catriona, 3, 4, 102
Motivation, 199–201, 203, 204. *See also* interest in science
Mueller, M. P., 92, 97
Muir, John, 91, 98
Multiculturalism, 68, 69, 81, 147, 148
Mun, K., 150
Muñoz, J. E., 49, 52
Murray, A., 73
Murray, Olivia, 12, 13
Museum of Medicine, Uppsala University, 31, 269–285
Museums. *See* children's museums; medical museums
Myers-Briggs Type Indicator®, 333
Mystery, 117, 121, 227, 228, 316
Mysticism, 315

N

Naming, 177, 181
Nash, P., 46
National Academy of Sciences, 327
National Association for Research in Science Teaching (NARST), 63, 67, 68, 82n5
National Organization of Gay and Lesbian Scientists and Technical Professionals (NOGLSTP), 132, 133

National Research Council, 191, 214, 301
National Science Education Standards (National Research Council), 77
Natural laws, 73, 138, 139
Naturalness, 102
Nature, 4, 76
  camping in, 289–303
  culture and, 130
  as gendered, sexualized, and racialized, 98
  interest in science and, 198, 199
  mysteries of, 117, 121
Natureculture, 3, 8, 34n2
"Nature's Queer Performativity" (Barad), 137, 138
Nature study, 92
Need, 119
Neoliberal agendas, 109–111, 265, 271, 323, 334
"Nerdy" scientist image, 114, 116, 156
Newcombe, N. S., 201
New materialism theory, 127, 128, 130, 133, 139, 140, 142n7, 289
Newton, Isaac, 132, 138, 139
Next Generation Science Standards (NGSS), 27–29, 150, 153, 155, 191–194, 199, 201–204, 210–218, 301
"Nigger," use of term, 42, 43, 55n
Nightingale, Florence, 275
Nobles, Connie, 82n5
Noddings, N., 83n18
Nomadic paths, 223–240
Nondominant student groups, 212–219
Nonformal education, 243–245
Nonhuman entities, 289, 291
Normal, 80, 81, 83n12, 172, 188, 189
Normal/abnormal binary, 90, 95, 96, 188
Normative/nonnormative binary, 12, 100, 172, 188
Normativity, 4, 8, 15–19, 35n17, 99, 170, 224. *See also* heteronormativity; homonormativity
  interior exclusions and, 4, 17, 73, 76
  mathematical, 309, 310, 315, 316
  mononormativity, 128, 129, 137, 142n4
  physics and, 128, 129
  in science education, 63
  sexuality and, 71
Nuance, 99, 102
Nursing
  gender and, 273–275, 278, 279, 281–283
  histories of, 31
Nursing students, 261, 269–271, 277–281
Nye, Bill, 156

O

Obama, Barack, 178
Obermark, L., 279
Objectivity, 21, 31, 34n4, 73, 139, 148, 151
Odi et amo, 51, 52
Ogbu, J. U., 43
Oldenziel, R., 322, 325
Olsen, T., 326
Olsson, L., 235, 238
One Flew Over the Cuckoo's Nest (film), 284
One Teacher in 10 (Jennings), 74
Onto-epistemology, 142n8
Ontology, 320
Oppression, 90, 93–95, 99, 101–104. *See also* discrimination
Oram, Alison, 271, 282
Orientation, 6, 7, 271–273, 275, 282–285. *See also* disorientation; sexual orientation
Osborne, J., 263

O'Shaughnessy, Tam Elizabeth, 131, 132
Other, 16–18, 53, 167, 168, 203, 277
Othering, 133
Oudshoorn, N., 260
Outdoor education, 92, 97, 98, 102
Outness, 83n11. *See also* coming out
Out to Innovate (NOGLSTP conference), 132

P

Pallotta-Chiarolli, M., 260
Palmer, J., 91
Pappas, Theoni, 316
Parsons, A., 283
Particle physics, 137
Passion, 3–5, 26, 33, 81, 109, 115, 131, 151, 153, 156, 157, 213, 279, 310, 314, 315. *See also* desire; interest in science
Patients, medical care and, 279–281
Patriarchal structures, 109–111
Patton, M. Q., 290
Pawley, A. L., 323, 326, 334
Payne, P., 91
Payne, R., 326
Pedagogy
  critical, 94, 95, 99–101
  critical queer environmental, 26, 89, 90, 94, 95, 99–101
  of desire, 11, 27, 230, 231
  ecopedagogy, 26, 95, 100
  engaged, 100
  inquiry, 77–79
  of the oppressed, 94, 100
  queer, 96, 99
  queer ecopedagogy, 23, 24, 26, 100
  queering, 77, 96, 97
  queer theories on, 22–24
  sexualized, 272
Pedagogy of the Oppressed (Freire), 94
Pekakirk, A. J., 165
Performativity, 27, 32, 48, 49, 67, 109, 116, 126–130, 140, 141, 272, 292
Perko, G., 129
Perseverance, 323, 327, 328
Personal responsibility, 328
Personhood, 320, 321, 331, 332, 334, 335
Peterson, C., 322
Pettersson, H., 128
Phelan, Shane, 43, 48, 75, 76
Phillips, C. M. L., 323, 326
Physicians. *See* doctors; medical students
Physics, teaching queering, 26, 27, 125–141
  classical mechanics and, 133, 138
  culture of physics, 133–136
  diversity and, 125, 127, 130–133
  knowledge and, 136–139
  quantum physics and, 127, 129, 130, 137, 138, 142n9
  textbooks and, 135, 136
  theoretical framework, 128–130
Physiology, 9, 261
Piaget, J., 233
Pickering, A., 137
Pieper, M., 128
Pinar, William F., 42, 43, 45, 48, 49, 55n, 73, 78, 82n9, 96, 97, 172, 216, 308, 314
Pittenger, D. J., 333
Platt, C., 51
Play, 30, 244–247, 250, 251
Pleasure, learning and, 247, 249, 290. *See also* desire; passion
Plecker, W. A., 53
Plushenko, Evgeni, 136
Political activism, 91, 92, 98, 99
Pond, G., 239
Popkewitz, T., 194, 199

Positionality
  of children's museums, 244
  environmental education and, 90, 103
  queer as, 13, 15, 16, 62, 95
Positivist sciences, 125, 126, 330
Postmodernism, 100
Poverty, 93, 323
Power
  doctor-patient relationships, 281
  in educational system, 94
  environmental education and, 89, 90, 100, 103, 104
  identity categories and, 192, 193, 198
  inequality and, 326
  intersectionality and, 96
  learning and, 172
  measurement and, 332
  neoliberal, 334
  queer theory and, 68
Powerful knowledge, 258–262, 265
Presley, M. L., 301
Prince, G., 290
Privilege
  ecotourism and, 93
  environmental education and, 89, 90, 102, 103
  merit and, 332
  strangeness and, 334
Problem-based learning (PBL), 77–79
Probyn, E., 111
Productivity, 325
Project-based learning, 244
Prusak, A., 156
Psychiatric hospitals, 275–277, 281, 283, 284
Purdy, M. A., 42
Putterman, Z., 230, 240

Q

Qian, X., 77
Quantum physics, 127, 129, 130, 137, 138, 142n9
Queer
  defined, 271
  as positionality, 13, 15, 16, 62
  as positivity, 66, 67
  use of term, 34n3, 289, 290
Queer curriculum theory, 307–310, 314–316
Queer ecopedagogy, 23, 24, 26, 100
Queer identity, 11–15, 42, 62, 63, 95, 96. *See also* gay men; identities; intersex identity; lesbians; LGBTQQI+ individuals; sexual orientation; transgender/trans* identity
Queer Inclusion in Teacher Education (Murray), 12, 13
Queering, 65, 69, 70, 95, 96. *See also* STEM education, queering
  competencies of, 126
  as critical practice, 14, 15, 290, 301
  as queer inquiry, 149, 150
  of straight teachers, 18
Queering Elementary Education (Letts & Sears), 64, 65, 68
Queering pedagogy, 77, 96, 97
Queering Science (Letts), 61
Queer inquiry, 149, 150
Queer pedagogy, 96, 99
"Queer Pedagogy" (Bryson & de Castell), 96, 97
Queer standpoints, 66, 67
Queer temporality, mental illness as, 283, 284
Queer theory
  affective turn, 111, 118
  defined, 4, 7, 8, 148, 149
  desire and, 8
  difference and, 237
  environmental education and, 89, 90
  gay and lesbian studies and, 23, 65, 68–70
  identity and, 11–15
  implication and, 7–11

museum research and, 271
normativity and, 4, 15–19, 35n17, 115, 116
orientation and, 6, 7, 34n10
physics and, 127, 128, 133, 140, 141
Questions, learning to ask, 182–186. *See also* inquiry
Quickstad, N. J., 195, 203
Quiet Crisis, The (Udall), 91

R

Race. *See also* black identity; ethnic minorities; whiteness
engineering and, 323, 335
environmental education and, 89, 90
interest in science and, 191–206
school science and, 42, 43, 48–55, 214
sexuality and, 269, 270
Racial difference, 28, 29, 194–199
Racial hierarchies, 197
Racism, 72, 93, 215, 326
Rakow, Steven, 82n5
Ramazanoglu, C., 256
Randall, Lisa, 131
Randall-Sundrum models, 131
Rands, K. (Kai), 22, 29, 32, 307, 308, 310, 311, 316
Rawls, John, 264
Read, S. J., 165
Reason, 31
Recapitulation theory, 195–197
Reid, A., 91
Reiss, Michael J., 8, 20, 21, 30, 31, 168, 255–259, 261, 263, 265
Religion
mathematics and, 315
science and, 73, 74, 117, 118, 183, 194, 195, 265
Rennie, L. J., 164, 166
Reparative orientations, 6, 7, 34n9
Representation, 67–69
Reproduction
conventional science education, 31, 35n29
gender and, 83n13, 83n21
physics and, 138
in schools, 295
in science textbooks, 20–22, 259, 260, 300
sexual orientation and, 73
single-sex classrooms and, 186
Rhizome, 247–250
Rich, Adrienne, 271, 279
Rich, M., 47
Richardson, H. A., 198
Ride, Sally, 131, 132
Riley, Donna M., 24, 30, 32, 132, 152, 203, 321, 323, 325, 328, 329, 333–335
Ringrose, J., 232
Robbins, I., 50
Roberts, C., 260
Roberts, John, 216
Rodéhn, Cecilia, 24, 31, 271
Rodriguez, A. J., 214
Rodriquez, Nelson, 13, 14
Rogoff, B., 212
Rogue, J., 89
Romero, M., 43, 46
Romesburg, D., 271
Rooke, A., 272
Rose, N., 201
Rosenberg, T., 271
Rosenblatt, L., 44
Rosenthal, B., 307
Ross, M. B., 49
Ross, S., 48
Roughgarden, J., 83n21
Rousseau, Jean-Jacques, 91, 98
Roy, K., 224, 227, 229, 232
Rudolph, J., 110
Ruffolo, David, 15, 17–19, 75, 99
Rumsfeld, Donald, 335
Rupp, L. J., 49

Russell, Constance L., 23, 89, 90, 98, 103
Russell, Joshua, 23, 24, 26, 89, 100, 102, 103
Russett, C. E., 195
Ryan, A., 24
Ryan, D. P., 218
Rylant, C., 47

S
Safe spaces, 336n5
Sandahl, C., 271, 276
Sand County Almanac, A (Leopold), 91
Sander, Scott, 11, 25, 30
Santavicca, Nicholas, 24, 32, 82n4
Sarick, Tema, 23, 89
Sawchyn, J., 24
Scantlebury, Kate, 69, 70, 140, 212
Scheich, E., 138
Schleter, W., 326
Scholer, Anne-Marie, 20, 82n5, 261
Schoof, A., 50
School science. *See also* curriculum; elementary school science; high school science; middle school science
  gender inclusiveness in, 214–216
  patriarchal and neoliberal agendas in, 109–111
  race and, 42, 43, 48–55, 214
Schroeder, U., 127
Schüklenk, U., 261
Schull, N. D., 324, 327
Schultz, K., 219
Schultz, S., 218
Schutz, A., 53
Schweingruber, H. A., 301
Science. *See also* STEM education, queering
  as chaotic, 186
  epistemologies of, 52
  ideals of, 113, 114, 121
  interest in. *See* interest in science
  kinds of content in, 9, 148
  as monster. *See* monsters/monstrousness
  ontologies of, 52
  religion and, 73, 74, 117, 118, 183, 194, 195, 265
  social construction of, 76
  as stranger. *See* strangeness
  as unbiased, 151
Science fair projects, 46, 47, 51, 151
"Science for all," 68
Science kits, 154, 155
Science literacy, 161, 262, 263
Scientific inquiry. *See* inquiry
Scientific practices, 150–153
Scientific process, 172
Scott, B., 166
Scribner, S., 162
Sears, James T., 41, 42, 154, 217, 262
Secondary education. *See* high school science
Sedgwick, Eve Kosofsky, 7, 8, 11, 33, 34n9, 64, 68, 234, 235, 303
Seidman, Steven, 8
Self, social construction of, 76, 165. *See also* identities
Self-actualization, 100, 101, 322
Self-discipline, 322, 325, 328
Self-efficacy, 322–324, 328
Sense of Wonder, The (Carson), 113
Separate Peace, A (Knowles), 43–50
Set theory, 311, 313
Sex
  environmental education and, 102
  gender and, 82n9, 210, 211. *See also* gender
  queer theory and, 68, 129, 149
  school science and, 30, 31, 255–265
  STEM education and, 3, 4, 8, 12
Sex difference, 28, 29, 194–199
Sex hormones, 260
Sexism, 64, 72, 326

Sex segregation, 28, 186–188, 279, 295
Sexual harassment, 274
Sexual identity. *See* sexual orientation
Sexualities, Silence, and Science Teacher Education (Letts & Fifield), 65, 66
Sexuality. *See also* desire; reproduction
  ecotourism and, 93
  elementary school science and, 27, 148, 153–157
  environmental education and, 89, 90
  nursing and, 275
  physics and, 136, 137
  queer theory and, 7–12, 68, 129, 149
  in schools, 119, 245–247
  school science and, 30, 31, 255–265, 294, 295, 300–303
  science knowledge and, 151–155
  secondary science teacher education and, 70, 71
  STEM education and, 3, 4, 7–12, 80, 81
  textbooks on, 20–22
Sexuality in School (Gilbert), 118, 119
Sexualized violence, 269, 270
Sexual orientation, 129, 214
  biological indicators of, 260, 261, 299
  disclosure of. *See* coming out
  genetics and, 73, 74, 260, 261, 299
Sexual strangers, 75, 76
Seymour, E., 131
Sfard, A., 156
Shame, 111, 114
Shamos, M. H., 263
Shannon, D., 89
Sharpe, S., 256
Sharples, M., 166
Sheldon, James, 22, 32, 309, 311, 316
Shields, S. A., 320
Shlasko, G. D., 154
Shouse, A. W., 162, 301
Shulevitz, J., 336n5
Sigman, M., 47
Silences, 285
Silent Spring (Carson), 91
Silin, J., 249
Silverman, M., 117
Simmel, G., 53
Simmons, B., 91
Simmons, N., 49
Simpson, T., 64
Single-sex classrooms. *See* sex segregation
Skills, 326–328
Slaton, Amy E., 24, 30, 32, 132, 152, 203, 319–323, 325, 326, 331, 332, 334
Small, E., 93
Smith, L., 271
Smith III, J. P., 82n9
Snow, C. P., 52
"Snow Queen," 41, 43, 48–55
Snyder, Vicky, 20, 21, 82n5, 110, 154, 259, 290
Social justice, 12, 109, 115, 147, 150, 194
  coming out and, 83n17
  science curriculum and, 262–264
Social learning, 166, 167
Social responsibility, 328
Social roles, learning and, 165, 169–173
Sohrabji, S., 132
Somerville, S. B., 193
Sontag, Susan, 309
South Carolina, 245
Standardized tests, 110, 204, 319, 323
Standards, educational, 110, 111, 150–153, 182–184, 239, 245–249, 301
Standpoint theory, 61, 62, 66, 67, 69
Starbird, M., 310
STEM (science, technology, engineering, and mathematics), 3–6

STEM education, queering, 4, 7, 8, 59, 60
children's museums and, 243–252
desire and, 8, 223–240
in elementary schools, 147–157
engineering and, 332–336
environmental education and, 89–91, 97–103
gender binaries and, 209–219
gender equity in, 70, 71
identity and, 11–15
implication and, 7–11
interest in science and, 191–206
learningscapes, 161, 168–173
literature mud map, 19–24
mathematics and, 307–317
in medical museums, 282–285
nature and, 289–303
"nonsexual" topics in, 9
normativity and, 4, 5, 15–19, 30, 35n17
orientation and, 6, 7, 34n10
physics and, 125–141
sex and, 255–265
sexuality and, 80, 81, 255–265
touchpoints, 60–80
transgressions, 30
wonder and, 26, 109–111, 113–119
Steorn, P., 271
Stevenson, R. B., 91, 92, 94
Stimpson, C. R., 35n29
St. Louis, K., 218
Stocklmayer, S. M., 164
Stolberg, T., 112, 113
Stoler, A. L., 193, 198
Storksdieck, M., 164
Straight identity, 99, 214
Strangeness, 4, 12, 16, 80, 81, 335. *See also* abnormality; monsters/monstrousness
exploration and, 226, 227
gender nonconformity and, 210
interest in science and, 205
personhoods and, 331
physics and, 136
privilege and, 334
race and, 43
sexual minorities and, 75, 76
Strauss, C., 161
StrengthsFinder®, 333
Strickland, Donna, 131
Strömberg, Augusta, 276, 277, 280
Structure of Scientific Revolutions, The (Kuhn), 45
Student-being, 246, 247
Students. *See* children; elementary school science; high school science; medical students; middle school science; nursing students; school science
Sturgis, P., 161
Subjectivation, 18, 19
Subjectivities, 90, 120, 289, 329
Subject positions, 65–67
Sublett, C., 24
Sucker Punch (2011), 284
Sullivan, Nikki, 8, 12, 34n7, 75, 149, 218
Sumara, Dennis, 18, 69, 73, 237, 272, 307, 308, 312, 313, 315
Sundrum, Raman, 131
Svensson, S., 274
Swain, Howard (Lee), 20, 21, 71–76, 320, 321, 330
Sweden, 22, 23, 275
Sykes, Heather, 21, 82n4, 110, 259, 290, 300, 302

T

Taboo topics, 126, 148, 247, 300
Talburt, Susan, 11, 14, 75
Talented Tenth, 50, 51
Tally, R., 226
Tamarak Farm, 101, 102
Tamboukou, M., 234
Taylor, D., 93

Taylor, V., 49, 166
Teacher-being, 246, 247
Teachers. *See also* curriculum; pedagogy; school science
  bias and, 94
  coming out, 71–76
  education programs for, 70, 71, 77–79, 147–157
  environmental educators, 101–103
  gendered norms for, 156, 157
  queering and, 177–189
  straight, queering of, 18
Temporal selves, 165
Tenacity, 322, 323
Terrall, M., 133, 136
Testosterone, 260
Tests, 182, 183, 330. *See also* standards, educational
  IQ, 199, 200, 328, 329
  standardized, 319, 323
Texas Essential Knowledge and Skills for Science, 181–183
Textbooks, 20–22
  biology, 259–261, 300–302
  gender in, 135, 136
  mathematics, 311–314
  physics, 135, 136
  on reproduction, 20–22, 259, 260, 300
"The Bargain" (Boulay), 59
Theory of Justice, A (Rawls), 264
Thermodynamics, 137
Thomas, Gail E., 191, 201
Thomas, S., 233
Thomas, W. I., 200
Thomson, R., 256
Thoreau, Henry David, 91
Thurstone, L. L., 200
Tierney, W. G. ("Bill"), 73, 80, 82n9, 83n10, 149, 320
Tippins, D. J., 97
Tolbert, Sara, 22
Tolerance, 99
Topkok, S. A., 47
Tough, A., 163
Townley, B., 324, 327
Toynton, R., 167
Transgender/trans* identity, 29, 134, 215, 219n, 270, 300. *See also* LGBTQQI+ individuals
Transgressions, 30, 244–247, 251, 252, 289, 307
Traweek, S., 128, 137
Trentemöller, S., 128, 133
Tribunella, E. L., 43
Trigger warnings, 336n5
Truths, 110, 115, 116, 238, 293–297
Tuana, N., 110
Turing, Alan, 116, 167
Turner, W. B., 34n7
Twain, Mark, 55n
Two spirit identity, 215
  defined, 219n
Tyburczy, J., 271
Tytler, R., 110

U

Udall, S. L., 91
Uden, L., 164
Ulleråker psychiatric hospital, 275–277, 281
Uncertainty, 25, 27, 110, 111, 113, 119–121, 155, 246, 249, 331. *See also* ambiguity; certainty
Uncomfortableness, 147, 148
Unhappiness, 119, 120
University courses, 20
University of Delaware, 70
  Lavender Scholars, 61
  Science Semester, 77–79
Unknown known, 335, 336
Unks, G., 264
Untamed thinking, 114, 120, 121
Unthought/unthinkable, 80, 244, 247, 330

Uppsala University, 272. *See also* Museum of Medicine, Uppsala University
U.S. Department of Education, 323, 328

V
Valero, P., 193
Vance, Carole, 35n29
Vandekerckhove, W., 53
van der Veen, J., 128
Vänersborg hospital, 282
Van Hooft, S., 53
Vanman, E. J., 165
Venkatraman, V., 132
Visibility, 167, 168, 214, 270
Volcano, A., 89
Vulnerability, 111, 117, 120, 121
Vygotsky, L. S., 166

W
Wagner, W., 54
Waidzunas, T., 24, 324–327, 332
Walker, W., 168, 172
Wallin, J., 235, 237
Wals, A. E. J., 91
Walter, M., 279
Wangsa, I. T., 164
Warner, M., 128
Warren, K., 98
Wasserman, J. A., 51
Waterhouse, R. H., 197
Watts, Alan, 230, 234, 240
Weeks, J., 166
Weeks, K., 333
Weinbach, H., 129
Weinstein, M., 117
Weld, J., 77
Wells, Ida B., 55n
Whalen, R., 326
What is the 'Science'? and Who are the 'All'? (Letts), 68
Whatley, Mariamne, 262
White, John, 256, 257
Whiteness. *See also* race
  ecotourism and privilege, 93
  normalization of, 99
  physics and, 130, 131
  science and, 41, 43, 48, 52, 214
  visibility and, 270
Whitlock, R. U., 47, 96, 99
Whitten, B. L., 131
Whittier, N., 166
Wickens, E., 218
Wickman, P., 112, 113
Widerström, Karolina, 275, 282, 283
Wiggles, 11, 230–234, 237, 240
Wilchins, R. A., 34n7, 89, 100
Wildness, of children, 119
Wilkins, J. L., 155
Williams, G., 131
Williams, J. P., 166
Williams, Raymond, 327
Williams, Z. R., 50, 51
Wilson, E. O., 115, 121
  Letters to a Young Scientist, 113, 114
Wilson, Z. V., 42
Witzig, Stephen, 24, 32, 82n4
Wolcott, H. F., 42
Women. *See also* femaleness; femininity; gender; girls
  gender equity and, 71, 115, 125, 139, 140
  nursing and, 31
  in physics, 130–133
Wonder, 151
  children and, 118, 119, 186, 229, 238, 251
  defined, 111, 112
  ideals of science and, 113, 114
  queer theory and, 26, 109–111, 113–119
  untamed thinking and, 114–117
Wong, D., 113
Woodhull, J. F., 29, 198

Woodson, C. G., 49, 50
Woolgar, S., 320
Working and Learning on the Margins (NARST symposium), 63
Worms, 297–300
Writing to Transgress (Boles-Friscia), 98, 99
Wylie, Alison, 67, 69
Wynter, S., 195

X
X, Malcolm, 49

Y
Youdell, D., 115
Young, Michael, 259
Yount, L., 131

Z
Zacharia, Z., 218
Zembylas, Michalinos, 3, 10, 34n8, 112, 230, 231, 234
Zermelo-Fraenkel set theory axioms, 313
Zimmerman, Timothy D., 24, 26
Zinsser, J. P., 132
Zitzelsberger, O., 127
Žižek, Slavoj, 335
Znaniecki, F., 200
Zongrone, Adrian, 30

www.ingramcontent.com/pod-product-compliance
Lightning Source LLC
Chambersburg PA
CBHW071414230926
51846CB00001B/90
9789004331051